ANNUAL EDITIONS

Environment 12/13

Thirty-First Edition

W9-CGK-929

Editor

Richard Eathorne
Northern Michigan University

Richard Eathorne is an Assistant Professor in the Department of Earth, Environmental, and Geographical Sciences at Northern Michigan University. His primary academic interest in the field of human geography with particular attention to the human–environment relationship helped create a new *Environmental Studies & Sustainability* major at the university for the 2011 academic year. *Annual Editions: Environment* 12/13 was a direct outgrowth of that effort.

Professor Eathorne lived for a decade in Alaska, where he spent most of his time living and learning the human–environment relationships of the Inupiaq Eskimos of the Unalakleet region of western Alaska, and the Gwich'in of Arctic Village in the central Brooks Range mountains of northern Alaska. He has also traveled extensively throughout Central America, the Galapagos Islands, the Ecuadorian rainforest, and highlands of Peru, exploring the man–land relationships of those regions as well. He brings his human geography experiences to the teaching of introductory classes in human geography and environmental science, as well as upper-level classes in economic geography, geography of Latin America, and geography of tourism.

Mc Graw Hill

Connect
Learn
Succeed™

ANNUAL EDITIONS: ENVIRONMENT, THIRTY-FIRST EDITION

Published by McGraw-Hill, a business unit of The McGraw-Hill Companies, Inc., 1221 Avenue
of the Americas, New York, NY 10020. Copyright © 2013 by The McGraw-Hill Companies, Inc.
All rights reserved. Previous edition(s) © 2012, 2011, 2009, 2008, and 2007. Printed in the United
States of America. No part of this publication may be reproduced or distributed in any form or by
any means, or stored in a database or retrieval system, without the prior written consent of The
McGraw-Hill Companies, Inc., including, but not limited to, in any network or other electronic
storage or transmission, or broadcast for distance learning.

Some ancillaries, including electronic and print components, may not be available to customers
outside the United States.

This book is printed on acid-free paper.

Annual Editions® is a registered trademark of The McGraw-Hill Companies, Inc.
Annual Editions is published by the **Contemporary Learning Series** group within the
McGraw-Hill Higher Education division.

1 2 3 4 5 6 7 8 9 0 QDB/QDB 1 0 9 8 7 6 5 4 3 2

ISBN: 978-0-07-3515618
MHID: 0-07-3515612
ISSN: 0272-9008 (print)
ISSN: 2159-1059 (online)

Managing Editor: *Larry Loeppke*
Developmental Editor: *Debra A. Henricks*
Senior Permissions Coordinator: *Shirley Lanners*
Senior Marketing Communications Specialist: *Mary Klein*
Senior Project Manager: *Joyce Watters*
Design Coordinator: *Margarite Reynolds*
Cover Designer: *Studio Montage, St. Louis, Missouri*
Buyer: *Susan K. Culbertson*
Media Project Manager: *Sridevi Palani*

Compositor: Laserwords Private Limited
Cover Image Credits: Atikokin High School students paddling the wilderness of Jesse Lake, Quetico
Provincial Park, Ontario, Canada, 2011 (inset); October morning solitude on Mountain Lake,
Sylvania Wilderness Area, Upper Peninsula, Michigan, 2008. (background); © Richard Eathorne

Preface

In publishing ANNUAL EDITIONS we recognize the enormous role played by the magazines, newspapers, and journals of the public press in providing current, first-rate educational information in a broad spectrum of interest areas. Many of these articles are appropriate for students, researchers, and professionals seeking accurate, current material to help bridge the gap between principles and theories and the real world. These articles, however, become more useful for study when those of lasting value are carefully collected, organized, indexed, and reproduced in a low-cost format, which provides easy and permanent access when the material is needed. That is the role played by ANNUAL EDITIONS.

Humankind has not woven the web of life. We are but one thread within it. Whatever we do to the web, we do to ourselves. All things are bound together. All things connect.

—*Chief Seattle*, 1855

Although never formally trained as an "environmental scientist," the eloquent simplicity of Chief Seattle's words express the fundamental bedrock from which the endeavors of environmental science emerge. And if the pursuit of understanding our relation between Human and Earth is not also grounded in that rock, in that understanding that "all things connect," then any efforts to achieve our quest of *sustainability* must end in vain as well.

This edition of *Annual Editions: Environment 12/13* is constructed first and foremost on the idea that approaching environmental issues is a threefold paradigm, a paradigm that implies connections and linkages. Figure 1 illustrates this idea. Whenever we address environmental issues, we cannot view their environmental aspects independent of the people aspect. Although this is generally understood, what we sometimes fail to recognize is that any particular issue must also be "placed" geographically. Only by doing so can we begin to uncover and appreciate the variable global patterns of environmental issues as well as how those variable patterns (of wealth, access to resources, poverty, political instability, etc.) are connected to finding resolutions and our success in implementing policy. For example, we understand the science of the hydrological cycle. But dealing with freshwater shortage/access is different for a wealthy person than a poor person. And the kind of policy implemented to improve access to freshwater will certainly be influenced by the "place": How degraded is the environment? Or, what kind of biome must we contend with?

Second, this edition draws on these connections to create a proactive learning vehicle for the reader. The Unit Overviews, Learning Outcomes, and Critical Thinking Questions all converge to encourage the reader to approach this text not for the consumption of more information, but to actively engage in the assessment of the articles' theses and content with regard to their linkages (or lack of) with the classroom, textbook, and reader's real-world experience. This kind of proactive reader approach requires going beyond analytical "reaction reading" (knowledge accumulation/demonstrating understanding).

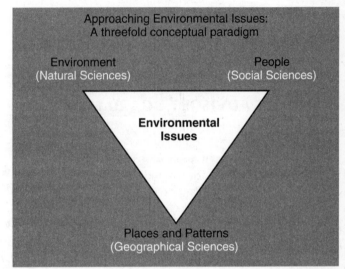

Figure 1 Addressing environmental issues requires a threefold approach that emphasizes the connections and linkages imbedded within any environmental problem. In addition, it is important to remember that the science applied (e.g., available technologies) and the people involved (e.g., cultural/ethnic, gender, religious factors) in a particular environmental problem can vary significantly by "place." Such place variability can significantly influence policy implementations, acceptance, appropriateness, and ultimately, success.

Critical-thinking environmental science reading seeks to generate questions and insights; help construct predictions and extrapolations; discover imbedded linkages and components of an environmental issue that can lead to resolutions; and encourage exploration of the *connections* between what a student is learning in class and what he/she is seeing outside of class. And finally, critical thinking of our environmental issues demands assessment of our purpose not only in environmental science, but also all scientific disciplines.

Distinguished economist and noted conservationist Barbara Ward reflected well this need for critical thinking and understanding of connections in environmental science when she noted, "For an increasing number of environmental issues, the difficulty is not to identify remedies. The remedies are well understood. The problem is to make them socially, economically, and politically acceptable. The solutions to environmental problems increasingly involve [seeing the connections between] human social systems as well as natural systems." And the National Science Foundation's Advisory Committee for Environmental Research and Education 2009 report underscores further that to live sustainably on the Earth, "greater priority must be given to advancing an integrated approach to Earth systems, and addressing the complexity of coupled natural and human systems from local to regional to global scales." As Chief Seattle said: "All things are bound together."

The *theme* of the 12/13 edition of *Environment* is the *Top Ten List of Environmental Issues* and is intended to provide a platform for "active learning." The list was compiled from dozens of reports and lists presented by national and international environmental organizations, agencies and environmental science curriculums. Although the combined lists certainly exceeded "10," many issues were similar and overlapped. Some issues were not included simply because they were not believed to be "top 10," although they are still very important. Articles were then selected to provide illustrations of each environmental issue and would provide a platform on which the reader could consider connections with other issues and components of the Human–Earth relationship. Several articles were selected because of their earlier publication date. The intent was to allow readers to compare the "hypotheses" regarding the future pattern of an environmental problem presented at that time of publication with the present situation. Were those predictions valid? If not, what "data" had not been included or was not available at the time? How might a better understanding of the reasons behind the accuracy or inaccuracy of these predictions help us in the construction of new hypotheses and new predictions leading to new knowledge for better planning for the future?

The thesis of the 12/13 edition of *Environment* is: Unit I—the issue of human population growth is directly connected to Unit 10—consuming the earth, and vice versa. And, the environmental issues presented in Units 2–9 are the direct result of the connection between Units 1 and 10. Furthermore, the United Nation's Millennium Development goals and the Millennium Ecosystem Assessment Report are argued to be inextricably linked to all the environmental issues presented, but connected none more so than to the present consumption patterns of humans.

How we presently consume the earth—transform its raw resources to meet our appetites—as well as the stress we put on the planet to attend to our waste may be perhaps our most critical environmental issue currently. But as the world's population grows, its societies and nations are destined to evolve socioeconomically as well. And herein may lay our greatest future challenge in the history of civilization: How do we change old eating habits and encourage new generations to take on the healthy diet lifestyles necessary for humanity to create a sustainable relationship with its planet?

Of course, answering this question lay not in the exclusive realm of environmental science, but will require insights and inputs from students and scholars of economics, geography, foreign policy, sociology, and others. And as such, I would like to express my gratitude to the McGraw-Hill Annual Editions Series and its entire editorial staff for providing a forum where scholars, authors, instructors and students can bring together knowledge, understanding and insights in an effort to resolve the challenges which lay ahead. I also would like to personally thank Larry Loeppke and David Welsh for providing me the opportunity to contribute to the series, and extend a very sincere thanks to Debra Henricks for her professional help with the manuscript, her patience, and her kindness.

Finally, feedback from readers, instructors and students is crucial if we hope to maintain the Annual Editions' spirit and intent of fostering critical thinking, intellectual inquiry and insight. Please feel free to share your opinions and suggestions regarding not only articles selected, but the *Annual Editions: Environment 12/13* thematic organization, pedagogic structure and construction of critical thinking questions.

Richard Eathorne
Editor

The Annual Editions Series

VOLUMES AVAILABLE

Adolescent Psychology

Aging

American Foreign Policy

American Government

Anthropology

Archaeology

Assessment and Evaluation

Business Ethics

Child Growth and Development

Comparative Politics

Criminal Justice

Developing World

Drugs, Society, and Behavior

Dying, Death, and Bereavement

Early Childhood Education

Economics

Educating Children with Exceptionalities

Education

Educational Psychology

Entrepreneurship

Environment

The Family

Gender

Geography

Global Issues

Health

Homeland Security

Human Development

Human Resources

Human Sexualities

International Business

Management

Marketing

Mass Media

Microbiology

Multicultural Education

Nursing

Nutrition

Physical Anthropology

Psychology

Race and Ethnic Relations

Social Problems

Sociology

State and Local Government

Sustainability

Technologies, Social Media, and Society

United States History, Volume 1

United States History, Volume 2

Urban Society

Violence and Terrorism

Western Civilization, Volume 1

World History, Volume 1

World History, Volume 2

World Politics

Contents

UNIT 1
Human Population Growth

Unit Overview   xxvi

UNIT 2
Global Development

Unit Overview   28

The concepts in bold italics are developed in the article. For further expansion, please refer to the Topic Guide.

UNIT 3
Feeding Humanity

The concepts in bold italics are developed in the article. For further expansion, please refer to the Topic Guide.

UNIT 4
A Thirsty Planet

UNIT 5
Changing Climate

The concepts in bold italics are developed in the article. For further expansion, please refer to the Topic Guide.

UNIT 6
Endangered Diversity

UNIT 7
Degrading Ecosystems

The concepts in bold italics are developed in the article. For further expansion, please refer to the Topic Guide.

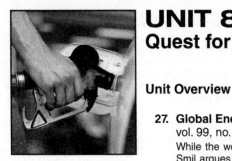

UNIT 8
Quest for Power

UNIT 9
A Global Environmental Ethic?

The concepts in bold italics are developed in the article. For further expansion, please refer to the Topic Guide.

UNIT 10
Consuming the Earth

The concepts in bold italics are developed in the article. For further expansion, please refer to the Topic Guide.

Correlation Guide

The *Annual Editions* series provides students with convenient, inexpensive access to current, carefully selected articles from the public press. **Annual Editions: Environment 12/13** is an easy-to-use reader that presents articles on important topics such as *consumption, economics, environmental ethics,* and many more. For more information on *Annual Editions* and other *McGraw-Hill Contemporary Learning Series* titles, visit www.mhhe.com/cls.

This convenient guide matches the units in **Annual Editions: Environment 12/13** with the corresponding chapters in three of our best-selling McGraw-Hill Environmental Science textbooks by Enger/Smith and Cunningham/Cunningham.

Annual Editions: Environment 12/13	Environmental Science: A Study of Interrelationships, 13/e by Enger/Smith	Environmental Science: A Global Concern, 12/e, by Cunningham/Cunningham	Principles of Environmental Science, 6/e, by Cunningham/Cunningham
Unit 1: Human Population Growth	**Chapter 2:** Environmental Ethics **Chapter 6:** Kinds of Ecosystems and Communities **Chapter 7:** Populations: Characteristics and Issues **Chapter 8:** Energy and Civilization **Chapter 19:** Environmental Policy and Decision Making	**Chapter 1:** Understanding Our Environment **Chapter 2:** Principles of Science and Systems **Chapter 4:** Evolution, Biological Communities, and Species Interactions **Chapter 6:** Population Biology **Chapter 7:** Human Populations **Chapter 22:** Urbanization and Sustainable Cites **Chapter 25:** What Then Shall We Do?	**Chapter 1:** Understanding Our Environment **Chapter 4:** Human Populations **Chapter 7:** Food and Agriculture **Chapter 14:** Economics and Urbanization **Chapter 15:** Environmental Policy and Sustainability
Unit 2: Global Development	**Chapter 1:** Environmental Interrelationships **Chapter 2:** Environmental Ethics **Chapter 3:** Environmental Risk: Economics, Assessment, and Management **Chapter 7:** Populations: Characteristics and Issues **Chapter 8:** Energy and Civilization **Chapter 9:** Energy Sources **Chapter 19:** Environmental Policy and Decision Making	**Chapter 5:** Biomes: Global Patterns of Life **Chapter 7:** Human Populations **Chapter 8:** Environmental Health and Toxicology **Chapter 9:** Food and Hunger **Chapter 12:** Biodiversity: Preserving Landscapes **Chapter 17:** Water Use and Management **Chapter 20:** Sustainable Energy **Chapter 22:** Urbanization and Sustainable Cites **Chapter 23:** Ecological Economics	**Chapter 4:** Human Populations **Chapter 6:** Environmental Conservation: Forests, Grasslands, Parks, and Nature Preserves **Chapter 9:** Air: Climate and Pollution **Chapter 10:** Water: Resources and Pollution **Chapter 14:** Economics and Urbanization **Chapter 15:** Environmental Policy and Sustainability
Unit 3: Feeding Humanity	**Chapter 1:** Environmental Interrelationships **Chapter 4:** Interrelated Scientific Principles: Matter, Energy, and Environment **Chapter 7:** Populations: Characteristics and Issues **Chapter 12:** Land-Use Planning **Chapter 14:** Agricultural Methods and Pest Management **Chapter 13:** Soil and Its Uses	**Chapter 3:** Matter, Energy, and Life **Chapter 9:** Food and Hunger **Chapter 7:** Human Populations **Chapter 10:** Farming: Conventional and Sustainable Practices **Chapter 17:** Water Use and Management **Chapter 23:** Ecological Economics	**Chapter 2:** Environmental Systems: Connections, Cycles, Flows, and Feedback Loops **Chapter 4:** Human Populations **Chapter 5:** Biomes and Biodiversity **Chapter 7:** Food and Agriculture **Chapter 15:** Environmental Policy and Sustainability

Correlation Guide

Annual Editions: Environment 12/13	Environmental Science: A Study of Interrelationships, 13/e by Enger/Smith	Environmental Science: A Global Concern, 12/e, by Cunningham/Cunningham	Principles of Environmental Science, 6/e, by Cunningham/Cunningham
Unit 4: A Thirsty Planet	**Chapter 1:** Environmental Interrelationships **Chapter 4:** Interrelated Scientific Principles: Matter, Energy, and Environment **Chapter 8:** Energy and Civilization **Chapter 11:** Biodiversity Issues **Chapter 12:** Land-Use Planning **Chapter 15:** Water Management **Chapter 19:** Environmental Policy and Decision Making	**Chapter 1:** Understanding Our Environment **Chapter 5:** Biomes: Global Patterns of Life **Chapter 7:** Human Populations **Chapter 10:** Farming: Conventional and Sustainable Practices **Chapter 12:** Biodiversity: Preserving Landscapes **Chapter 17:** Water Use and Management **Chapter 18:** Water Pollution **Chapter 21:** Solid, Toxic, and Hazardous Waste	**Chapter 2:** Environmental Systems: Connections, Cycles, Flows, and Feedback Loops **Chapter 4:** Human Populations **Chapter 10:** Water: Resources and Pollution **Chapter 13:** Solid and Hazardous Waste **Chapter 14:** Economics and Urbanization
Unit 5: Changing Climate	**Chapter 2:** Environmental Ethics **Chapter 3:** Environmental Risk: Economics, Assessment, and Management **Chapter 9:** Energy Sources **Chapter 16:** Air Quality Issues **Chapter 10:** Nuclear Energy	**Chapter 1:** Understanding Our Environment **Chapter 3:** Matter, Energy, and Life **Chapter 5:** Biomes: Global Patterns of Life **Chapter 15:** Air, Weather, and Climate **Chapter 16:** Air Pollution **Chapter 20:** Sustainable Energy	**Chapter 1:** Understanding Our Environment **Chapter 8:** Environmental Health and Toxicology **Chapter 9:** Air: Climate and Pollution **Chapter 15:** Environmental Policy and Sustainability
Unit 6: Endangered Diversity	**Chapter 3:** Environmental Risk: Economics, Assessment, and Management **Chapter 5:** Interactions: Environments and Organisms **Chapter 6:** Kinds of Ecosystems and Communities **Chapter 11:** Biodiversity Issues	**Chapter 4:** Evolution, Biological Communities, and Species Interactions **Chapter 5:** Biomes: Global Patterns of Life **Chapter 6:** Population Biology **Chapter 11:** Biodiversity: Preserving Species **Chapter 12:** Biodiversity: Preserving Landscapes **Chapter 13:** Restoration Ecology **Chapter 23:** Ecological Economics	**Chapter 1:** Understanding Our Environment **Chapter 5:** Biomes and Biodiversity **Chapter 15:** Environmental Policy and Sustainability **Chapter 6:** Environmental Conservation: Forests, Grasslands, Parks, and Nature Preserves
Unit 7: Degrading Ecosystems	**Chapter 1:** Environmental Interrelationships **Chapter 2:** Environmental Ethics **Chapter 5:** Interactions: Environments and Organisms **Chapter 11:** Biodiversity Issues **Chapter 16:** Air Quality Issues	**Chapter 2:** Principles of Science and Systems **Chapter 3:** Matter, Energy, and Life **Chapter 5:** Biomes: Global Patterns of Life **Chapter 8:** Environmental Health and Toxicology **Chapter 12:** Biodiversity: Preserving Landscapes **Chapter 13:** Restoration Ecology **Chapter 17:** Water Use and Management **Chapter 21:** Solid, Toxic, and Hazardous Waste	**Chapter 1:** Understanding Our Environment **Chapter 2:** Environmental Systems: Connections, Cycles, Flows, and Feedback Loops **Chapter 5:** Biomes and Biodiversity **Chapter 6:** Environmental Conservation: Forests, Grasslands, Parks, and Nature Preserves **Chapter 15:** Environmental Policy and Sustainability

Correlation Guide

Annual Editions: Environment 12/13	Environmental Science: A Study of Interrelationships, 13/e by Enger/Smith	Environmental Science: A Global Concern, 12/e, by Cunningham/Cunningham	Principles of Environmental Science, 6/e, by Cunningham/Cunningham
Unit 8: Quest for Power	**Chapter 8:** Energy and Civilization **Chapter 9:** Energy Sources **Chapter 10:** Nuclear Energy **Chapter 16:** Air Quality Issues **Chapter 19:** Environmental Policy and Decision Making	**Chapter 3:** Matter, Energy, and Life **Chapter 19:** Conventional Energy **Chapter 20:** Sustainable Energy **Chapter 23:** Ecological Economics **Chapter 24:** Environmental Policy, Law, and Planning	**Chapter 1:** Understanding Our Environment **Chapter 7:** Food and Agriculture **Chapter 12:** Energy **Chapter 15:** Environmental Policy and Sustainability
Unit 9: A Global Environmental Ethic?	**Chapter 2:** Environmental Ethics **Chapter 8:** Energy and Civilization **Chapter 11:** Biodiversity Issues **Chapter 19:** Environmental Policy and Decision Making	**Chapter 1:** Understanding Our Environment **Chapter 11:** Biodiversity: Preserving Species **Chapter 23:** Ecological Economics **Chapter 24:** Environmental Policy, Law, and Planning	**Chapter 1:** Understanding Our Environment **Chapter 4:** Human Populations **Chapter 7:** Food and Agriculture **Chapter 8:** Environmental Health and Toxicology **Chapter 14:** Economics and Urbanization **Chapter 15:** Environmental Policy and Sustainability
Unit 10: Consuming the Earth	**Chapter 2:** Environmental Ethics **Chapter 3:** Environmental Risk: Economics, Assessment, and Management **Chapter 8:** Energy and Civilization **Chapter 11:** Biodiversity Issues **Chapter 19:** Environmental Policy and Decision Making	**Chapter 8:** Environmental Health and Toxicology **Chapter 9:** Food and Hunger **Chapter 12:** Biodiversity: Preserving Landscapes **Chapter 16:** Air Pollution **Chapter 17:** Water Use and Management **Chapter 25:** What Then Shall We Do?	**Chapter 1:** Understanding Our Environment **Chapter 10:** Water: Resources and Pollution **Chapter 12:** Energy **Chapter 14:** Economics and Urbanization **Chapter 15:** Environmental Policy and Sustainability

Topic Guide

This topic guide suggests how the selections in this book relate to the subjects covered in your course. You may want to use the topics listed on these pages to search the Web more easily.

On the following pages a number of websites have been gathered specifically for this book. They are arranged to reflect the units of this Annual Editions reader. You can link to these sites by going to www.mhhe.com/cls.

All the articles that relate to each topic are listed below the bold-faced term.

Agriculture
(*small-scale, large scale, science, re-localization*)
7. Promises and Poverty
9. Radically Rethinking Agriculture for the 21st Century
10. The Politics of Hunger: How Illusion and Greed Fan the Food Crisis
11. Across the Globe, Empty Bellies Bring Rising Anger
12. How to Feed 8 Billion People
13. Where Oil and Water Do Mix: Environmental Scarcity and Future Conflict in the Middle East and North Arica
14. The Big Melt
15. The World's Water Challenge
17. The Last Straw
19. Global Warming Battlefields: How Climate Change Threatens Security
24. Ecosystems and Human Well-Being
26. Ecosystems Services: How People Benefit from Nature
38. How Much Should a Person Consume?

Climate change
(*disturbances, induced conflict, global warming*)
1. Population, Human Resources, Health, and the Environment: Getting the Balance Right
9. Radically Rethinking Agriculture for the 21st Century
12. How to Feed 8 Billion People
14. The Big Melt
15. The World's Water Challenge
16. Climate Change
17. The Last Straw
18. How to Stop Climate Change: The Easy Way
19. Global Warming Battlefields: How Climate Change Threatens Security
24. Ecosystems and Human Well-Being
26. Ecosystems Services: How People Benefit from Nature
27. Global Energy: The Latest Infatuations

Consumption
(*consumer society: gas consumption, trends, consumption gap, carbon emissions*)
1. Population, Human Resources, Health, and the Environment: Getting the Balance Right
3. Consumer Trends in Three Different "Worlds"
9. Radically Rethinking Agriculture for the 21st Century
18. How to Stop Climate Change: The Easy Way
24. Ecosystems and Human Well-Being
27. Global Energy: The Latest Infatuations
31. Gas Costs Squeeze Daily Life: Survey Reveals How High Prices Have Pushed Us into New Routines
32. Do Global Attitudes and Behaviors Support Sustainable Development?
33. The Ethics of Respect for Nature
36. Consumption, Not Population Is Our Main Environmental Threat
37. Consumption and Consumerism
38. How Much Should a Person Consume?
39. Reversal of Fortune

Degradation
(*ecosystems, environmental, variable consequences, food production systems*)
1. Population, Human Resources, Health, and the Environment: Getting the Balance Right

4. The New Population Bomb: The Four Megatrends That Will Change the World
7. Promises and Poverty
9. Radically Rethinking Agriculture for the 21st Century
12. How to Feed 8 Billion People
14. The Big Melt
17. The Last Straw
19. Global Warming Battlefields: How Climate Change Threatens Security
20. Executive Summary from Secretariat of the Convention on Biological Diversity
21. When Diversity Vanishes
24. Ecosystems and Human Well-Being
26. Ecosystems Services: How People Benefit from Nature
33. The Ethics of Respect for Nature
39. Reversal of Fortune

Demographics
(*trends, transition*)
2. Booms, Busts, and Echoes
3. Consumer Trends in Three Different "Worlds"
4. The New Population Bomb: The Four Megatrends That Will Change the World
6. The World Is Spiky
12. How to Feed 8 Billion People
32. Do Global Attitudes and Behaviors Support Sustainable Development?
36. Consumption, Not Population Is Our Main Environmental Threat
38. How Much Should a Person Consume?

Development
(*global trends, patterns, eco-friendly, innovations, failed sates, flat world, spiky world, millennium development goals, happiness*)
1. Population, Human Resources, Health, and the Environment: Getting the Balance Right
3. Consumer Trends in Three Different "Worlds"
4. The New Population Bomb: The Four Megatrends That Will Change the World
5. It's a Flat World, After All
6. The World Is Spiky
7. Promises and Poverty
15. The World's Water Challenge
17. The Last Straw
21. When Diversity Vanishes
30. It's Still the One
37. Consumption and Consumerism

Diversity
(*biological, cultural maintenance, reduction, collapse, protection, threats*)
4. The New Population Bomb: The Four Megatrends That Will Change the World
20. Executive Summary from Secretariat of the Convention on Biological Diversity
21. When Diversity Vanishes
22. When Good Lizards Go Bad: Komodo Dragons Take Violent Turn
23. Cry of the Wild
33. The Ethics of Respect for Nature

Internet References

The following Internet sites have been selected to support the articles found in this reader. These sites were available at the time of publication. However, because websites often change their structure and content, the information listed may no longer be available. We invite you to visit www.mhhe.com/cls for easy access to these sites.

Annual Editions: Environment 12/13

General Sources

About: Geography
http://geography.about.com

This website, created by the About network, contains hyperlinks to many specific areas of geography, including cartography, population, country facts, historic maps, physical geography, topographic maps, and many others.

The Association of American Geographers (AAG)
www.aag.org

Surf this site of the Association of American Geographers to learn about AAG projects and publications, careers in geography, and information about related organizations.

Britannica's Internet Guide
www.britannica.com

This site presents extensive links to material on world geography and culture, encompassing material on wildlife, human lifestyles, and the environment.

CO$_2$ Calculator
www.conservation.org/CarbonCalculator

This site is a fun link to illustrate where we use energy in our daily lives and how we can reduce our CO$_2$ footprint. Graphical interface.

Debunking Climate Skeptics
http://scholarsandrogues.wordpress.com/2007/07/23/anti-global-heating-claims-a-reasonably-thorough-debunking

Anti-global heating claims—a reasonably thorough debunking.

Earth Science Enterprise
www.earth.nasa.gov

Information about NASA's Mission to Planet Earth program, its Science of the Earth program, and its Science of the Earth System can be found here. Surf to learn about satellites, El Niño, and even "strategic visions" of interest to environmentalists.

EnviroLink
www.envirolink.org

One of the world's largest environmental information clearinghouses, EnviroLink is a grassroots nonprofit organization that unites organization volunteers around the world and provides up-to-date information and resources.

EPA
www.epa.gov/wastes/conserve/index.htm

EPA's website concerning recycling.

Global Climate Change, NASA's Eyes on Earth
www.climate.nasa.gov

A remarkably informative and graphical discussion about climate change.

IPPC (Internation Panel on Climate Change)
www.ipcc-wg2.gov

Link to the Nobel Prize–winning report on climate change. An incredible resource. Very detailed, but with a short informative Summary for Policymakers.

Library of Congress
www.loc.gov

Examine this extensive website to learn about resource tools, library services/resources, exhibitions, and databases in many different subfields of environmental studies.

National Geographic Society
www.nationalgeographic.com

Links to National Geographic's huge archive are provided here. There is a great deal of material related to the atmosphere, the oceans, and other environmental topics.

The New York Times
www.nytimes.com

Browsing through the archives of The New York Times will provide a wide array of articles and information related to the different subfields of the environment.

RealClimate
www.realclimate.org

Climate Science from Climate Scientists. Now that makes sense.

Research and Reference (Library of Congress)
http://lcweb.loc.gov/rr

This research and reference of the Library of Congress will lead to invaluable information on different countries. It provides links to numerous publications, bibliographies, and guides in area studies that can be of great help to environmentalists.

SocioSite: Sociological Subject Areas
www.pscw.uva.nl/sociosite/TOPICS

This huge sociological site from the University of Amsterdam provides many discussion and references of interest to students of the environment, such as the links to information on ecology and consumerism.

Solstice: Documents and Databases
http://solstice.crest.org/index.html

In this online source for sustainable energy information, the Center for Renewable Energy and sustainable Technology (CREST) offers documents and databases on renewable energy, energy efficiency, and sustainable living. The site also offers related websites, case studies, and policy issues.

U.S. Geological Survey
www.usgs.gov

This site and its many links are replete with information and resources in environmental studies, from explanations of El Niño to discussion of concerns about water resources.

U.S. Global Change Research Program
www.usgcrp.gov

This government program supports research on the interaction of natural and human-induced changes on the global environment and their implication for study. Find details on the atmosphere, climate change, global carbon and water cycles, ecosystems, and land use plus human contributions and responses.

Internet References

Union of Concerned Scientists

www.ucsusa.org

Citizens and scientists for practical solutions to environmental problems.

Wikipedia

www.wikipedia.org

A free encyclopedia with millions of articles contributed collaboratively using Wiki software. Remarkable resource created by users the world over.

Unit 1: Human Population Growth

Can Cities Save the Future?

www.huduser.org/publications/econdev/habitat/prep2.html

This press release about the second session of the Preparatory Committee for Habitat II is an excellent discussion of the question of global urbanization.

Linkages on Environmental Issues and Development

www.iisd.ca/linkages

Linkages is a site provided by the International Institute for Sustainable Development. It is designed to be an electronic clearinghouse for information on past and upcoming international meetings related to both environmental issues and economic development in the developing world.

People & Planet

www.peopleandplanet.org

People & Planet is an organization of student groups at universities and colleges across the United Kingdom. Organized in 1969 by students at Oxford University, it is now an independent pressure group campaigning on world poverty, human rights, and the environment.

Population Action International

www.populationaction.org

According to its mission statement, Population Action International is dedicated to advancing policies and programs that slow population growth to enhance the quality of life for all people.

Population Reference Bureau

www.prb.org

This site provides data on the world population and census information.

SocioSite: Sociological Subject Areas

www.pscw.uva.nl/sociosite/TOPICS

This huge site provides many references of interest to those interested in global issues, such as links to information on ecology and the impact of consumerism.

World Health Organization (WHO)

www.who.ch

The WHO's objective, according to its website, is the attainment by all peoples of the highest possible level of health. Health, as defined in the WHO constitution, is a state of complete physical, mental, and social well-being and not merely the absence of disease or infirmity.

World Resources Institute

www.wri.org

The World Resources Institute provides information and practical proposals for policy and institutional change that will foster environmentally sound, socially equitable development.

WWW Virtual Library: Demography & Population Studies

http://demography.anu.edu.au/VirtualLibrary

A definitive guide to demography and population studies can be found at this site. It contains a multitude of important links to information about global poverty and hunger.

Unit 2: Global Development

Earth Renewal

www.earthrenewal.org/global_economics.htm

Listing of various excellent sites that address economic environmental policies and solutions.

Geography and Socioeconomic Development

www.ksg.harvard.edu/cid/andes/Documents/Background%20Papers/Geography&Socioeconomic%20Development.pdf

John I. Gallup wrote this 19-page background paper examining the state of the Andean region. He explains the strong and pervasive effects geography has on economic and social development.

Global Footprint Network

footprints@footprintnetwork.org

This website has a link to Africa Factbook 2009, an electronic set of some of the most up-to-date charts and recent statistics about human and environmental trends in Africa. The Factbook is compiled by The Global Footprint Network, the Swiss Agency for Development, and other sponsors.

Global Trends Project

www.globaltrendsproject.org

The Center for Strategic and International Studies explores the coming global trends and challenges of the new millennium. Read their summary report at his website. Also access Enterprises for the Environment, Global Information Infrastructure Commission, and Americas at this site.

Graphs Comparing Countries

http://humandevelopment.bu.edu/use_existing_index/start_comp_graph.cfm

This site allows you to compare the statistics of various countries and nation-states using a visual tool.

IISDnet

www.iisd.org

The International Institute for Sustainable Development, a Canadian organization, presents information through gateways entitled Business, Climate Change, Measurement and Assessment, and Natural Resources. IISD Linkages is its multimedia resource for environment and development policymakers.

United Nations

www.unsystem.org

Visit this official website locator for the United Nations System of Organizations to get a sense of the scope of international environmental inquiry today. Various UN organizations concern themselves with everything from maritime law to habitat protection to agriculture.

United Nations Environment Programme (UNEP)

www.unep.ch

Consult this home page of UNEP for links to critical topics of concern to environmentalists, including desertification, migratory species, and the impact of trade on the environment. The site will direct you to useful databases and global resource information.

Internet References

World Resources Institute (WRI)
www.wri.org

The World Resources Institute is committed to change for a sustainable world and believes that change in human behavior is urgently needed to halt the accelerating rate of environmental deterioration in some areas. It sponsors not only the general website already listed but also The Environmental Information Portal (www.earthtrends.wri.org) that provides a rich database on interaction between human disease, pollution, and large-scale environmental, development, and demographic issues.

Unit 3: Feeding Humanity

The Hunger Project
www.thp.org

Browse through this nonprofit organization's site, whose goal is the sustainable end to global hunger through leadership at all levels of society. The Hunger Project contends that the persistence of hunger is at the heart of the major security issues threatening our planet.

National Geographic Society
www.nationalgeographic.com

This site provides links to National Geographic's huge archive of maps, articles, and other documents. There is a great deal of material related to social and cultural topics that will be of great value to those interested in the study of cultural pluralism.

Penn Library: Resources by Subject
www.library.upenn.edu/cgi-bin/res/sr.cgi

This vast site is rich in links to information about subjects of interest to students of global issues. Its extensive population and demography resources address such concerns as migration, family planning, and health and nutrition in various world regions.

Sustainable Development.Org
www.sustainabledevelopment.org

Extensive links at this site lead to such sustainable development categories as agriculture, energy, environment, finance, health, micro enterprise, public policy, and technologies.

Unit 4: A Thirsty Planet

Freshwater Society
www.freshwater.org

The mission of the Freshwater Society is to promote the conservation, protection and restoration of all freshwater resources.

EnviroLink
www.envirolink.org

One of the world's largest environmental information clearinghouses, EnviroLink is a grassroots nonprofit organization that unites organization volunteers around the world and provides up-to-date information and resources. There are many links to freshwater issues found here.

National Geographic Society
www.nationalgeographic.com

National Geographic is developing a strategic approach to freshwater issues with the goal of motivating people across the world to care about and conserve freshwater. Links to National Geographic's huge archive are provided here. There is a great deal of material related to freshwater.

The World's Water
www.worldwater.org

For more than a decade, the biennial report The World's Water has provided key data and expert insights into our most pressing freshwater issues.

United Nations Environment Program (UNEP)
www.enep.ch

Consult this home page of UNEP for links to critical topics of concern related to fresh water.

Unit 5: Changing Climate

The Climate Project
www.theclimateprojectus.org

The Climate Project United States is the American branch of Nobel Laureate and former vice president Al Gore's climate change leadership program founded in June 2006.

Global Climate Change
www.puc.state.oh.us/consumer/gcc/index.html

The goal of this PUCO (Public Utilities Commission of Ohio) site is to serve as a clearinghouse of information related to global climate change. Its extensive links provide an explanation of the science and chronology of global climate change, acronyms, definitions, and more.

Global Climate Change Facts: The Truth, The Consensus, and the Skeptics
www.climatechangefacts.info

NASA Global Change site
www.climate.nasa.gov

Displays vital signs of the planet and other valuable information.

National Oceanic and Atmospheric Administration (NOAA)
www.noaa.gov

Through this home page of NOAA, part of the U.S. Department of Commerce, you can find information about coastal issues, fisheries, climate, and more. The site provides many links to research and to other Web resources.

National Operational Hydrologic Remote Sensing Center (NOHRSC)
www.nohrsc.nws.gov

Flood images are available at this site of the NOHRSC, which works with the U.S. National Weather Service to track weather-related information.

National Security and the Threat of Climate Change
www.npr.org/documents/2007/apr/security_climate.pdf

Report issued by eleven retired senior military officials in 2007 to the Center for Naval Analysis warning that climate change will affect all aspects of the United States' defense readiness.

RealClimate
www.realclimate.org

A blog for climate science with information from climate scientists. This is a one-stop link for resources that people can use to get up to speed on the issue of climate change.

Unit 6: Endangered Diversity

Endangered Species
www.endangeredspecie.com

This site provides a wealth of information on endangered species anywhere in the world. Links providing data on the causes,

Internet References

interesting facts, law issues, case studies, and other issues on endangered species are available.

Friends of the Earth
www.foe.co.uk/index.html

Friends of the Earth, a nonprofit organization based in the United Kingdom, pursues a number of campaigns to protect the Earth and its living creatures. This site has links to many important environmental sites, covering such broad topics as ozone depletion, soil erosion, and biodiversity.

Greenpeace
www.greenpeace.org

Greenpeace is an international NGO (nongovernmental organization) that is devoted to environmental protection.

Natural Resources Defense Council
http://nrdc.org

The Natural Resources Defense Council (NRDC) uses law, science, and the support of more than 1 million members and activists to protect the planet's wildlife, plants, water, soils, and other resources. The site provides abundant information on global issues and political responses.

Smithsonian Institution Website
www.si.edu

Looking through this site, which will provide access to many of the enormous resources of the Smithsonian, offers a sense of the biological diversity that is threatened by humans' unsound environmental policies and practices.

World Wildlife Federation (WWF)
www.wwf.org

This home page of the WWF leads to an extensive array of information links about endangered species, wildlife management and preservation, and more. It provides many suggestions for how to take an active part in protecting the biosphere.

Unit 7: Degrading Ecosystems

National Center for Electronics Recycling
www.electronicsrecycling.org/Public/default.aspx

This site is dedicated to the development and enhancement of a national infrastructure for the recycling of used electronics in the United States.

Terrestrial Sciences
www.cgd.ucar.edu/tss

The Terrestrial Sciences Section (TSS) is part of the Climate and Global Dynamics (CGD) Division at the National Center for Atmospheric Research (NCAR) in Boulder, Colorado. Scientists in the section study land–atmosphere interactions, in particular surface forcing of the atmosphere, through model development, application, and observational analyses. Here, you'll find a link to VEMAP, The Vegetation/Ecosystem Modeling and Analysis Project.

World Health Organization
www.who.int

This home page of the World Health Organization will provide you with links to a wealth of statistical and analytical information about health and the environment in the developing world.

The World Watch Institute
www.worldwatch.org

The Worldwatch Institute advocates environmental protection and sustainable development.

Unit 8: Quest for Power

Alliance for Global Sustainability (AGS)
http://globalsustainability.org

The AGS is a cooperative venture seeking solutions, to today's urgent and complex environmental problems. Research teams from four universities study large-scale, multidisciplinary environmental problems that are faced by the world's ecosystems, economies, and societies.

Alternative Energy Institute (AEI)
www.altenergy.org

The AEI will continue to monitor the transition from today's energy forms to the future in a "surprising journey of twists and turns." This site is the beginning of an incredible journey. On this site created by a nonprofit organization, discover how the use of conventional fuels affects the environment. Also learn about research work on new forms of energy.

Arctic Climate Impact Assessment
www.acia.uaf.edu

An international project of the Arctic Council and the International Arctic Science Committee (IASC), to evaluate and synthesize knowledge on climate variability, climate change, and increased ultraviolet radiation and their consequences.

Department of Energy—Energy Efficiency and Renewable Energy
www.eere.energy.gov

The U.S. government's official site regarding energy, its use, and government policy.

Energy and the Environment: Resources for a Networked World
http://zebu.uregon.edu/energy.html

An extensive array of materials having to do with energy sources—both renewable and nonrenewable—as well as other topics of interest to students of the environment is found on this site.

Fuel Economy (Department of Energy)
www.fueleconomy.gov

Gas mileage (mpg), greenhouse gas emissions, air pollution ratings, and safety information for new and used cars and trucks. Also discusses alternative fuel vehicles.

Global Climate Change
http://climate.jpl.nasa.gov

JPL's graphic-rich site discussing all aspects of climate change. A must see.

Institute for Global Communication/EcoNet
www.igc.org

This environmentally friendly site provides links to dozens of governmental, organizational, and commercial sites having to do with energy sources. Resources address energy efficiency, renewable generating sources, global warming, and more.

Nuclear Power Introduction
http://library.thinkquest.org/17658/pdfs/nucintro.pdf

Information regarding alternative energy forms can be accessed here. There is a brief introduction to nuclear power and a link to maps that show where nuclear power plants exist.

Internet References

U.S. Department of Energy
www.energy.gov

Scrolling through the links provided by this Department of Energy home page will lead to information about fossil fuels and a variety of sustainable/renewable energy sources.

Unit 9: A Global Environmental Ethic?

The Earth Institute at Columbia University
www.earth.columbia.edu

The Earth Institute at Columbia University, led by Professor Jeffery D. Sachs, is dedicated to addressing a number of complex issues related to sustainable development and the needs of the world's poor.

Earth Pledge Foundation
www.earthpledge.org

The Earth Pledge Foundation promotes the principles and practices of sustainable development—the need to balance the desire for economic growth with the necessity of environmental protection.

Energy Justice Network
www.energyjustice.net/peak

Advocates for "a clean energy, zero-emission, zero-waste future for all."

EnviroLink
http://envirolink.org

EnviroLink is committed to promoting a sustainable society by connecting individuals and organizations through the use of the Internet.

Grass-roots.org
www.grass-roots.org

This site describes innovative grassroots programs in the United States that have helped people better their communities.

Poverty Mapping
www.povertymap.net

Poverty maps can quickly provide information on the spatial distribution of poverty. Here you will find maps, graphics, data, publications, news, and links that provide the public with poverty mapping from the global to the subnational level.

Resources
www.etown.edu/vl

Surf this site and its extensive links to learn about specific countries and regions, to research various think tanks and international organizations, and to study such vital topics as international law, development, the international economy, human rights, and peacekeeping.

Unit 10: Consuming the Earth

Alliance for Global Sustainability (AGS)
www.global-sustainability.org

The AGS is a cooperative venture seeking solutions to today's urgent and complex environmental problems. Research teams from four universities study large-scale, multidisciplinary environmental problems that are faced by the world's ecosystems, economies, and societies.

The Dismal Scientist
www.dismal.com

Often referred to as the "best free lunch on the Web," this is an excellent site with many interactive features. It provides access to economic data, briefings on the current state of the economy, and original articles on economic issues.

Global Footprint Network
http://footprints@footprintnetwork.org

At Global Footprint Network programs are designed to influence decision makers at all levels of society and to create a critical mass of powerful institutions using the Footprint to put an end to ecological overshoot and get economies back into balance.

U.S. Information Agency (USIA)
www.america.gov

USIA's home page provides definitions, related documentation, and discussions of topics of concern to students of global issues. The site addresses today's hot topics as well as ongoing issues that form the foundation of the field.

World Population and Demographic Data
http://geography.about.com/cs/worldpopulation

On this site you will find information about world population and addition a demographic data for all the countries of the world.

World Map

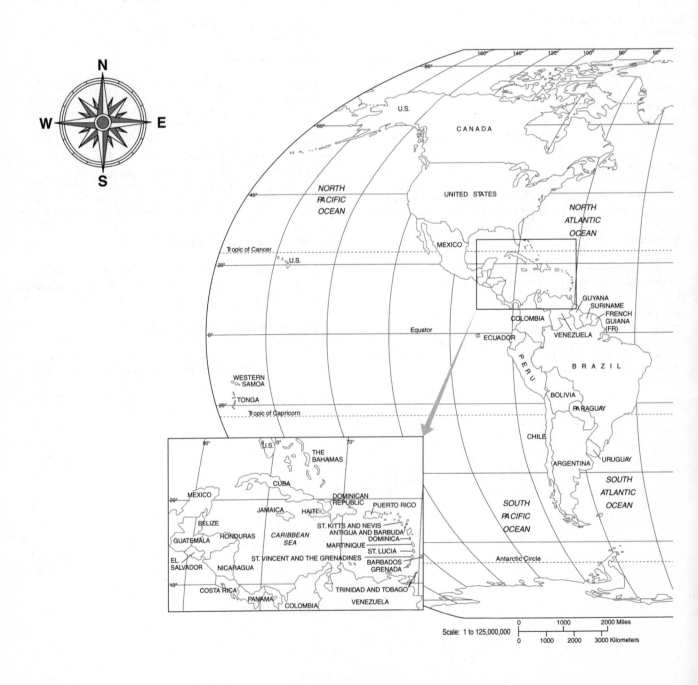

N
W E
S

160° 140° 120° 100° 80° 60°
80°
U.S.
60°
CANADA

NORTH
PACIFIC
OCEAN

UNITED STATES

NORTH
ATLANTIC
OCEAN

Tropic of Cancer
20°
U.S.
MEXICO

GUYANA
SURINAME
FRENCH
GUIANA
(FR)
COLOMBIA
VENEZUELA

Equator
0°
ECUADOR

P E R U

B R A Z I L

WESTERN
SAMOA

BOLIVIA

PARAGUAY

TONGA
20°
Tropic of Capricorn

CHILE

ARGENTINA URUGUAY

SOUTH
ATLANTIC
OCEAN

SOUTH
PACIFIC
OCEAN

Antarctic Circle

90° U.S. 0° 70°
THE
BAHAMAS
CUBA
MEXICO
DOMINICAN
REPUBLIC PUERTO RICO
20°
JAMAICA HAITI
ST. KITTS AND NEVIS
BELIZE ANTIGUA AND BARBUDA
DOMINICA
GUATEMALA HONDURAS CARIBBEAN MARTINIQUE
SEA ST. LUCIA
EL ST. VINCENT AND THE GRENADINES BARBADOS
SALVADOR NICARAGUA GRENADA
10°
COSTA RICA TRINIDAD AND TOBAGO
PANAMA
COLOMBIA VENEZUELA

Scale: 1 to 125,000,000

| 0 | 1000 | 2000 Miles |
| 0 | 1000 | 2000 | 3000 Kilometers |

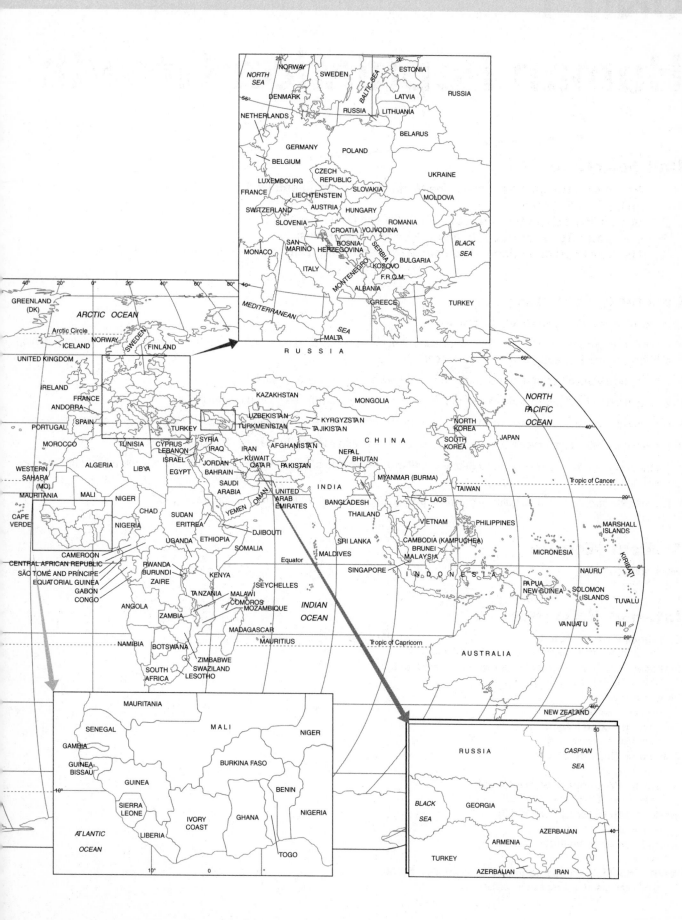

UNIT 1

Human Population Growth

Unit Selections

1. **Population, Human Resources, Health, and the Environment: Getting the Balance Right,** Anthony J. McMichael
2. **Booms, Busts, and Echoes,** David E. Bloom and David Canning
3. **Consumer Trends in Three Different "Worlds,"** Andy Hines
4. **The New Population Bomb: The Four Megatrends That Will Change the World,** Jack A. Goldstone

Learning Outcomes

After reading this Unit, you should be able to:

- Discuss linkages between the environmental issue of human population growth and global development patterns, demographic changes, consumer trends, and the environment.

- Discuss how human health issues are affected by the health of natural environments.

- Summarize the relationship trends in human capital, population growth, health, and the environment.

- Offer insights on what the effects of changing population demographics might have on global development and natural resources.

- Identify underlying connections between consumption trends and possible future environmental issues.

- Elaborate on the four demographic megatrends that are predicted to change the world and potential environmental impacts they may imply.

- Discuss the implications of population growth in poor countries, urbanization, and the consequences.

Student Website

www.mhhe.com/cls

Internet References

Can Cities Save the Future?
www.huduser.org/publications/econdev/habitat/prep2.html

Linkages on Environmental Issues and Development
www.iisd.ca/linkages

People & Planet
www.peopleandplanet.org

Population Action International
www.populationaction.org

Population Reference Bureau
www.prb.org

SocioSite: Sociological Subject Areas
www.pscw.uva.nl/sociosite/TOPICS

World Health Organization (WHO)
www.who.ch

World Resources Institute
www.wri.org

WWW Virtual Library: Demography & Population Studies
http://demography.anu.edu.au/VirtualLibrary

The command "Be fruitful and multiply" was promulgated, according to our authorities, when the population of the world consisted of two people.

—William Ralph Inge,
More Lay Thoughts of a Dean, 1931

There can be no true discourse on today's leading environmental issues without including the primary driving agent of each and every one of our environmental problems—*Top Ten Environmental Issue 1:* Human Population Growth. If Planet Earth were devoid of humans, would Mother Nature be fighting for her life? But it's not simply our presence here that threatens the planet. It's our population growth *patterns* and the *demographic characteristics* of those patterns that are creating the potential threats to the future of our environment, and our own survival as a species. Many scholars have suggested that the earth can support a human population of 9, 10, even 11 billion people. However, the earth's ability to support "quantity" of life is not the issue. It's quality of life. How do we want our billions to live? To begin to answer this, we first need to recognize, and admit, that humans must consume the Earth—transform raw resources into a form we can use to ensure our well-being as a species, or ensure our survival. Second, we need to realize that our current human consumption patterns are proving to be unsustainable and damaging to the ecosystems on which all living things must depend. Third, and most importantly, we need to better understand what possible future human population growth patterns—geographic, demographic, social, and economic—may unfold and how those patterns may affect our ability to ensure *quality of life* for all humans and maintain a *sustainable relationship* with our planet. Science is about being able to make accurate predictions. If we can predict outcomes better, we can plan better, and if we can plan better, we can often change the inevitability of unpleasant surprises and even fatal mistakes. Unit 1 aims to stimulate our thinking about the characteristics, patterns, and subsequent environmental issues to consider regarding the future of human population growth. Such critical thinking now may help us to develop better plans of action for addressing myriad potential future consequences of our population growth.

Article 1, "Population, Human Resources, Health and the Environment: Getting the Balance Right," provides an overview of the connection between the trends of our growing population, maintaining the health of our environment, and our own health patterns in the future. Both the 1987 Brundtland Report and the 2000 Millennium Development Goals emphasize the close connection between building a sustainable relationship with our environment and achieving our development goals for humanity. According to author McMichael, the challenge presented in Article 1—getting the balance right between sustainability, human health, and population growth—will require keeping a close eye on not only

© Flat Earth Images

where population is growing now, but also how the patterns of its earth consumption may unfold in the future.

Article 2, "Booms, Busts, and Echoes," examines what patterns in human population growth we may see in the coming decades. There is considerable agreement that much of the growth will take place in the developing world, and with it will come substantial demographic changes. Such changes are predicted to also impact the geopolitics and economies for these emerging nations. Can serious environmental changes be inferred from these predictions as well? The reader is encouraged to note the date of the article's publication and assess our current global demographic patterns today in 2012. And then, consider the accuracy of our predictions then, and what we may surmise regarding our predictions now for what the future may hold.

Changing patterns of population growth and demographics are predicted to create changing global consumer trends in three different geopolitical regions of the world, as well, according to Andy Hines, author of Article 3, "Consumer Trends in Three Different Worlds." Given the need for humans to consume the earth, what implications may these new resource consumption patterns have regarding future environmental impacts? Although the author provides insights into primarily the business implications of these possible trends, the reader is asked to compare the article's 2008 predictions and critically assess what changes have or have not unfolded, and what the future may hold.

Finally, Article 4's author, Jack Goldstone, offers some current predictions (2010) regarding what he believes to be the next "population bomb" and the global megatrends it will set in motion in his article, "The New Population Bomb: The Four Megatrends That Will Change the World." The trends of global population growth in the world's poorest countries, continued urbanization of our population, the changing demographics of the industrialized nations' populations, and the subsequent political economy consequences are generally well agreed on as the articles in this unit suggest. However, it is also imperative that we expand our thinking on the issues presented in these articles and assess the potential future scenarios in light of both place—local and regional—impacts within nations, and the potential environmental impacts (which will vary considerably by place) resulting from these changing human earth consumption patterns. In terms of human population growth we must constantly be cognizant of the question: Who is currently consuming the Earth and at what rate? And who will be consuming the Earth in the future, and at what rate? Sustainably or unsustainably?

Population, Human Resources, Health, and the Environment

Getting the Balance Right

ANTHONY J. McMICHAEL

The UN's World Commission on Environment and Development (WCED)—the "Brundtland Commission," chaired by Gro Harlem Brundtland—released its seminal report *Our Common Future* in 1987.[1] Much has changed on the global environment front since then, only some of which was (or could have been) anticipated by that report. As human population continues to grow and as human societies, cultures, and economies become more interconnected against the background crescendo of "globalization" in recent decades, the collective human impact on the biosphere has increasingly assumed a global and systemic dimension. While issues like climate change, freshwater deficits, and degradation of food-producing systems and ocean fisheries were appearing on the horizon in 1987, they have now moved to the foreground. Today, it is evident that these momentous changes pose threats not only to economic systems, environmental assets, infrastructural integrity, tourism, and iconic nature, but also to the stability, health, and survival of human communities. This realization—along with the fact that human-induced global environmental changes impinge unequally on human groups—heightens the rationale for seeking sustainable development.

While the WCED report explored the rationale and the path toward sustainable development, the extent of subsequent large-scale environmental problems arising from the scale and the energy and materials intensity of prevailing modes of development could not have been fully anticipated in 1987. Indeed, paradoxically, concern over world population growth had temporarily receded in the mid-1980s, reflecting the prevailing mix of politics and optimism. The optimism derived from the apparent alleviation of hunger that had been achieved by the Green Revolution of the 1970s and 1980s in much of the developing world, and from the downturn in fertility rates in at least some developing regions. Today, however, the population issue is reemerging in public discussion, reflecting renewed recognition that population growth, along with rising consumption levels, is exacerbating climate change and other global environmental changes.[2]

If the commission's assessment were re-run this decade, its updated terms of reference would necessarily focus more attention on the social and health dimensions of the "development" process, both as inputs and, importantly, as outcomes. The charge to the commission, which focused on the often-conflicted relationship between economic activity and environmental sustainability, was framed at a time when the orthodox Rostovian view (that economic development occurs in five basic stages from "traditional society" to "age of high mass consumption") still remained influential.[3] Today, human capital and social capital—both of which were first properly understood and factored into the development calculus in the 1990s, along with the need for sound governance—are better recognized as prerequisites for environmentally sustainable development. At the same time, realization is growing that the attainment of positive human experience is the core objective of human societies.[4] In contrast, the commission's primary mandated focus was on how to reconcile environmental sustainability with social-economic development. That orientation afforded little stimulus to considering why, in human experiential terms, achieving such a balance is not an end in itself, but is a prerequisite for attaining human security, well-being, health, and survival. Why else do we seek sustainability?

People, Resources, Environment, and Development

The UN General Assembly Resolution A/38/161 of 1983 establishing the WCED specified that the commission would "take account of the interrelationships between people, resources, environment and development."[5] The full text of the resolution emphasized—as did the commission's name—the dual need for long-sighted environmental management strategies and greater cooperation among countries in seeking a sustainable development path to the common future. Two words in the quoted phrase are of particular interest: "people" and "resources."

Reference to "people," rather than to "populations," seems to emphasize the *human* dimension. However, it also distracts from issues of fertility and population size—a distraction that probably reflected two prevailing circumstances. In the 1980s—when world population growth was at its historic high—the United States's conservative Reagan Administration withheld international aid from family planning because of its perceived links with abortion counseling. This ill-informed and culturally high-handed approach, coming from a powerful country with great financial influence over UN policies, was complemented by the fact that many low-income countries considered that issues of fertility and population size were their own business. Nevertheless, and to its credit, the WCED report directly addressed the question of population size and its environmental consequences, urging lower fertility rates as a prerequisite for both poverty alleviation and environmental sustainability.

The word "resources" is ambiguous; it could be taken to refer to natural environmental resources or to human resources (human capital, including education and health status). To what extent did the WCED consider human well-being and health in relation to changing environmental conditions, population size, and resources? "Many such changes are accompanied by life-threatening hazards," stated the WCED in its overview of the report,[6] suggesting that the report would indeed explore how the state of the natural environment, our basic habitat, sets limits on human well-being, health, and survival, both now—and of particular relevance to sustainability—in future. Indeed, in launching the report in Oslo, on 20 March 1987, Chair Brundtland said:

> Our message is directed towards people, whose wellbeing is the ultimate goal of all environment and development policies. . . . If we do not succeed in putting our message of urgency through to today's parents and decision makers, we risk undermining our children's fundamental right to a healthy, life-enhancing environment.[7]

Despite these promising statements, the report itself gave only limited attention to considering how environmental degradation and ecological disruption affect the foundations of human population health. The report focused primarily on the prospects for achieving an "ecologically sustainable" form of social and economic development that conserves the natural environmental resource base for future human needs. It paid little attention to the fact that the conditions of the world's natural environment signify much more than assets for production, consumption, and economic development in general; the biosphere and its component ecosystems and biophysical processes provide the functions and flows that maintain life processes and therefore good health. Indeed, all extant forms of life have evolved via an exquisite dependency on environmental conditions.

This somewhat restricted vision on the part of the WCED is not surprising. Indeed, such a perspective has been reflected often in subsequent forays of UN agencies into the rationale and objectives of sustainable development—forays that have consistently overlooked or sometimes trivialized the role of sustainable development as a precondition to attaining well-being, health, and survival (see the box on for the example of the UN's Millennium Development Goals).[8] In defense of the report, however, it does state:

It is misleading and an injustice to the human condition to see people merely as consumers. Their well-being and security—old age security, declining child mortality, health care, and so on—are the goal of development.[9]

In the 1980s and early 1990s, there was little evidence and understanding of the relationship between environmental conditions, ecological systems, and human health. For example, the First Assessment Report of the Intergovernmental Panel on Climate Change (IPCC), released in 1991, contained only passing reference to how global climate change would affect human health.[10] The IPCC report reviewed in detail the risks to farms, forests, fisheries, feathered and furry animals, to settlements, coastal zones, and energy generation systems. In contrast, it glossed cursorily over the risks to human health (and gave undue emphasis to solar ultraviolet exposure and skin cancer, which is very marginal to the climate change and health topic).

There was, then, only a rudimentary awareness that the profile and scale of environmental hazards to human health were undergoing a profound transformation. For instance, the human health risks due to stratospheric ozone depletion, first recognized during the late 1970s and early 1980s, had been easily understood. They belonged to the familiar category of direct-acting hazardous environmental exposures. An increase in ambient levels of ultraviolet radiation at Earth's surface would increase the risks of skin damage and skin cancer and would affect eye health (for example, cataract formation). Recognition of this straightforward risk to human biology facilitated the ready international adoption of the Montreal Protocol in 1987, requiring national governments to eliminate release of ozone-destroying gases (mostly chlorofluorocarbons, nitrous oxide, and methyl bromide).

In contrast, the great diversity of (mostly) less direct-acting but potentially more profound risks to human health from changes to Earth's climate system, agroecosystems, ocean fisheries, freshwater flows, and general ecosystem functioning (such as pollination, nutrient cycling, and soil formation) were only dimly perceived in the 1980s. Those health risks received relatively little attention in the WCED report, which focused instead on health hazards related to inadequate water supply and sanitation, malnutrition, drug addiction, and exposure to carcinogens and other toxins in homes and the workplace.

An Incomplete Model of Health Determinants

In discussing population health, the WCED report took a largely utilitarian view, discussing good health as an input to economic development and, specifically, as stimulus to the reduction of fertility and poverty. In this respect it was in good company: both the pioneering sanitary revolution of nineteenth-century England and World Health Organization's International Commission on Macroeconomics and Health, established in 2000, espoused the same rationale: good health fosters national wealth. To the extent that the WCED report addressed the determinants of population health, it focused mainly on the contributions of

Millennium Development Goals: How Much Progress Has Been Made?

By coincidence, the 20-year anniversary of the Brundtland report nearly coincides with the halfway mark of another UN project, the Millennium Development Goals (MDGs), 2000–2015.[1] The MDGs were launched in 2000 against a backdrop of increasing attention on what was termed "ecologically sustainable development" in large part stimulated by the WCED report. They encompass eight goals (each with associated targets): to eradicate extreme poverty and hunger; achieve universal primary education; promote gender equality and empower women; reduce child mortality; improve maternal health; combat HIV/AIDS, malaria, and other diseases; ensure environmental sustainability; and develop a global partnership for development.

Achievement of the MDGs is becoming increasingly improbable as time passes. Some headway has been made in relation to poverty reduction and child school enrollment. But there has been little alleviation of hunger and malnutrition, maternal mortality, and infant-child death rates (which have declined by around one sixth in poorer countries, well short of the two-thirds reduction target).

Inevitably, progress toward the goals has varied between regions and countries. China, for example, has made social and health advances on many fronts, albeit at the cost of increasingly serious environmental degradation. In contrast, in sub-Saharan Africa, no country is coming close to halving poverty, providing universal primary education, or stemming the devastating HIV/AIDS epidemic. More than 40 percent of persons in sub-Saharan Africa live in extreme poverty.

One quarter of the world's children aged less than 5 are underfed and underweight. This, as a proportion, is an improvement on the figure of one third in 1990. However, in sub-Saharan Africa and South Asia, nearly half the children remain underweight, and gains are minimal.

The total number of people living with AIDS has increased by nearly 7-million since 2001, to a total now of 40 million. Neither malaria nor tuberculosis is being effectively curtailed, with the attempt to reduce tuberculosis being threatened further by the recent emergence of strains with more extreme forms of antimicrobial resistance.

Perhaps this lack of progress is in part reflected in the UN's failure to explore and emphasize the primary interconnected role of Goal 7 for the achievement of the MDGs overall. Goal 7 seeks "environmental sustainability"—and achieving this particular goal is the bedrock for attaining most of the targets of the other seven goals. Without an intact and productive natural environment and its life-supporting global and regional systems and processes (such as climatic conditions, ocean vitality, ecosystem functioning, and freshwater circulation), the prospects are diminished for food production, safe drinking water adequate household and community energy sources, stability of infectious disease agents, and protection from natural environmental disasters.

The subsequent treatment by the UN of Goal 7 in relation to its health implications has been rather superficial, and mostly in relation to familiar, localized, environmental health hazards. For example, the UN's 2007 report on the MDGs focuses particularly on how Goal 7 relates to child diarrhoeal diseases. It states:

> The health, economic and social repercussions of open defecation, poor hygiene and lack of safe drinking water are well documented. Together they contribute to about 88 per cent of the deaths due to diarrhoeal diseases—more than 1.5 million—in children under age five. Infestation of intestinal worms caused by open defecation affects hundreds of millions of predominantly school-aged children, resulting in reduced physical growth, weakened physical fitness and impaired cognitive functions. Poor nutrition contributes to these effects.[2]

More encouraging is the recent, wider-visioned approach taken by the UN Millennium Project, undertaken for the Commission on Sustainable Development.[3] This project's definition of "environmental sustainability" refers explicitly to the health impacts of environmental changes, and states as follows:

> Achieving environmental sustainability requires carefully balancing human development activities while maintaining a stable environment that predictably and regularly provides resources such as freshwater, food, clean air, wood, fisheries and productive soils and that protects people from floods, droughts, pest infestations and disease.[4]

Notes

1. UN Secretary General, Millennium Development Goals (New York. United Nations, 2000), www.un.org/millenniumgoals/goals.html (accessed 23 August 2007).

2. United Nations, *The Millennium Development Goals Report 2007* (New York: United Nations, 2007).

3. J. Sachs and J. McArthur, "The Millennium Project: A Plan for Meeting the Millennium Development Goals," *Lancer* 365, no. 9456 (2005): 347–53.

4. Y. K. Nayarro, J. McNeely, D. Melnick, R. R. Sears, and G. Schmidt-Traub, *Environment and Human Wellbeing: A Practical Strategy* (New York: UN Millennium Project Task Force on Environmental Sustainability, 2005).

economic development, health care systems, and public health programs—and not on the fundamental health-supporting role of the natural environment and its ecosystem services.

The report noted the success of some relatively poor nations and provinces, such as China, Sri Lanka, and Kerala State in India, in lowering infant mortality and improving population health by investing in education (especially for girls), establishing primary health clinics, and enacting other health-care programs. The report extended this analysis, citing the history of the well-documented mortality decline in the industrial world—which preceded the advent of modern drugs and medical care, deriving instead from betterment of nutrition, housing, and hygiene. Progressive policies, strong social institutions, and innovative health care and public health protection (especially against infectious diseases), without generalized gains in national wealth, the report's authors said, can be sufficient to raise population health markedly.

This important insight, though, makes no explicit reference to the role of wider environmental conditions. While the control of mosquito populations with window-screens and insecticides certainly confers some health protection, for example, land-use practices, surface water management, biodiversity (frogs and birds eat mosquitoes), and climatic conditions can affect mosquito ecology and mosquito-borne disease transmission more profoundly. The issue must be tackled at both levels.

In fairness, understanding the patterns and determinants of human population health within a wider ecological frame has been impeded by strong cultural and intellectual undercurrents. The rise of modern western science and medicine, in concert with the contemporary ascendancy of neo-liberalism and individualism, has recast our views of health and disease in primarily personal terms. The Christian biblical notion from two thousand years ago of the Four Horsemen of the Apocalypse as the major scourges of population health and survival—war, conquest, famine, and pestilence—has been replaced by today's prevailing model of health and disease as predominantly a function of individual-level consumer behaviors, genetic susceptibility, and access to modern health care technologies.

In addition to this cultural misshaping of our understanding, our increasing technological sophistication has created the illusion that we no longer depend on nature's "goods and services" for life's basic necessities. In this first decade of the twenty-first century, however, we are being forcibly reminded of that fundamental dependence. Hence, a repeat WCED report, written now, would give much higher priority to the relationship between biosphere, environmental processes, human biological health, and survival.

Footprints, Environmental Conditions, and Human Well-being

It is interesting that the WCED report was being drafted at about the time when, according to recent assessments, the demands and pressures of the global human population were first over-reaching the planet's carrying capacity.[11]

In the time since the publication of the report, the "ecological footprint" has become a familiar concept. For any grouping of persons, it measures the amount of Earth's surface required to provide their materials and food and to absorb their wastes. Collectively, humankind reached a point in the mid-1980s when it began to exceed the limit of what Earth could supply and absorb on a sustainable basis. Since then, the human population has moved from having a precariously balanced environmental budget that left nothing in reserve to a situation today in which we are attempting to survive on a substantial, growing,

overdraft: our global standard-of-living is estimated to be at the level that requires approximately 1.3 Earths (see Table).[12] We are therefore consuming and depleting natural environmental capital. This explains the accruing evidence of climate change, loss of fertile soil, freshwater shortages, declining fisheries, biodiversity losses and extinctions. This is not a sustainable trajectory, and it is what, generically, the WCED report exhorted the world to avoid.

In the 1980s, there was more ambivalence about the population component of the "footprint" concept. The absolute annual increments in human numbers were at a historical high, and many demographers and some enlightened policymakers were concerned that population growth needed constraining. That view faced an emergent western political ethos that eschewed family planning, abortion counseling, and governmental intervention. In the upshot, population growth has begun to slow in a majority of countries. Meanwhile, this is being offset by the rapid rise in wealth and consumption in many larger developing countries, including China, India, Brazil, and Mexico.

This planet simply cannot support a human population of 8 to 10 billion living at the level of today's high-income country citizens. Each of those citizens, depending on their particular country, needs 4 to 9 hectares of Earth's surface to provide materials for their lifestyle and to absorb their wastes. Meanwhile, India's population of 1.2 billion has to get by with less than 1 hectare per person. With an anticipated world population of 8 to 10 billion living within Earth's limits, there would be no more than about 1.5 hectares of ecological footprint per average-person—and this arithmetic would limit the resources available for other species. To comply equitably with this limit will necessitate radical changes in value systems and social institutions everywhere.

Global Environment: Emerging Evidence

The Brundtland Commission foresaw at least some of the impending serious erosion of large-scale environmental resources and systems. Indeed, the WCED report judged that by early in the twenty-first century, climate change might have increased average global temperatures sufficiently to displace agricultural production areas, raise sea levels (and perhaps flood coastal cities), and disrupt national economies. This apparently has not yet happened, although very recent scientific reports point strongly to an acceleration in the climate change process,[13] as the global emissions of carbon dioxide from fossil-fuel combustion and of other greenhouse gases from industrial and agricultural activities alter the global climate faster than previously expected.

Several other adverse environmental trends have emerged since 1987. Accessible oil stocks may now be declining—thereby stimulating an (ill-judged) scramble to divert food-grain production into biofuel production as an alternative source of liquid energy.[14] It has also become apparent that human actions are transforming the global cycles of various elements other than carbon, particularly nitrogen, phosphorus, and sulfur.[15] Human agricultural and industrial activity now generates as much biologically activated nitrogen (nitrogenous

Table 1   Changes in Key Global Indicators of Environment and Population Health (1987–2007)

	1987 (1985–1989)	2007 (2005–2009)	Comments
World population size	4.9 billion	6.7 billion	Slight reduction in absolute annual increment
Annual population growth rate	1.7%	1.2%	
Fertility rate (births/woman)	3.4	2.4	
Percent over age 65 years	6%	8%	Low-income countries have increased from 4% to 5.5%
Life expectancy, years	65	68	
Maternal mortality (per 100,000 births)	430	400	
Under 5 mortality, per 1,000 births	115	70	
Infant mortality, per 1,000 births	68	48	
Primary schooling	~60%	82%	See also Figure 1
Malnutrition prevalence	870 million	850 million	Recent increase, relative to the turn of century (~820 million)
Child stunting, less than age 5, prevalence	~30%	25%	Down from 35% circa 1950, but a persistent and serious problem in sub-Saharan Africa (highest prevalence) and South Asia
HIV/AIDS, prevalent cases	10 million	40 million	
AIDS deaths per year	~0.2 million	3.2 million	
Lack safe drinking water	1.3 billion (27%)	1.1 billion (15%)	Percent of world population shown in parenthesis
Lack sanitation	2.7 billion (54%)	2.6 billion (40%)	Percent of world population shown in parenthesis
CO_2 atmospheric concentration	325 parts per million	385 parts per million	Approx 0.5% rise per year, currently accelerating. (Pre-industrial concentration 275 parts per million)
Increase in average global temperature relative to 1961–1990 baseline	0.1 degrees Celsius	0.5 degrees Celsius	Warming faster at high latitude, especially in northern hemisphere
Global ecological footprint	1.0 planet Earths	1.3 planet Earths	Estimate of number of planet Earths needed to supply, sustainably, the world population's energy, materials and waste disposal needs

Source: Compiled from various international agency reports, databases, and scientific papers.

compounds such as ammonia) as do lightning, volcanic activity, and nitrogen-fixation on the roots of wild plants. Meanwhile, worldwide land degradation, freshwater shortages, and biodiversity losses are increasing. Those environmental problems were all becoming evident in the mid-1980s and were duly referred to in the WCED report, albeit without particular connection to considerations of human health.

Some other large-scale environmental stresses, however, were not evident in the 1980s. The scientific community had not anticipated the acidification of the world's oceans caused by absorption of increasingly abundant atmospheric carbon dioxide. This acidification—global average ocean pH has declined by a little over 0.1 points during the past several decades—endangers

the calcification processes in the tiny creatures at the base of the marine food web. Nor was much attention paid to the prospect of loss of key species in ecosystems, such as pollinating insects (especially bees). Both those processes are now demonstrably happening, further jeopardizing human capital development, poverty alleviation, and good health.

During 2001–2006, the Millennium Eco-system Assessment (MA) was conducted as a comprehensive international scientific assessment with processes similar to those of IPCC. The MA documented the extent to which recent human pressures have accelerated the decline of stocks of many environmental assets, including changes to ecosystems.[15] The MA also projected likely future trends. This assessment documented how several other globally significant environmental graphs peaked in the mid-1980s. On land, the annual per capita production of cereal grains peaked and has subsequently drifted sideways and, recently, downwards. The harvest from the world's ocean fisheries also peaked at that time and has subsequently declined slowly—albeit with compensatory gains from aquaculture. These emergent negative trends in food-producing capacity jeopardize attempts to reduce hunger, malnutrition, and child stunting—a key target area of the Millennium Development Goals (see the box on next page).

The WCED report, if rewritten today, would presumably take a more integrative and systems-oriented approach to the topic of environmental sustainability and would incorporate greater awareness of the risks posed to human well-being and health.

Trends in Human Capital and Population Health

As discussed, the original UN resolution calling for the WCED report referred ambiguously to "resources." Within the overarching environmental context of the commission, the intended reference of that word may well have been to environmental resources (such as oil, strategic and precious metals, water supplies, etc.). Interestingly, the WCED treated the word as referring primarily to *human* resources in chapter 4, titled "Population and Human Resources."

The global population was 4.9 billion at the time the WCED report was published, and now exceeds 6.7 billion. It continues to increase by more than 70 million persons annually. Because overall fertility rates have declined a little faster than was previously expected, the current "medium" UN projection for population growth by 2050 is for a total of approximately 9.1 billion.[16] Most of that increase will occur in the low-income countries, predominantly in rapidly expanding cities.

Population growth necessarily increases demands on the local environment. But as the WCED report correctly argued, "the population issue is not solely about numbers."[17] Population size, density, and movement are part of a larger set of pressures on the environment. In some regions, resource degradation occurs because of the combination of poverty and the farming of thinly populated drylands and forests. Elsewhere, per-person levels of consumption and waste generation are the critical drivers of environmental stress. Extrapolation of current global economic trends foreshadows a potential five- to tenfold increase in economic activity by 2050. But this looks increasingly unachievable without radical changes in world technological choices and economic practices. The current experience of China is salutary in this regard: that country's rapid economic growth is engendering huge problems of freshwater supply, air quality, environmental toxins in food, desertification of western provinces—and, now, the world's largest national contribution to greenhouse gas emissions.

Is there an upside to population? "People," stated the WCED report, "are the ultimate resource. Improvements in education, health, and nutrition allow them to better use the resources they command, to stretch them further."[18] How have we progressed since 1987 in providing these improvements?

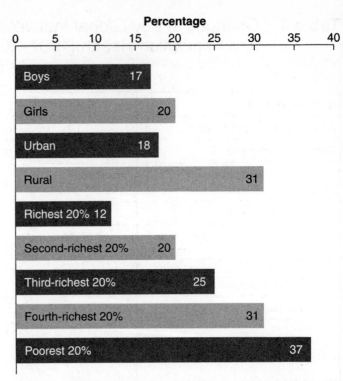

Figure 1 Children of primary school age not in school, by sex, place of residence, and household wealth, 2005

Source: United Nations, *Millennium Development Goals Report 2007* (New York: United Nations, 2007), www.un.org/millenniumgoals/goals.html (accessed 23 August, 2007).

Access to primary schooling has increased since 1987 (see Table 1). In particular, the proportion of young girls completing primary school has increased (starting from a lower base than for boys). Figure 1 on this page shows current proportions of the world's children not in primary schooling by key categories. Impediments persist in the form of poverty, parental illiteracy, civil war, and Islamic extremism (banning female education).

Beyond environmental stresses and deficits, the task of improving population health faces other, systemic difficulties. As my colleague C. D. Butler and I wrote last year:

> The gap between rich and poor, both domestically and internationally, has increased substantially in recent decades. Inequality between countries has weakened the United Nations and other global organisations and institutions. Foreign aid has declined, replaced by claims that market forces and the removal of trade-distorting subsidies will reduce poverty and provide public goods, including health care and environmental stability.[19]

Hunger and malnutrition persist at high levels (see box). Famines in Africa remain frequent, and 300 million people in India are undernourished. Further, the almost 50 percent prevalence of underweight children in sub-Saharan Africa and South Asia causes widespread stunting of growth, intellectual development, and energy levels. Yet elsewhere, hundreds of millions of people in all continents are overfed and, via obesity, at increased risk of diabetes and heart disease.

Over the past two decades, demographic and epidemiological transitions have become less orderly than was anticipated

Recent Trends in Population Health

Human health experienced unprecedented gains last century. Globally, average life expectancy approximately doubled from around 35 years to almost 70 years.[1] Rises in life expectancy have slowed a little in recent years in high-income countries. Meanwhile, rises are continuing (from a lower base) in much of the rest of the world. However, the regional picture is very uneven, and some divergence has occurred. The rise in life expectancy has stalled in much of sub-Saharan Africa, various ex-Soviet countries, North Korea, and Iraq (see the figure on the next page). Meanwhile, health inequalities persist both between and within countries and reflect, variously, differences in economic circumstance, literacy, social institutions, and political regimen.

Improved food supply is the likely cause of much of the health gain in modern western populations. The second agricultural revolution, which began in eighteenth-century Europe, brought mechanization, new cultivars, and, eventually, fossil fuel power. Consequently, the millennia-old pattern of subsistence crises diminished and then disappeared. The greater security and abundance of food apparently explains why adult males in northern European countries have grown around 10 centimeters taller and 20–30 kilograms heavier than their eighteenth-century predecessors.[2] Others have argued that improved food quality and safety raised the resistance of better-nourished persons to infectious diseases.[3]

Despite these gains, an estimated 850 million persons remain malnourished. In absolute terms, that figure has grown since the time of the WCED report, including over the past decade.[4] Meanwhile, it has become increasingly evident in both high-income and lower-income countries that an abundance of food energy, especially in the form of refined and selectively produced energy-dense (high fat, high sugar) foods, poses various serious risks to health.

In the 1980s, the general assumption was that these non-communicable diseases appear in the later stages of economic development and would increase with further gains in wealth and modernity. However it has become clear in the past two decades that these diseases, particularly heart disease, hypertensive stroke, and type 2 diabetes, are increasing markedly in lower-income populations as they undergo urbanization, and dietary change. The burden of cardiovascular disease—which accounts for around 30 percent of all deaths in today's world—will continue this shift to low- and middle-income countries. This, plus the persistent infectious disease burden, particularly in poorer subpopulations, will further increase global health inequalities.[5]

Notes

1. A. J. McMichael, M. McKee, V. Shkolnikov, and T. Vaikonen, "Mortality Trends and Setbacks: Global Convergence or Divergence?" *Lancet* 363, no. 9415, (2004): 1155–59.

2. R. W. Fogel, *The Escape from Hunger and Premature Death, 1700–2100: Europe, American and the Third World* (Cambridge: Cambridge University Press, 2004).

3. T. McKeown, R. G. Brown, and R. Record, "An Interpretation of the Modern Rise of Population in Europe," *Population Studies* 26 no. 3 (1972): 345–82.

4. Food and Agriculture Organization of the United Nations (FAO), *The State of Food Insecurity in the World 2004* (Rome: FAO, 2005).

5. M. Ezzati et al., "Rethinking the 'Diseases of Affluence' Paradigm: Global Patterns of Nutritional Risks in Relation to Economic Development." *PLoS Medicine* 2, no. 5 (2005), e133.doi.10.1371/journal.pmed.0020133.

---

by conventional demographic models. There has been considerable divergence between countries in trends in death rates (life expectancy) and fertility rates. National health trends (see box), particularly in poor and vulnerable populations, are falling increasingly under the shadow of climate change and other adverse environmental trends.

In many, but not all low-income countries, fertility rates have declined faster than might have been predicted. However, in some countries (such as East Timor, Nigeria, and Pakistan) fertility remains high (4–7 children per woman). In some regions, the fertility decline has led to an economically and socially unbalanced age structure, especially in China, where in the wake of their "one-child policy," the impending dependency ratio is remarkably high—many fewer young adults will have to provide economic support for an older, longer-living generation.

In some other countries, population growth has declined substantially because of rapid falls in life expectancy.[20] Russia and parts of sub-Saharan Africa have very different demographic characteristics, and yet common elements may underlie their downward trends in life expectancy. Both regions lack public goods for health.[21] In Russia there is a lack of equality, safety, and public health services—and many men have lost status and authority following the collapse of the Communist party structure. Meanwhile, in a number of sub-Saharan African countries, there is serious corruption in government, deficient governance structures, food insecurity, and inadequate public health services.

The conventional assumption, also evident in the WCED report, has been that a health dividend will flow from poverty alleviation. However, it is becoming clear that those anticipated health gains are likely to be lower because of the now-worldwide rise of various non-communicable diseases, including those due to obesity, dietary imbalances, tobacco use, and urban air pollution.[22]

Conclusion

The WCED was commissioned to examine critically the relationship between environmental resource use and sustainable development and to propose solutions for the tensions between

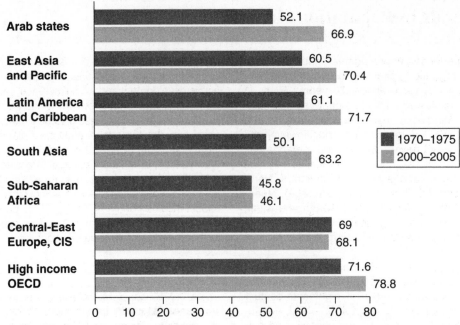

Changes in life expectancy by region, over the past three decades

Note: Differences are marked between regions—especially the lack of gains in sub-Saharan Africa and the central and eastern European (including ex-Soviet) countries.

Source: Based on M. Marmot, "Health in an Unequal World," *Lancet* 368, no. 9552 (2006): 2081–94.

environment (including the pressures of population growth and urbanization) and development. A prime task for the commission was to formulate a "global agenda for change" within the frame of ecologically sustainable development, while recognizing the aspirational goals of people and communities everywhere.[23]

During the time that the commission was developing its report, a widely held view, reinforced by the Green Revolution successes of the 1970s and early 1980s, was that continuing population growth need not have adverse environmental impacts. The commission was hesitant to embrace that view, which has recently been re-evaluated,[24] with renewed recognition of the adverse effects of rapid population growth, especially in developing countries, on both social and environmental conditions.[25]

In the 1980s, national governments and multilateral agencies began to see that economic development issues could not be separated from environment issues. Many forms of development erode the environmental resource base (including forests, fertile soils, and coastal zones) necessary for sustained development. And conversely, environmental degradation can jeopardize economic development. The WCED report rightly emphasized the futility of addressing environmental problems without alleviating poverty and international inequality. The report also recognized the needs for stronger social structures and legal processes to deal with tensions over environmental commons, and for more enlightened public agency structures at the international level to address these issues. It advocated partnerships with the private sector—a sector in which there is now a growing recognition that business-as-usual is no longer an option.

Those formulations remain important and valid, but they are an incomplete basis for future strategic policy. They overlook the fundamental role that sustaining an intact biosphere and its component systems plays in enabling the social and human developmental processes that can reduce poverty, undernutrition, unsafe drinking water, and exposures to endemic and epidemic infectious diseases. The report, if updated today, would seek a better balance between these sets of relationships.

The idea of "ecologically sustainable development" was, in the latter 1980s, ahead of its time. We had, then, neither the evidence nor the insight to know just how fundamental that framework was to achieving the other human goals that would be embraced over the next two decades. Today, the ongoing growth of the global population and—with economic development and rising consumer expectations—the increasingly great environmental impact of that population means that we may be less than one generation away from exhausting much of the biosphere's environmental buffering capacity.[13] Unless we can constrain our excessive demands on the natural world, the demographic and epidemiological transitions (faltering in some regions) will be further affected and human fulfillment will thus be eroded.

Twenty years on from the report of the World Commission on Environment and Development, we can see additional layers to the environment challenge that were little understood in the 1980s. Clearly, some fundamental changes are needed in how we live, generate energy, consume materials, and dispose of wastes. Population arithmetic will impose a further dimension of challenge: 4.8 billion in 1987; 6.7 billion in 2007; perhaps 8 billion by 2027. Beyond that, the numbers and outcomes will be influenced by what current and future "Brundtland reports" formulate, and how seriously and urgently we and our governments take their formulations and recommendations.

Notes

1. World Commission on Environment and Development, *Our Common Future* (Cambridge, UK, and New York: Cambridge University Press, 1987).

2. A. C. Kelley, "The Population Debate in Historical Perspective: Revisionism Revised," in N. Birdsall, A. C. Kelley, and S. W. Sinding, eds., *Population Matters: Demographic Change, Economic Growth, and Poverty in the Developing World* (Oxford, UK: Oxford University Press, 2001), 24–54.

3. P. McMichael, *Development and Social Change: A Global Perspective* (Thousand Oaks, California: Pine Forge Press, 2004).

4. A. J. McMichael, M. McKee, V. Shkolnikov, and T. Valkonen, "Mortality Trends and Setbacks: Global Convergence or Divergence?" *Lancet* 363, no. 9415 (2004): 1155–59; and R. Eckersley, "Is Modern Western Culture a Health Hazard?" *International Journal of Epidemiology* 35, no. 5 (2006): 252–58.

5. United Nations, "Process of Preparation of the Environmental Perspective to the Year 2000 and Beyond," General Assembly Resolution 38/161, 19 December 1983.

6. WCED, note 1 above, page 1.

7. G. H. Brundtland, speech given at the launch of the WCED report, Oslo, Norway, 20 March 1987.

8. D. G. Victor, "Recovering Sustainable Development," *Foreign Affairs 85*, no. 1 (January/February 2006): 91–103.

9. WCED, note 1 above, page 98.

10. Intergovernmental Panel on Climate Change, *Climate Change. The IPCC Scientific Assessment* (Cambridge, UK: Cambridge University Press, 1990).

11. Ibid.; and C. M. Wackernagel et al., "Tracking the Ecological Overshoot of the Human Economy," *Proceedings of the National Academy of Sciences* 99, no. 14 (2002): 9266–71.

12. Worldwide Fund for Nature International (WWF), *Living Planet Report 2006* (Gland, Switzerland: WWF, 2006), http://assets.panda.org/dowloads/living_planet_report.pdf (accessed 23 Aug 2007).

13. S. Rahmstorf et al., "Recent Climate Observations Compared to Projections," *Science* 316, no. 5825 (4 May 2007): 709.

14. See R. L. Naylor et al., "The Ripple Effect: Biofuels, Food Security, and the Environment," *Environment* 49, no. 9 (November 2007): 30–43.

15. Millennium Ecosystem Assessment, *Ecosystems and Human Wellbeing. Synthesis* (Washington, DC: Island Press, 2005).

16. UN Department of Economic and Social Affairs, Population Division: http://esa.un.org/unpp/p2k0data.asp (accessed Nov 1, 2007).

17. WCED, note 1 above, page 95.

18. WCED, note 1 above, page 95. This statement has faint resonance with the ideas of the late U.S. economist Julian Simon, whose book *The Ultimate Resource* made the tendentious argument that the more people on Earth the greater the probability of occurrence of important new ideas. J. L. Simon, *The Ultimate Resource* (Princeton, NJ: Princeton University Press, 1981).

19. A. J. McMichael and C. D. Butler, "Emerging Health Issues: The Widening Challenge for Population Health Promotion," *Health Promotion International* 21, no. 1 (2006): 15–24.

20. McMichael, McKee, Shkolnikov, and Valkonen, note 4 above.

21. R. Smith, R. Beaglehole, D. Woodward, and N. Drager, eds., *Global Public Goods for Health* (Oxford: Oxford University Press, 2003).

22. M. Ezzati et al., "Rethinking the 'Diseases of Affluence' Paradigm: Global Patterns of Nutritional Risks in Relation to Economic Development," *Plos Medicine* 2, no. 5 (2005): e133.

23. Brundtland, note 7 above.

24. Kelley, note 2 above.

25. M. Campbell, J. Cleland, A. Ezeh, and N. Prata, "Return of the Population Growth Factor," *Science* 315, no. 5818 (2 February 2007): 1501–2.

Critical Thinking

1. Define the terms: *development, sustainable environment,* and *human well-being.* How may the definitions of these terms vary for different people in different regions of the world?

2. How might maintaining the lifestyles of people in industrialized, wealthy nations interfere with achieving the MDGs?

3. Can any of MDG goals be obtained if Goal 7: Environmental Sustainability is not considered the most important goal of all? Explain why or why not.

4. If you had to pick one fundamental reason why we have not gotten "the balance right" with regard to humans' relationship to their environment, what would that reason be? Explain.

5. Do some research on the Internet sites provided; after 4 years, have we moved closer to "getting the balance right"?

6. Identify in this article three key terms, concepts, or principles that are used in your textbook (environmental science, economics, sociology, history, geography, etc.) or employed in the discipline you are currently studying. (Note: The terms, concepts, or principles may be implicit, explicit, implied, or inferred.)

ANTHONY J. MCMICHAEL is a professor at the National Centre for Epidemiology and Population Health (NCEPH) at Australia National University in Canberra. From 2001 to 2007, he was director of NCEPH, where he has led the development of a program of epidemiological research on the environmental influences on immune disorders, particularly autoimmune diseases such as multiple sclerosis. Meanwhile, he has continued his pioneering research on the health risks of global climate change, developed in conjunction with his central role in the assessment of health risks for the Intergovernmental Panel on Climate Change. His work on climate and environmental change, along with longstanding interests in social and cultural influences on patterns of health and disease, also underlie his interests in understanding the determinants of the emergence and spread of infectious diseases in this seemingly "renaissant" microbial era. He may be contacted at Tony.McMichael@anu.edu.au.

Booms, Busts, and Echoes

How the biggest demographic upheaval in history is affecting global development.

DAVID E. BLOOM AND DAVID CANNING

For much (and perhaps most) of human history, demographic patterns were fairly stable: the human population grew slowly, and age structures, birth rates, and death rates changed very little. The slow long-run growth in population was interrupted periodically by epidemics and pandemics that could sharply reduce population numbers, but these events had little bearing on long-term trends.

Over the past 140 years, however, this picture has given way to the biggest demographic upheaval in history, an upheaval that is still running its course. Since 1870 death rates and birth rates have been declining in developed countries. This long-term trend toward lower fertility was interrupted by a sharp, post–World War II rise in fertility, which was followed by an equally sharp fall (a "bust"), defining the "baby boom." The aging of this generation and continued declines in fertility are shifting the population balance in developed countries from young to old. In the developing world, reductions in mortality resulting from improved nutrition, public health infrastructure, and medical care were followed by reductions in birth rates. Once they began, these declines proceeded much more rapidly than they did in the developed countries. The fact that death rates decline before birth rates has led to a population explosion in developing countries over the past 50 years.

Even if the underlying causes of rapid population growth were to suddenly disappear, humanity would continue to experience demographic change for some time to come. Rapid increases in the global population over the past few decades have resulted in large numbers of people of childbearing age (whose children form an "echo" generation). This creates "population momentum," where the populations of most countries, even those with falling birth rates, will grow for many years, particularly in developing countries.

These changes have huge implications for the pace of economic development. Economic analysis has tended to focus on the issue of population numbers and growth rates as factors that can put pressure on scarce resources, dilute the capital-labor ratio, or lead to economies of scale. However, demographic change has important additional dimensions. Increasing average life expectancy can change life-cycle behavior affecting education, retirement, and savings decisions—potentially boosting the financial capital on which investors draw and the human capital that strengthens economies. Demographic change also affects population age structure, altering the ratio of workers to dependents. This issue of *F&D* looks at many facets of the impact of demographic change on the global economy and examines the policy adjustments needed in both the developed and the developing world.

Sharp Rise in Global Population

The global population, which stood at just over 2.5 billion in 1950, has risen to 6.5 billion today, with 76 million new inhabitants added each year (representing the difference, in 2005, for example, between 134 million births and 58 million deaths). Although this growth is slowing, middle-ground projections suggest the world will have 9.1 billion inhabitants by 2050.

These past and projected additions to world population have been, and will increasingly be, distributed unevenly across the world. Today, 95 percent of population growth occurs in developing countries. The populations of the world's 50 least developed countries are expected to more than double by the middle of this century, with several poor countries tripling their populations over the period. By contrast, the population of the developed world is expected to remain steady at about 1.2 billion, with declines in some wealthy countries.

The disparity in population growth between developed and developing countries reflects the considerable heterogeneity in birth, death, and migration processes, both over time and across national populations, races, and ethnic groups. The disparity has also coincided with changes in the age composition of populations. An overview of these factors illuminates the mechanisms of population growth and change around the world.

Total fertility rate. The total world fertility rate, that is, the number of children born per woman, fell from about 5 in 1950 to a little over 2.5 in 2006 (see Figure 1). This number is projected to fall to about 2 by 2050. This decrease is attributable largely to changes in fertility in the developing world and can be ascribed to a number of factors, including declines in infant

(total fertility rate; children per woman)

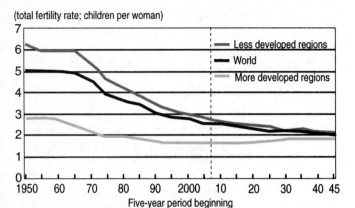

Figure 1 Smaller Families. Fertility rates are tending to converge at lower levels after earlier sharp declines.

Source: United Nations, *World Population Prospects,* 2004.

(life expectancy in years)

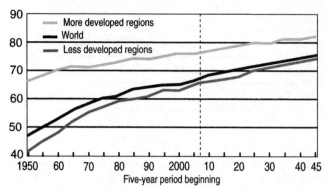

Figure 2 Living Longer. Life expectancy is continuing to rise, but there are big differences between rates in well-off and poorer countries.

Source: United Nations, *World Population Prospects,* 2004.

mortality rates, greater levels of female education and increased labor market opportunities, and the provision of family-planning services.

Infant and child mortality decline. The developing world has seen significant reductions in infant and child mortality over the past 50 years. These gains are primarily the result of improved nutrition, public health interventions related to water and sanitation, and medical advances, such as the use of vaccines and antibiotics. Infant mortality (death prior to age 1) in developing countries has dropped from 180 to about 57 deaths per 1,000 live births. It is projected to decline to fewer than 30 by 2050. By contrast, developed countries have seen infant mortality decline from 59 deaths per 1,000 live births to 7 since 1950, and this is projected to decline further still, to 4 by 2050. Child mortality (death prior to age 5) has also fallen in both developed and developing countries.

Life expectancy and longevity. For the world as a whole, life expectancy increased from 47 years in 1950–55 to 65 years in 2005. It is projected to rise to 75 years by the middle of this century, with considerable disparities between the wealthy industrial countries, at 82 years, and the less developed countries, at 74 years (see Figure 2). (Two major exceptions to the upward trend are sub-Saharan Africa, where the AIDS epidemic has drastically lowered life expectancy, and some of the countries of the former Soviet Union, where economic dislocations have led to significant health problems.) As a result of the global decline in fertility, and because people are living longer, the proportion of the elderly in the total population is rising sharply. The number of people over the age of 60, currently about half the number of those aged 15 to 24, is expected to reach 1 billion (overtaking the 15–24 age group) by 2020 and almost 2 billion by 2050. The proportion of individuals aged 80 or over is projected to rise from 1 percent to 4 percent of the global population by 2050.

A new UN report says that in 2007 the worldwide balance will tip and more than half of all people will be living in urban areas.

Age distribution: working-age population. Baby booms have altered the demographic landscape in many countries. As the experiences of several regions during the past century show, an initial fall in mortality rates creates a boom generation in which high survival rates lead to more people at young ages than in earlier generations. Fertility rates fall over time, as parents realize they do not need to give birth to as many children to reach their desired family size, or as desired family size contracts for other reasons. When fertility falls and the baby boom stops, the age structure of the population then shows a "bulge" or baby-boom age cohort created by the nonsynchronous falls in mortality and fertility. As this generation moves through the population age structure, it constitutes a share of the population larger than the cohorts that precede or follow. This creates particular challenges and opportunities for countries, such as a large youth cohort to be educated, followed by an unusually large working-age (approximately ages 15–64) population, with the prospect of a "demographic dividend," and characterized eventually by a large elderly population, which may burden the health and pension systems (see Figure 3).

Migration. Migration also alters population patterns. Globally, 191 million people live in countries other than the one in which they were born. On average during the next 45 years, the United Nations projects that over 2.2 million individuals will migrate annually from developing to developed countries. It also projects that the United States will receive by far the largest number of immigrants (1.1 million a year), and China, Mexico, India, the Philippines, and Indonesia will be the main sources of emigrants.

Urbanization. In both developed and developing countries, there has been huge movement from rural to urban areas since 1950. Less developed regions, in aggregate, have seen their population shift from 18 percent to 44 percent urban, while the corresponding figures for developed countries are 52 percent to 75 percent. A new UN report says that in 2007 the worldwide balance will tip and more than half of all people will be living in urban areas. This shift—and the concomitant urbanization of areas that were formerly peri-urban or rural—is consistent with the shift in most countries away from agriculturally based economies.

(ratio of working-age to non-working-age population)

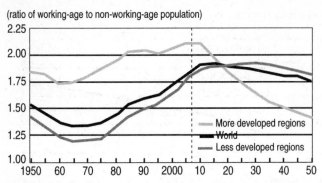

Figure 3 Tracking the Bulge. Developing countries are nearing the peak of their opportunity to benefit from a high ratio of workers to dependents.

Source: United Nations, *World Population Prospects,* 2004.

(world population, aged 80+; millions)

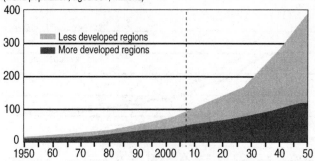

Figure 4 Retiree Boom. The number of people living past 80 is projected to rise sharply, but labor shortages could drive up living costs for retirees.

Source: United Nations, *World Population Prospects,* 2004.

The existence and growth of megacities (that is, those with 10 million or more residents) is a late-20th-century phenomenon that has brought with it special problems. There were 20 such cities in 2003, 15 in developing countries. Tokyo is by far the largest, with 35 million people, followed by (in descending order) Mexico City, New York, São Paulo, and Mumbai (all with 17 to 19 million). Cities in general allow for economies of scale—and, most often, for a salutary mix of activities, resources, and people—that make them centers of economic growth and activity and account, in some measure, for their attractiveness. As continued movement to urban areas leads to megacities, however, these economies of scale and of agglomeration seem to be countered, to some extent, by problems that arise in transportation, housing, air pollution, and waste management. In some instances, socioeconomic disparities are particularly exacerbated in megacities.

What Is the Impact on Economies?

The economic consequences of population growth have long been the subject of debate. Early views on the topic, pioneered by Thomas Malthus, held that population growth would lead to the exhaustion of resources. In the 1960s, it was proposed that population growth aided economic development by spurring technological and institutional innovation and increasing the supply of human ingenuity. Toward the end of the 1960s, a neo-Malthusian view, focusing again on the dangers of population growth, became popular. Population control policies in China and India, while differing greatly from each other, can be seen in this light. Population neutralism, a middle-ground view, based on empirical analysis of the link between population growth and economic performance, has held sway for the past two decades. According to this view, the net impact of population growth on economic growth is negligible.

Population neutralism is only recently giving way to a more fine-grained view of the effects of population dynamics in which demographic change does contribute to or detract from economic development. To make their case, economists and demographers point to both the "arithmetic accounting" effects of age structure change and the effects of behavioral change caused by longer life spans (see box).

Arithmetic accounting effects. These effects assume constant behavior within age and sex groups, but allow for changes in the relative size of those groups to influence overall outcomes. For example, holding age- and sex-specific labor force participation rates constant, a change in age structure affects total labor supply.

As a country's baby-boom generation gets older, for a time it constitutes a large cohort of working-age individuals and, later, a large cohort of elderly people. The span of years represented by the boom generation (which determines how quickly this cohort moves through the age structure) and the size of the population bulge vary greatly from one country to another. In all circumstances, there are reasons to think that this very dynamic age structure will have economic consequences. A historically high proportion of working-age individuals in a population means that, potentially, there are more workers per dependent than previously. Production can therefore increase relative to consumption, and GDP per capita can receive a boost.

Life cycle patterns in savings also come into play as a population's age structure changes. People save more during their working-age years, and if the working-age cohort is much larger than other age groups, savings per capita will increase.

Behavioral effects. Declining rates of adult mortality and the movement of large cohorts through the global population pyramid will lead to a massive expansion in the proportion of elderly in the world population (see the projections for 2050 in Figure 4). Some simple economic projections show catastrophic effects of this aging. But such projections tend to be based on an "accounting" approach, which assumes that age-specific behavior remains unchanged and ignores the potentially significant effects of behavior change.

The aging of the baby-boom generation potentially promotes labor shortages, creating upward pressure on wages and downward pressure on the real incomes of retirees. In response, people may adjust their behavior, resulting in increased labor force participation, the immigration of workers from developing countries, and longer working lives. Child mortality declines can also have behavioral effects, particularly for women, who tend to be the primary caregivers for children. When the reduced fertility effect of a decrease in child mortality is in place, more women participate in the workforce, further boosting the labor supply.

How Much Can the Human Life Span Be Stretched?

In most of the world, children born today can expect to live for many decades longer than their ancestors born in the 19th or early 20th centuries. In Japan, life expectancy at birth is now 82 years, and other regions have also made great progress as medical and public health advances, improved nutrition, and behavioral changes encouraged by improved education have combined to reduce the risk of death at all ages. But how far can these increases in longevity go?

Continuing increases in life expectancy in low-mortality populations have led some demographers to forecast further gains. Kenneth Manton, Eric Stallard, and H. Dennis Tolley, for example, estimate that populations with extremely healthy lifestyles—that is, with an absence or near-absence of risk factors such as infectious disease, smoking, alcohol abuse, and obesity, and the presence of health-promoting behaviors such as a healthy diet and exercise—could achieve a life expectancy of between 95 and 100 years.

But others have reached different conclusions. Nan Li and Ronald Lee estimate that life expectancy in the United States will rise from a 1996 figure of 76.3 to 84.9 by 2050, with that in Japan rising from 80.5 to 88.1. S. Jay Olshansky, Bruce Carnes, and Aline Desesquelles predicted in 1990 that life expectancy at birth would not surpass 85 years, even in low-mortality settings. Death rates, they argued, would not fall sufficiently for life expectancy to rise rapidly, and earlier increases were driven largely by dramatic reductions in infant and child mortality, which could not recur (Samuel Preston, on the other hand, observes that 60 percent of the life expectancy increase in the United States since 1950 is due to mortality declines in people over the age of 50). Perhaps more important, they saw no reason why the future should necessarily mirror the past—new threats to health such as influenza pandemics, antibiotic resistance, and obesity could reverse gains made in recent decades; technological improvements could stall and the drugs needed to counter the diseases of aging might not be found; and environmental disasters, economic collapse, or war could derail health systems at the same time that they weaken individuals' ability to protect their own health.

The Missing Link

Demographic effects are a key missing link in many macroeconomic analyses that aim to explain cross-country differences in economic growth and poverty reduction. Several empirical studies show the importance of demographics in explaining economic development.

East Asia's baby boom. East Asia's remarkable economic growth in the past half century coincided closely with demographic change in the region. As infant mortality fell from 181 to 34 per 1,000 births between 1950 and 2000, fertility fell from six to two children per woman. The lag between falls in mortal-ity and fertility created a baby-boom generation: between 1965 and 1990, the region's working-age population grew nearly four times faster than the dependent population. Several studies have estimated that this demographic shift was responsible for one-third of East Asia's economic growth during the period (a welcome demographic dividend).

Labor supply and the Celtic Tiger. From 1960 to 1990, the growth rate of income per capita in Ireland was approximately 3.5 percent a year. In the 1990s, it jumped to 5.8 percent, well in excess of any other European economy. Demographic change contributed to the country's economic surge. In the decade following the legalization of contraceptives in 1979, Ireland saw a sharp fall in the crude birth rate. This led to decreasing youth dependency and a rise in the working-age share of the total population. By the mid-1990s, the dependency burden in Ireland had dropped to a level below that in the United Kingdom.

Two additional demography-based factors also helped fuel economic growth by increasing labor supply per capita. First, while male labor force participation rates remained fairly static, the period 1980–2000 saw a substantial increase in female labor force participation rates, particularly among those aged between 25 and 40. Second, Ireland historically had high emigration levels among young adults (about 1 percent of the population a year) because its economy was unable to absorb the large number of young workers created by its high fertility rate. The loss of these young workers exacerbated the problem of the high youth dependency rate. The decline in youth cohort sizes and rapid economic growth of the 1990s led to a reversal of this flow, resulting in net in-migration of workers, made up partly of return migrants and also, for the first time, substantial numbers of foreign immigrants.

Continued high fertility in sub-Saharan Africa. Demographic change of a very different type can account for slow economic development. Much of sub-Saharan Africa remains stalled at the first stage of a demographic transition. Fertility rates actually increased a bit from the 1950s through the 1970s and only recently have begun a slow fall. As swollen youth cohorts have entered the labor force, an inadequate policy and economic environment in most countries has prevented many young people from being able to engage in productive employment. The existence of large dependent populations (in this case, of children) has kept the proportion of working-age people low, making it more difficult for these economies to rise out of poverty.

Looking to the Future

Based on the indicators that are available, we can make a few important points:

- *All signs point to continued but slowing population growth.* This growth will result in the addition of roughly 2.5 billion people to the world population, before it stabilizes around 2050 at about 9 billion. Managing this increase will be an enormous challenge, and the economic consequences of failing to do so could be severe.
- *The world's population is aging rapidly.* The United Nations predicts that 31 percent of China's population in 2050—432 million people—will be age 60 or older.

The corresponding figures for India are 21 percent and 330 million. No longer can aging be thought of as just a developed-world phenomenon.

- *International migration will continue, but the extent is unclear.* The pressures that encourage people to migrate—above all, the lure of greater economic well-being in the developed countries—will undoubtedly persist, but the strength of countervailing policy restrictions that could substantially stanch the flow of migrants is impossible to predict.
- *Urbanization will continue, but the pace is also hard to predict.* Greater economic opportunities in the cities will surely continue to attract migrants from rural areas, but environmental and social problems may stymie growth.

Getting the Focus Right

Rapid and significant demographic change places new demands on national and international policymaking. Transitions from high mortality and fertility to low mortality and fertility can be beneficial to economies as large baby-boom cohorts enter the workforce and save for retirement. Rising longevity also tends to increase the incentives to save for old age.

The ability of countries to realize the potential benefits of the demographic transition and to mitigate the negative effects of aging depends crucially on the policy and institutional environment. Focusing on the following areas is likely to be key:

Health and nutrition. Although it has long been known that increased income leads to improved health, recent evidence indicates that good health may also be an important factor in economic development. Good nutrition in children is essential for brain development and for allowing them to become productive members of society. Health improvements—especially among infants and children—often lead to declines in fertility, above and beyond the heightened quality of life they imply. Focusing on the diseases of childhood can therefore increase the likelihood of creating a boom generation and certain positive economic effects. Countries wishing to accelerate fertility declines may benefit from focusing on access to family-planning services and education about fertility decisions.

The ability of countries to realize the potential benefits of the demographic transition and to mitigate the negative effects of aging depends crucially on the policy and institutional environment.

Education. Children are better able to contribute to economic growth as they enter the workforce if they have received an effective education. East Asia capitalized on its baby boom by giving its children a high-quality education, including both general schooling and technical skills, that equipped them to meet the demands of an ever-changing labor market. Ireland also profited from its baby boomers by introducing free secondary schooling and expanding tertiary education.

Labor market institutions. Restrictive labor laws can limit a country's ability to benefit from demographic change, particularly when they make it unduly difficult to hire and fire workers or to work part-time. International outsourcing, another controversial subject, may become an increasingly important means of meeting the demand for labor.

Trade. One way that East Asian countries provided their baby-boom cohorts with productive opportunities was by carefully opening up to international trade. By providing a new avenue for selling the region's output, this opening helped countries avoid the unemployment that could have arisen. We have found that open economies benefit much more from demographic change than the average, and that closed economies do not derive any statistically significant benefit from age structure changes.

Retirement. Population aging will require increased savings to finance longer retirements. This will likely affect financial markets, rates of return, and investment. In addition, as more people move into old age, health care costs will tend to increase, with the expansion of health care systems and growth in long-term care for the elderly. As nontradable, labor-intensive sectors with a low rate of technical progress, health care and elder care may slow economic growth. The ability of individuals to contribute to the financing of their retirement may be hampered by existing social security systems, many of which effectively penalize individuals who work beyond a fixed retirement age.

Although demographic changes are generally easier to predict than economic changes, the big picture outlook is nonetheless unclear. Indeed, many forces that affect the world's demographic profile are highly unpredictable. Will an outbreak of avian flu or another disease become pandemic, killing many millions and decimating economies? What happens if these diseases are, or become, drug-resistant? Conversely, scientific advances in areas such as genomics, contraceptive methods, or vaccines for diseases such as AIDS or malaria could save and improve millions of lives. Global warming and other environmental change could completely alter the context of demographic and economic predictions. Or—to take things to extremes—wars could result in massive premature mortality, thereby rendering irrelevant most predictions about demographic and related economic changes.

References

Bloom, David E., and David Canning, 2004, "Global Demographic Change: Dimensions and Economic Significance," in *Global Demographic Change: Economic Impacts and Policy Challenges,* proceedings of a symposium, sponsored by the Federal Reserve Bank of Kansas City, Jackson Hole, Wyoming, August 26–28, pp. 9–56.

Lee, Ronald, 2003, "The Demographic Transition: Three Centuries of Fundamental Change," *Journal of Economic Perspectives,* Vol. 17 (Fall), pp. 167–90.

National Research Council, 1986, *Population Growth and Economic Development: Policy Questions* (Washington: National Academies Press).

Critical Thinking

1. What are the potential environmental consequences of a world population of 9 billion?

2. Will there be enough earth resources, and will the earth's ecosystems be able to allow there to be 9 billion people to live like the average American? Why or why not?

3. Describe what associations smaller families, living longer, international migration, and continuing urbanization will have on the environment?

4. Underline the number of times the word *environment* is referred to in this article. Do you think in discussion about demographic upheaval it is important to include reference to "environment"? Why or why not?

5. The article was written in 2006. Do some research on the Internet sites provided. What updates to the author's predictions can you offer?

6. Identify in this article three key terms, concepts, or principles that are used in your textbook (environmental science, economics, sociology, history, geography, etc.) or employed in the discipline you are currently studying. (Note: The terms, concepts, or principles may be implicit, explicit, implied, or inferred.)

DAVID E. BLOOM is Professor of Economics and Demography and **DAVID CANNING** is Professor of Economics and International Health at the Harvard School of Public Health.

Consumer Trends in Three Different "Worlds"

In the first of a two-part series, a professional futurist and business-trend watcher looks at the big trends in demography, money, and consumerism that will shape the world in the next decade.

ANDY HINES

What are the top 20 trends affecting consumer life around the world? The futurists and analysts at Social Technologies have done a number of deep dives into the topic and produced a list of "durable" trends that are sure to be important over the next decade and beyond. We grouped these trends into five categories: demography, rising wealth, culture, infrastructure, and values, and we suggest business implications for each of these trends.

Throughout this analysis, we refer to World 1 (or W1) World 2 (or W2), and World 3 (or W3) as our three "worlds" based on an index that rates a country's economic and social development and technological capability, then groups them among peers. W1 consists of fewer than a billion people in the affluent countries, including the United States, Western Europe, Japan, South Korea, and Australia. W3 consists of the 1 to 2 billion people who are in dire straits, including most of Africa, Bangladesh, and Haiti. The W2 world is the large segment of 3 to 4 billion in the middle, in nations that are relatively balanced in terms of needs and resources, though still vulnerable to setbacks. Some W2 nations, such as India and China, are growing particularly rapidly, while nations in other regions, such as in Latin America and eastern Europe, less so.

Look to these trends to be influential, assuming relative economic stability, over the next decade.

Demographic Trends

Demography is a solid starting point for our trend analysis, because demographic numbers are fairly predictable. However, when we look at the future of demography, a rule of thumb is that the *exception* is becoming the rule.

Trend 1: Aging Population

We are all familiar with the notion of an aging population. Imagine the world resembled Florida or Italy, where the median age will be 50 by mid-century. The populations of the world's most developed, W1, countries are skewing toward more older people and fewer younger people ("aging," in other words). W2 is just a generation behind. While those in the United States might complain about the "broken" Social Security system, most of W2 and W3 don't even *have* a social security system. Of course, opportunity emerges from need, and the financial services in these countries are poised for explosive growth, particularly in W1 countries.

An interesting aspect of aging in nations like the United States is the increasing obsolescence of retirement. U.S. baby boomers are choosing post-work lifestyles that don't resemble the stereotype of the quaint, restful senior citizen. Surveys reveal that about three-fourths of boomers plan to keep working beyond retirement age. Boomers are going to live longer, generally be in better health, and have a better education than previous generations. Why not work longer?

A futurist who joined Social Technologies about a year ago is a poster child for the post-work boomer. She retired in 2006 after 30 years with a large multinational corporation. She works for us part time because she finds futures work interesting and is willing to accept less money than she made in the corporate world. She also teaches a class at a university near her home—where she is working on another degree for herself. She is active in her church, goes to boot camp fitness training at 5 A.M. *every* day, takes motorcycle trips with her boomer husband, and is currently planning to run the Rome Marathon. I jokingly tell her she is going to have to go back to work to get some rest!

Business implications. In their quest to discover new activities and have new experiences, non-retiring seniors will look for ways to stay active despite diminishing physical abilities. Companies will surely want to tap their valuable expertise and experience. However, many seniors will insist on working on their own terms, setting their own working conditions or even establishing home-based freelance businesses. The urban renewal trend under way in cities like Baltimore and San Francisco—where empty-nesters leave the suburbs and good school districts to relocate to high-rise condos in downtown areas—is also part of this trend.

Trend 2: Changing Families

It's time to put the traditional family on the endangered species list in W1. "Alternative" family arrangements, on the fringe a generation ago, are now mainstream. Consider that 51% of adult women are now without a spouse and 37% of babies are born out of wedlock in the United States. Couples are delaying childbirth to pursue career goals and then generally having smaller families.

Single-person households are the fastest-growing type of household in the United States—second only to multigeneration households, in which parents move in with their kids who still have kids of their own at home. This isn't surprising considering the cost of raising and schooling children in many W1 countries. In March 2008, the U.S. Department of Agriculture deduced that the cost of clothing, housing, and educating a child in the United States until the age of 18 is $204,060.

Kids now grow up faster and, ironically, adults are "growing up" more slowly. Children absorb adult messages from media and are moving through childhood more quickly. Meanwhile, many adults are seeking to recapture their lost childhoods—these youth-seeking grown-ups have been called "rejuveniles" or "kidults." Those who come back to live with their parents have been dubbed "boomerangers."

An indicator of this demographic boundary blurring is the formation of adult kickball leagues and participation in other activities formerly associated with children. At the same time, adults are going back to school in record numbers—in many cases outnumbering youths in the classroom.

Business implications. Given all these changes in the nature of families, organizations will need to reconsider how they think about families and how they communicate with them. These shifts are already driving market researchers crazy—at least those who still rely on traditional age-based segmentations. Age is simply not a reliable indicator of people's actions or interests. The choices open to people across life stages have expanded, and aging populations are taking full advantage of them.

Trend 3: Migration

It's logistically easier now than ever before to relocate from one area to another—be it across a country or around the world. This phenomenon is raising challenging social and political issues. In W1, the debate continues over whether immigrants are a net benefit or drain on the recipient economy, and it has become a perennially popular campaign issue for some politicians, as well. Do we let them in? Do we keep them out? Several European nations are confronting these issues, made particularly acute as their native-born population growth has stopped and in some cases declined. Incorporating immigrants into national social/entitlement systems can present major political challenges.

It is likely that politicians will continue to make a lot of noise about "cracking down" on illegal immigration, but in the end, we forecast, the net benefits that immigrants bring by doing both manual and high-value-added knowledge work will keep the flows of migrants coming. One of the traditional strengths of the U.S. economy, for example, has been its willingness to embrace new people with new ideas. Many argue that it has been a key competitive advantage.

Business implications. Organizations struggling to deal with the issue of growing diversity are particularly affected by this trend. In the future, more organizations will find themselves confronted with having to diversify their workforces in the face of a global economy that demands an international perspective to succeed.

Trend 4: Population Growth

About 10 to 15 years ago, a series of books came out sounding the alarm that population growth threatens Earth's carrying capacity. The most recent statistics from the U.S. Census Bureau suggest that global population is going to top off at somewhere between 9 billion and 10 billion people (we're closing in on 7 billion today).

W2 and W3 have been proceeding through the demographic transition more quickly than some had anticipated, important because 95% of population growth will come from these worlds. As these nations develop economically, large families are confronting disincentives in the form of the high cost of raising children. Previously, more children meant more labor to work the farm. But the process of industrialization has diminished the economic advantages of having kids in the developed world. Most European countries and Japan are already below replacement-level fertility, and the projection is that global population growth will reach this mark by 2050.

Business implications. Ironically, a newly emerging "worry" is that soon W1 will have too few people and too few workers. This concern results from the common misconception that economies and organizations must continue to grow. The challenge brought by slowing population increases is that some of the "automatic" growth that businesses have experienced due to ever-higher GDP (fueled by population increases) will gradually decline. This is already happening in the United States, Europe, and parts of Asia. W2 and W3 are rapidly catching up.

Trend 5: Urbanization

The majority of the world's population is now living in cities. In W1, for instance, urbanization rates typically hover around 75%. W2 and W3 are in the process of rapid urbanization, as many of those countries are transitioning from an agricultural to an industrial economy. When that happens, people tend to settle in huge, multimillion-person megalopolises, often around national capital cities; one exception is the Nigerian mega-city of Lagos.

Thanks to technology, anyone armed with a laptop or PDA can connect to the world, and the workplace, via the Internet. As a result, some interesting choices will emerge about where to live and work. For decades, urbanization has been agglomerating people together, grouping them in big brick or glass tubes called office buildings after they fight a daunting commute and joust for expensive parking spots. Many of these people go up 30 floors to sit in a cubicle and use a computer or phone all day—the typical activities of a knowledge worker. Why fight this battle in an age of telecommunications? Why not work at home, the local coffee shop, or any venue where you can plug in and stay connected, going to the office only when it's necessary to physically meet with people?

Business implications. Managing virtual workforces will present challenges to organizations. In the past, measuring the "good" workers by how many hours they spent at their desks was easy. Organizations will now need to devise new ways to measure productivity based on output or value created, rather than hours spent in the office.

Rising Wealth Trends

We characterize this economically based set of consumer trends as "rising wealth" to reflect the world's rising standard of living—particularly in W2 nations such as China. Interestingly, this trend is challenging the fundamental assumption of scarcity, defined as the assumption that, in order for there to be wealthy people, there must also be poor people, and that a certain amount of lethal poverty is inevitable. Rising wealth suggests a need for an economic system based on a new assumption of abundance. The global economy produces more than enough wealth to support a comfortable standard of living for all three worlds, but it still hasn't figured out an equitable way to distribute that wealth or produce it sustainably.

The open-source movement (where computer developers post code they've developed online for free to be used and added to by other developers) is an example of how an abundance economy might work: Contributors give away their expertise to have a better product; in the process, they enhance their reputations and build networking connections that may later pay off in a "real" project or job. The open-source idea is spreading beyond software development. There is even a Web site for open-source prosthetics.

Another indicator among young people, particularly in W1, is that they are less inclined to follow traditional paths to earning a living. Some, for example, may spend their day blogging and not getting paid, because they believe that if they provide value it will pay off at some point, sometime, somewhere. And they are often right. Somebody picks up one of their stories, which builds their reputation and eventually leads to a paid gig. We might call it the "karma economy."

Trend 6: Asia Rising

The rise of Asia became real for me personally when I was chatting with a U.S. representative from a large Korean conglomerate. He was telling me that his company was interested in entering the U.S. market, but that their big push in the next five to 10 years was going be in China, because they saw it as a bigger opportunity.

There are various estimates of when China will be the world's largest economy, and there will be many debates about whether India might be the winner. Both China and India are ascending the economic ladder of development; workers are becoming more educated and skilled as legitimate players in the global knowledge economy. Singapore, for example, is a small island of 5 million that has ascended from W3 to W1 in just a generation. It has become one of the richest countries in the world thanks to the technological and knowledge-based prowess of its people.

Business implications. The twenty-first century can be characterized as the Asian century. Asia's economic and political clout is clearly rising. Many in the United States (the number one economy for some time now) are frightened by the trend. But it's good news for the global economy. These W2 and W3 nations will be more economically and militarily competitive, but they will also offer large and lucrative markets for W1 organizations.

Trend 7: Consumerism

For the first time, hundreds of millions of people in W2 and, to a lesser extent, W3 will be living in a consumer economy. As these nations become wealthier, the work and lifestyle patterns of their people will shift—and soon they will be able to purchase more goods and services as they build economies of specialized labor. Over time, the purchase of consumer goods becomes an increasingly important expression of one's values and identity. For instance, being able to afford luxury branded goods is a way to signal to one's peers that one has "made it."

In W1, on the other hand, there are growing questions about consumerism. One could argue that a primary assumption of our capitalist system is that consumption will make you happy. The more stuff one accumulates, the happier one should be. Data convincingly suggests that this is a dangerously flawed assumption. Yes, there is a correlation between income growth and happiness as nations modernize and citizens move beyond a survival mode. But in affluent, developed-world countries, studies show that more money does not equal more happiness.

Although consumerism is still alive and well in the developed world, some early indicators suggest it may be changing. For example, there is the story of 10 San Francisco friends who vowed not to purchase a single new thing in 2006 except food and some bare necessities such as toilet paper, brake fluid, and underwear. People in more-developed economies will have to make choices about how to pursue happiness in the future.

Business implications. The evolution of consumerism is likely to play out in different ways around the world, and companies would be well-advised to develop different strategies for approaching it in the different "worlds." For example, whereas material goods are likely to thrive in W2 nations, having rich personal experiences is increasingly important for people in more-developed countries. And an ongoing need for W3 is to devise strategies to reach those at the "bottom of the pyramid."

Trend 8: Middle-Class Growth

In affluent countries, particularly the United States, the middle class is shrinking as the upper and lower classes expand. But rather than haves and have-nots, it is more accurate to say the divisions are between haves and have-lots: The standard of living at the bottom of W1 is still higher than in much of W2 and W3.

While the size of the growing W2 middle class in countries like China is difficult to nail down, it is growing along a similar trajectory to the one that places like Europe, the United States, and later Japan underwent a few generations ago, which created large numbers of "first-time buyers." These markets will be the places to look for sales volumes to increase. Given that many affluent-nation populations and their middle classes are shrinking, the automatic growth that many businesses have become accustomed to will decline in these places. Thus, attention will turn to W2 and eventually W3 for growing the top line.

Business implications. In the bifurcating W1 markets, a pattern of status-conscious shopping is emerging. Increasingly, consumers are willing to spend lots of money on "identity-related" purchases, while other less-important purchases are based solely on cost. Thus, we have people shopping at Gucci for a handbag and at Wal-Mart for dishwasher detergent and a new lawnmower. This suggests a need for business models where manufacturers and retailers can profit by making small amounts of customized products—especially those identity-based products.

Trend 9: Time Pressure

As never before, consumers in wealthy W1 nations consistently say they never have enough time to do everything on their to-do lists. Of course, the perception is often worse than the reality, as some of the pressure is self-imposed. A time-use study of the U.S. population by sociologist John Robinson at the University of Maryland suggests that people have more free time now than in the past—they just fill it up as soon as they get it. The complaint goes something like: "Oh, I don't have to go to choir practice. Great, now I can do boot camp. . . . I'm so busy!" A popular lament in some psychotherapy sessions is that people are becoming human "doings" rather than human beings, because they constantly feel the need to be *doing* something. Robinson's research also suggests that free time is now available in smaller chunks—convenient for checking e-mail, but inconvenient for reading a book.

In W2 and to a lesser extent W3, free time is more likely to actually decrease as people spend more time working in their growing economies. This will be even more the case as those economies experience rapid growth.

Business implications. Time will continue to be at a premium. Don't believe it? Survey your colleagues and ask whether they would prefer a week of unfettered vacation—not the kind where it means double the work upon return—or the equivalent sum in pay. Don't be surprised when vacation wins hands-down. The time shortage means that businesses may benefit if they provide time-saving products. It also serves as a reminder of the value of customer service. An automated call-response system that costs less than flesh-and-blood phone operators isn't necessarily saving you money. If it's a hassle to use, it isn't just annoying your customers, it's robbing them.

Trend 10: Personal Outsourcing

Consumers used to save money by doing their own chores, but today's time shortage is leading consumers to outsource services they used to provide for themselves. Thus, a new breed of time-saving products and services is emerging. In W1, for example, pet sitters and personal concierges are both growth jobs. In W2, labor-saving devices such as refrigerators and dishwashers are as sought after and trendy as they were in the 1950s in the United States. These purchases free up time for participation in the paid workforce.

The growing value of time-saving technologies and services will bring up some interesting choices for consumers. Is gardening a hobby or a chore to be outsourced? Should I mow my own lawn? Is having a housekeeper a practical choice or an unnecessary indulgence? In many cases, it may make economic sense to outsource an activity.

Business implications. Some people may feel guilty about spending money to save time, but as time pressure mounts, statistics suggest more people will get over their reservations. This values shift will provide great opportunities for an increasing number of businesses to earn money by offering convenience, performing tasks that people could do for themselves but haven't the patience for any longer.

This partial list is intended to provide a primer on the global trends in play today and how they might develop and evolve into the next decade. It is a starting point for analysis, and we'll follow up in the next issue with trends related to culture, social mobility, and women's equality. We've suggested some of the business implications of these trends, but all of them also carry consequences for individuals, organizations, and nations.

Understanding established trends such as these provides a foundation for thinking in a productive way about the future. The next step is to apply the knowledge by uncovering new opportunities, detecting threats, crafting strategy, guiding policy, and exploring new markets, products, and services.

Critical Thinking

1. The author describes 10 consumer trends and then their "business implications." Discuss the "environmental implications" of those trends.

2. The article analyzes these trends via three different worlds. How are these three worlds defined? In view of these three categories, compare/contrast the characteristics of the natural environments of each country. What implications do you see regarding their consumer trends?

3. The article was written in 2008; what environmental issues exist in the world now (2012) that may profoundly alter the future of these predictions?

4. Describe the demographic scenario you see for the world of year 2050, and include the environmental changes that may be part of that scenario.

5. Identify in this article three key terms, concepts, or principles that are used in your textbook (environmental science, economics, sociology, history, geography, etc.) or employed in the discipline you are currently studying. (Note: The terms, concepts, or principles may be implicit, explicit, implied, or inferred.)

ANDY HINES is the director of Custom Projects at Social Technologies. He co-founded and is currently chair of the Association of Professional Futurists. *Thinking About the Future,* his third book, co-edited with Peter Bishop, was published by Social Technologies in 2006.

This article focuses on some the Top 20 trends affecting consumer life around the world, as judged by the futurists of Social Technologies. The list was developed for the sponsors of Social Technologies Global Lifestyles research program. The second part of this list, detailing trends in culture, gender equity, etc., will appear in the September–October 2008 edition of THE FUTURIST. For more information, visit www.socialtechnologies.com.

The New Population Bomb: The Four Megatrends That Will Change the World

JACK A. GOLDSTONE

Forty-two years ago, the biologist Paul Ehrlich warned in The Population Bomb that mass starvation would strike in the 1970s and 1980s, with the world's population growth outpacing the production of food and other critical resources. Thanks to innovations and efforts such as the "green revolution" in farming and the widespread adoption of family planning, Ehrlich's worst fears did not come to pass. In fact, since the 1970s, global economic output has increased and fertility has fallen dramatically, especially in developing countries.

The United Nations Population Division now projects that global population growth will nearly halt by 2050. By that date, the world's population will have stabilized at 9.15 billion people, according to the "medium growth" variant of the UN's authoritative population database World Population Prospects: The 2008 Revision. (Today's global population is 6.83 billion.) Barring a cataclysmic climate crisis or a complete failure to recover from the current economic malaise, global economic output is expected to increase by two to three percent per year, meaning that global income will increase far more than population over the next four decades.

But twenty-first-century international security will depend less on how many people inhabit the world than on how the global population is composed and distributed: where populations are declining and where they are growing, which countries are relatively older and which are more youthful, and how demographics will influence population movements across regions.

These elements are not well recognized or widely understood. A recent article in The Economist, for example, cheered the decline in global fertility without noting other vital demographic developments. Indeed, the same UN data cited by The Economist reveal four historic shifts that will fundamentally alter the world's population over the next four decades: the relative demographic weight of the world's developed countries will drop by nearly 25 percent, shifting economic power to the developing nations; the developed countries' labor forces will substantially age and decline, constraining economic growth in the developed world and raising the demand for immigrant workers; most of the world's expected population growth will increasingly be concentrated in today's poorest, youngest, and most heavily Muslim countries, which have a dangerous lack of quality education, capital, and employment opportunities; and, for the first time in history, most of the world's population will become urbanized, with the largest urban centers being in the world's poorest countries, where policing, sanitation, and health care are often scarce. Taken together, these trends will pose challenges every bit as alarming as those noted by Ehrlich. Coping with them will require nothing less than a major reconsideration of the world's basic global governance structures.

Europe's Reversal of Fortunes

At the beginning of the eighteenth century, approximately 20 percent of the world's inhabitants lived in Europe (including Russia). Then, with the Industrial Revolution, Europe's population boomed, and streams of European emigrants set off for the Americas. By the eve of World War I, Europe's population had more than quadrupled. In 1913, Europe had more people than China, and the proportion of the world's population living in Europe and the former European colonies of North America had risen to over 33 percent. But this trend reversed after World War I, as basic health care and sanitation began to spread to poorer countries. In Asia, Africa, and Latin America, people began to live longer, and birthrates remained high or fell only slowly. By 2003, the combined populations of Europe, the United States, and Canada accounted for just 17 percent of the global population. In 2050, this figure is expected to be just 12 percent—far less than it was in 1700. (These projections, moreover, might even understate the reality because they reflect the "medium growth" projection of the UN forecasts, which assumes that the fertility rates of developing countries will decline while those of developed countries will increase. In fact, many developed countries show no evidence of increasing fertility rates.) The West's relative decline is even more dramatic if one also considers changes in income. The Industrial Revolution made Europeans not only more numerous than they had been but also considerably richer per capita than others worldwide. According to the economic historian Angus Maddison, Europe, the United States, and Canada together produced about 32 percent of the world's GDP at the beginning of the nineteenth century. By 1950, that proportion had increased to a remarkable 68 percent of the world's total output (adjusted to reflect purchasing power parity).

This trend, too, is headed for a sharp reversal. The proportion of global GDP produced by Europe, the United States, and Canada fell from 68 percent in 1950 to 47 percent in 2003 and will decline even more steeply in the future. If the growth rate of per capita income (again, adjusted for purchasing power parity) between 2003 and 2050 remains as it was between 1973 and 2003—averaging 1.68 percent annually in Europe, the United States, and Canada and 2.47 percent annually in the rest of the world—then the combined GDP of Europe, the United States, and Canada will roughly double by 2050, whereas the GDP of the rest of the world will grow by a factor of five. The portion of global GDP produced by Europe, the United States, and Canada in 2050 will then be less than 30 percent—smaller than it was in 1820.

These figures also imply that an overwhelming proportion of the world's GDP growth between 2003 and 2050—nearly 80 percent—will occur outside of Europe, the United States, and Canada. By the middle of this century, the global middle class— those capable of purchasing durable consumer products, such as cars, appliances, and electronics—will increasingly be found in what is now considered the developing world. The World Bank has predicted that by 2030 the number of middle-class people in the developing world will be 1.2 billion—a rise of 200 percent since 2005. This means that the developing world's middle class alone will be larger than the total populations of Europe, Japan, and the United States combined. From now on, therefore, the main driver of global economic expansion will be the economic growth of newly industrialized countries, such as Brazil, China, India, Indonesia, Mexico, and Turkey.

Aging Pains

Part of the reason developed countries will be less economically dynamic in the coming decades is that their populations will become substantially older. The European countries, Canada, the United States, Japan, South Korea, and even China are aging at unprecedented rates. Today, the proportion of people aged 60 or older in China and South Korea is 12–15 percent. It is 15–22 percent in the European Union, Canada, and the United States and 30 percent in Japan. With baby boomers aging and life expectancy increasing, these numbers will increase dramatically. In 2050, approximately 30 percent of Americans, Canadians, Chinese, and Europeans will be over 60, as will more than 40 percent of Japanese and South Koreans.

Over the next decades, therefore, these countries will have increasingly large proportions of retirees and increasingly small proportions of workers. As workers born during the baby boom of 1945–65 are retiring, they are not being replaced by a new cohort of citizens of prime working age (15–59 years old).

Industrialized countries are experiencing a drop in their working-age populations that is even more severe than the over-all slowdown in their population growth. South Korea represents the most extreme example. Even as its total population is projected to decline by almost 9 percent by 2050 (from 48.3 million to 44.1 million), the population of working-age South Koreans is expected to drop by 36 percent (from 32.9 million to 21.1 million), and the number of South Koreans aged 60 and

older will increase by almost 150 percent (from 7.3 million to 18 million). By 2050, in other words, the entire working-age population will barely exceed the 60-and-older population. Although South Korea's case is extreme, it represents an increasingly common fate for developed countries. Europe is expected to lose 24 percent of its prime working-age population (about 120 million workers) by 2050, and its 60-and-older population is expected to increase by 47 percent. In the United States, where higher fertility and more immigration are expected than in Europe, the working-age population will grow by 15 percent over the next four decades—a steep decline from its growth of 62 percent between 1950 and 2010. And by 2050, the United States' 60-and-older population is expected to double.

All this will have a dramatic impact on economic growth, health care, and military strength in the developed world. The forces that fueled economic growth in industrialized countries during the second half of the twentieth century—increased productivity due to better education, the movement of women into the labor force, and innovations in technology—will all likely weaken in the coming decades. College enrollment boomed after World War II, a trend that is not likely to recur in the twenty-first century; the extensive movement of women into the labor force also was a one-time social change; and the technological change of the time resulted from innovators who created new products and leading-edge consumers who were willing to try them out—two groups that are thinning out as the industrialized world's population ages.

Overall economic growth will also be hampered by a decline in the number of new consumers and new households. When developed countries' labor forces were growing by 0.5–1.0 percent per year, as they did until 2005, even annual increases in real output per worker of just 1.7 percent meant that annual economic growth totaled 2.2–2.7 percent per year. But with the labor forces of many developed countries (such as Germany, Hungary, Japan, Russia, and the Baltic states) now shrinking by 0.2 percent per year and those of other countries (including Austria, the Czech Republic, Denmark, Greece, and Italy) growing by less than 0.2 percent per year, the same 1.7 percent increase in real output per worker yields only 1.5–1.9 percent annual overall growth. Moreover, developed countries will be lucky to keep productivity growth at even that level; in many developed countries, productivity is more likely to decline as the population ages.

A further strain on industrialized economies will be rising medical costs: as populations age, they will demand more health care for longer periods of time. Public pension schemes for aging populations are already being reformed in various industrialized countries—often prompting heated debate. In theory, at least, pensions might be kept solvent by increasing the retirement age, raising taxes modestly, and phasing out benefits for the wealthy. Regardless, the number of 80- and 90-year-olds—who are unlikely to work and highly likely to require nursing-home and other expensive care—will rise dramatically. And even if 60- and 70-year-olds remain active and employed, they will require procedures and medications—hip replacements, kidney transplants, blood-pressure treatments— to sustain their health in old age.

All this means that just as aging developed countries will have proportionally fewer workers, innovators, and consumerist young households, a large portion of those countries' remaining economic growth will have to be diverted to pay for the medical bills and pensions of their growing elderly populations. Basic services, meanwhile, will be increasingly costly because fewer young workers will be available for strenuous and labor-intensive jobs. Unfortunately, policymakers seldom reckon with these potentially disruptive effects of otherwise welcome developments, such as higher life expectancy.

Youth and Islam in the Developing World

Even as the industrialized countries of Europe, North America, and Northeast Asia will experience unprecedented aging this century, fast-growing countries in Africa, Latin America, the Middle East, and Southeast Asia will have exceptionally youthful populations. Today, roughly nine out of ten children under the age of 15 live in developing countries. And these are the countries that will continue to have the world's highest birthrates. Indeed, over 70 percent of the world's population growth between now and 2050 will occur in 24 countries, all of which are classified by the World Bank as low income or lower-middle income, with an average per capita income of under $3,855 in 2008.

Many developing countries have few ways of providing employment to their young, fast-growing populations. Would-be laborers, therefore, will be increasingly attracted to the labor markets of the aging developed countries of Europe, North America, and Northeast Asia. Youthful immigrants from nearby regions with high unemployment—Central America, North Africa, and Southeast Asia, for example—will be drawn to those vital entry-level and manual-labor jobs that sustain advanced economies: janitors, nursing-home aides, bus drivers, plumbers, security guards, farm workers, and the like. Current levels of immigration from developing to developed countries are paltry compared to those that the forces of supply and demand might soon create across the world.

These forces will act strongly on the Muslim world, where many economically weak countries will continue to experience dramatic population growth in the decades ahead. In 1950, Bangladesh, Egypt, Indonesia, Nigeria, Pakistan, and Turkey had a combined population of 242 million. By 2009, those six countries were the world's most populous Muslim-majority countries and had a combined population of 886 million. Their populations are continuing to grow and indeed are expected to increase by 475 million between now and 2050—during which time, by comparison, the six most populous developed countries are projected to gain only 44 million inhabitants. Worldwide, of the 48 fastest-growing countries today—those with annual population growth of two percent or more—28 are majority Muslim or have Muslim minorities of 33 percent or more.

It is therefore imperative to improve relations between Muslim and Western societies. This will be difficult given that many Muslims live in poor communities vulnerable to radical appeals and many see the West as antagonistic and militaristic.

In the 2009 Pew Global Attitudes Project survey, for example, whereas 69 percent of those Indonesians and Nigerians surveyed reported viewing the United States favorably, just 18 percent of those polled in Egypt, Jordan, Pakistan, and Turkey (all U.S. allies) did. And in 2006, when the Pew survey last asked detailed questions about Muslim-Western relations, more than half of the respondents in Muslim countries characterized those relations as bad and blamed the West for this state of affairs.

But improving relations is all the more important because of the growing demographic weight of poor Muslim countries and the attendant increase in Muslim immigration, especially to Europe from North Africa and the Middle East. (To be sure, forecasts that Muslims will soon dominate Europe are outlandish: Muslims compose just three to ten percent of the population in the major European countries today, and this proportion will at most double by midcentury.) Strategists worldwide must consider that the world's young are becoming concentrated in those countries least prepared to educate and employ them, including some Muslim states. Any resulting poverty, social tension, or ideological radicalization could have disruptive effects in many corners of the world. But this need not be the case; the healthy immigration of workers to the developed world and the movement of capital to the developing world, among other things, could lead to better results.

Urban Sprawl

Exacerbating twenty-first-century risks will be the fact that the world is urbanizing to an unprecedented degree. The year 2010 will likely be the first time in history that a majority of the world's people live in cities rather than in the countryside. Whereas less than 30 percent of the world's population was urban in 1950, according to UN projections, more than 70 percent will be by 2050.

Lower-income countries in Asia and Africa are urbanizing especially rapidly, as agriculture becomes less labor intensive and as employment opportunities shift to the industrial and service sectors. Already, most of the world's urban agglomerations—Mumbai (population 20.1 million), Mexico City (19.5 million), New Delhi (17 million), Shanghai (15.8 million), Calcutta (15.6 million), Karachi (13.1 million), Cairo (12.5 million), Manila (11.7 million), Lagos (10.6 million), Jakarta (9.7 million)—are found in low-income countries. Many of these countries have multiple cities with over one million residents each: Pakistan has eight, Mexico 12, and China more than 100. The UN projects that the urbanized proportion of sub-Saharan Africa will nearly double between 2005 and 2050, from 35 percent (300 million people) to over 67 percent (1 billion). China, which is roughly 40 percent urbanized today, is expected to be 73 percent urbanized by 2050; India, which is less than 30 percent urbanized today, is expected to be 55 percent urbanized by 2050. Overall, the world's urban population is expected to grow by 3 billion people by 2050.

This urbanization may prove destabilizing. Developing countries that urbanize in the twenty-first century will have far lower per capita incomes than did many industrial countries when they first urbanized. The United States, for example, did not reach 65 percent urbanization until 1950, when per capita income was nearly $13,000 (in 2005 dollars). By contrast, Nigeria, Pakistan, and the Philippines, which are approaching similar levels of urbanization, currently have per capita incomes of just $1,800–$4,000 (in 2005 dollars).

According to the research of Richard Cincotta and other political demographers, countries with younger populations are especially prone to civil unrest and are less able to create or sustain democratic institutions. And the more heavily urbanized, the more such countries are likely to experience Dickensian poverty and anarchic violence. In good times, a thriving economy might keep urban residents employed and governments flush with sufficient resources to meet their needs. More often, however, sprawling and impoverished cities are vulnerable to crime lords, gangs, and petty rebellions. Thus, the rapid urbanization of the developing world in the decades ahead might bring, in exaggerated form, problems similar to those that urbanization brought to nineteenth-century Europe. Back then, cyclical employment, inadequate policing, and limited sanitation and education often spawned widespread labor strife, periodic violence, and sometimes—as in the 1820s, the 1830s, and 1848—even revolutions.

International terrorism might also originate in fast-urbanizing developing countries (even more than it already does). With their neighborhood networks, access to the Internet and digital communications technology, and concentration of valuable targets, sprawling cities offer excellent opportunities for recruiting, maintaining, and hiding terrorist networks.

Defusing the Bomb

Averting this century's potential dangers will require sweeping measures. Three major global efforts defused the population bomb of Ehrlich's day: a commitment by governments and nongovernmental organizations to control reproduction rates; agricultural advances, such as the green revolution and the spread of new technology; and a vast increase in international trade, which globalized markets and thus allowed developing countries to export foodstuffs in exchange for seeds, fertilizers, and machinery, which in turn helped them boost production. But today's population bomb is the product less of absolute growth in the world's population than of changes in its age and distribution. Policymakers must therefore adapt today's global governance institutions to the new realities of the aging of the industrialized world, the concentration of the world's economic and population growth in developing countries, and the increase in international immigration.

During the Cold War, Western strategists divided the world into a "First World," of democratic industrialized countries; a "Second World," of communist industrialized countries; and a "Third World," of developing countries. These strategists focused chiefly on deterring or managing conflict between the First and the Second Worlds and on launching proxy wars and diplomatic initiatives to attract Third World countries into the First World's camp. Since the end of the Cold War, strategists have largely abandoned this three-group division and have tended to believe either that the United States, as the sole superpower, would maintain a Pax Americana or that the world would become multipolar, with the United States, Europe, and China playing major roles.

Unfortunately, because they ignore current global demographic trends, these views will be obsolete within a few decades. A better approach would be to consider a different three-world order, with a new First World of the aging industrialized nations of North America, Europe, and Asia's Pacific Rim (including Japan, Singapore, South Korea, and Taiwan, as well as China after 2030, by which point the one-child policy will have produced significant aging); a Second World comprising fast-growing and economically dynamic countries with a healthy mix of young and old inhabitants (such as Brazil, Iran, Mexico, Thailand, Turkey, and Vietnam, as well as China until 2030); and a Third World of fast-growing, very young, and increasingly urbanized countries with poorer economies and often weak governments. To cope with the instability that will likely arise from the new Third World's urbanization, economic strife, lawlessness, and potential terrorist activity, the aging industrialized nations of the new First World must build effective alliances with the growing powers of the new Second World and together reach out to Third World nations. Second World powers will be pivotal in the twenty-first century not just because they will drive economic growth and consume technologies and other products engineered in the First World; they will also be central to international security and cooperation. The realities of religion, culture, and geographic proximity mean that any peaceful and productive engagement by the First World of Third World countries will have to include the open cooperation of Second World countries.

Strategists, therefore, must fundamentally reconsider the structure of various current global institutions. The G-8, for example, will likely become obsolete as a body for making global economic policy. The G-20 is already becoming increasingly important, and this is less a short-term consequence of the ongoing global financial crisis than the beginning of the necessary recognition that Brazil, China, India, Indonesia, Mexico, Turkey, and others are becoming global economic powers. International institutions will not retain their legitimacy if they exclude the world's fastest-growing and most economically dynamic countries. It is essential, therefore, despite European concerns about the potential effects on immigration, to take steps such as admitting Turkey into the European Union. This would add youth and economic dynamism to the EU—and would prove that Muslims are welcome to join Europeans as equals in shaping a free and prosperous future. On the other hand, excluding Turkey from the EU could lead to hostility not only on the part of Turkish citizens, who are expected to number 100 million by 2050, but also on the part of Muslim populations worldwide.

NATO must also adapt. The alliance today is composed almost entirely of countries with aging, shrinking populations and relatively slow-growing economies. It is oriented toward the Northern Hemisphere and holds on to a Cold War structure that cannot adequately respond to contemporary threats. The young and increasingly populous countries of Africa, the Middle East, Central Asia, and South Asia could mobilize insurgents much more easily than NATO could mobilize the troops it would need if it were called on to stabilize those countries. Long-standing

NATO members should, therefore—although it would require atypical creativity and flexibility—consider the logistical and demographic advantages of inviting into the alliance countries such as Brazil and Morocco, rather than countries such as Albania. That this seems far-fetched does not minimize the imperative that First World countries begin including large and strategic Second and Third World powers in formal international alliances.

The case of Afghanistan—a country whose population is growing fast and where NATO is currently engaged—illustrates the importance of building effective global institutions. Today, there are 28 million Afghans; by 2025, there will be 45 million; and by 2050, there will be close to 75 million. As nearly 20 million additional Afghans are born over the next 15 years, NATO will have an opportunity to help Afghanistan become reasonably stable, self-governing, and prosperous. If NATO's efforts fail and the Afghans judge that NATO intervention harmed their interests, tens of millions of young Afghans will become more hostile to the West. But if they come to think that NATO's involvement benefited their society, the West will have tens of millions of new friends. The example might then motivate the approximately one billion other young Muslims growing up in low-income countries over the next four decades to look more kindly on relations between their countries and the countries of the industrialized West.

Creative Reforms at Home

The aging industrialized countries can also take various steps at home to promote stability in light of the coming demographic trends. First, they should encourage families to have more children. France and Sweden have had success providing child care, generous leave time, and financial allowances to families with young children. Yet there is no consensus among policymakers—and certainly not among demographers—about what policies best encourage fertility.

More important than unproven tactics for increasing family size is immigration. Correctly managed, population movement can benefit developed and developing countries alike. Given the dangers of young, underemployed, and unstable populations in developing countries, immigration to developed countries can provide economic opportunities for the ambitious and serve as a safety valve for all. Countries that embrace immigrants, such as the United States, gain economically by having willing laborers and greater entrepreneurial spirit. And countries with high levels of emigration (but not so much that they experience so-called brain drains) also benefit because emigrants often send remittances home or return to their native countries with valuable education and work experience.

One somewhat daring approach to immigration would be to encourage a reverse flow of older immigrants from developed to developing countries. If older residents of developed countries took their retirements along the southern coast of the Mediterranean or in Latin America or Africa, it would greatly reduce the strain on their home countries' public entitlement systems. The developing countries involved, meanwhile, would

benefit because caring for the elderly and providing retirement and leisure services is highly labor intensive. Relocating a portion of these activities to developing countries would provide employment and valuable training to the young, growing populations of the Second and Third Worlds.

This would require developing residential and medical facilities of First World quality in Second and Third World countries. Yet even this difficult task would be preferable to the status quo, by which low wages and poor facilities lead to a steady drain of medical and nursing talent from developing to developed countries. Many residents of developed countries who desire cheaper medical procedures already practice medical tourism today, with India, Singapore, and Thailand being the most common destinations. (For example, the international consulting firm Deloitte estimated that 750,000 Americans traveled abroad for care in 2008.)

Never since 1800 has a majority of the world's economic growth occurred outside of Europe, the United States, and Canada. Never have so many people in those regions been over 60 years old. And never have low-income countries' populations been so young and so urbanized. But such will be the world's demography in the twenty-first century. The strategic and economic policies of the twentieth century are obsolete, and it is time to find new ones.

Reference

Goldstone, Jack A. "The new population bomb: the four megatrends that will change the world." *Foreign Affair*s 89.1 (2010): 31. *General OneFile*. Web. 23 Jan. 2010. http://0-find.galegroup.com.www .consuls.org/gps/start.do?proId=IPS& userGroupName=a30wc.

Critical Thinking

1. What does the author contend will be the characteristics of future population growth?

2. The article argues that future population trends will have significant political and economic consequences. What do you see as the "environmental consequences"?

3. How might these environmental consequences vary around the world for different populations?

4. Summarize briefly the major thesis of each Article 2, 3, and 4, and list the similarities and differences (if any) regarding their demographic trend predictions.

5. Use your analyses of question 4 to discuss what consequences these trends may have regarding the achievment of MDG goal #7: environmental sustainability discussed in Article 1.

6. Identify in this article three key terms, concepts, or principles that are used in your textbook (environmental science, economics, sociology, history, geography, etc.) or employed in the discipline you are currently studying. (Note: The terms, concepts, or principles may be implicit, explicit, implied, or inferred.)

UNIT 2
Global Development

Unit Selections

5. **It's a Flat World, After All,** Thomas L. Friedman
6. **The World Is Spiky,** Richard Florida
7. **Promises and Poverty,** Tom Knudson
8. **A User's Guide to the Century,** Jeffrey D. Sachs

Learning Outcomes

After reading this Unit, you should be able to:

- Describe the global variability of environmental issue #2: global development, and assess the role that environmental resources play in that development.

- Evaluate the importance of access to natural and human capital with regard to economic development.

- Describe the global pattern of the "geography of innovation" and the environmental characteristics of the nodes.

- Discuss what assumptions need to be addressed regarding new technological trends of the growing global economy and potential environmental impacts.

- Explain what impacts "flattening" the world may have on achieving a sustainable relationship with the planet.

- Offer insights on why discourse about economic development cannot be discussed without also identifying its connections to environmental externalities.

- Articulate how the kinds of new global foreign policies we construct will impact directly the success of our achieving a sustainable relationship with the planet and its peoples.

Student Website

www.mhhe.com/cls

Internet References

Earth Renewal
www.earthrenewal.org/global_economics.htm

Geography and Socioeconomic Development
www.ksg.harvard.edu/cid/andes/Documents/Background%20Papers/Geography&Socioeconomic%20Development.pdf

Global Footprint Network
footprints@footprintnetwork.org

Global Trends Project
www.globaltrendsproject.org

Graphs Comparing Countries
http://humandevelopment.bu.edu/use_existing_index/start_comp_graph.cfm

IISDnet
www.iisd.org

United Nations
www.unsystem.org

United Nations Environment Programme (UNEP)
www.unep.ch

World Resources Institute (WRI)
www.wri.org

God forbid that India should ever take to industrialism after the manner of the west . . . keeping the world in chains. If [our nation] took to similar economic exploitation, it would strip the world bare like locusts.
— Mahatma Gandhi

If human population growth is posited as the first and foremost driver influencing the success of our ability to achieve a sustainable relationship with the Earth, the way we approach *Top Ten Environmental Issue 2,* Global Development, may most certainly determine the ultimate outcome of our success. Why? Because at present, political economy philosophies generally fall into one of two camps. For one group, the concept of development infers principles such as economic growth, gross national product (GNP), material wealth, and expanding consumption. Here, such growth is the result of innovations, technologies, and economic growth, and development is measured in terms of raising standards of living (e.g., having bottled water available at an affordable price in biodegradable bottles). For the other group, development means provision of and opportunities for basic sustenance, health, shelter, human self-esteem, freedom from tyranny and oppression, protection of human rights, education, and freedom of expression, to name a few (e.g., all people having affordable access to freshwater sources and maintaining sustainable levels of water consumption). Here, development is measured in terms of improving quality of life. Both development philosophies will require innovations, technologies, resource development, energy consumption, economic activities, production functions, and so on. However, although there is not necessarily a "right" or "wrong" way for a nation to "develop," the social, economic, and environmental consequences of those development inputs (innovations, technologies, etc.) can result in vastly different global endgames, depending on which development approach is focused on by any nation or society. We must remember that, like population growth, development requires human consumption of the Earth. And, we must constantly ask ourselves, "What are the endgames of our innovations, technologies, and economic growth? Freshwater is freshwater, regardless of whether it's served in a crystal glass, plastic bottle, or hand-carved gourd.

The four articles selected for **Unit 2** are intended to stimulate the reader's critical thinking about these endgames. For instance, what is the price to our environment of these technologies? What does a "spiky" versus "flat" socioeconomic world mean for our environment? What are the socioeconomic externalities that can occur even in an eco-friendly approach to the production functions required of development? What relationships can we uncover about the future use of our environment, and the impacts of that use, by examining the geo-economic patterns of the essential elements necessary for development (innovations, technology, science, economy, etc.)? What will humans do with their new technologies and innovation as we progress (develop)? What kind of global development foreign policies must we construct to ensure a future sustainable relationship with our planet and its peoples?

© Brand X Pictures/PunchStock

Unit 2 opens with Article 5, "It's a Flat World, After All." Thomas Friedman argues that certain technological trends (e.g., the information technology revolution) and globalization are leveling the global economic plying field and creating what he refers to as a "flat world." New patterns of global development are emerging, according to Friedman, due to new technology based geo-economics. There is no doubt that the evidence of economic development in regions such as Southeast Asia, China, and India support the argument. But is this the kind of development that is contributing to raising standards of living (material wealth) or improving quality of life? What are the environmental resource requirements of this leveling of the playing fields? Or can such leveling development help humanity achieve environmental sustainability?

Richard Florida disagrees with Thomas Friedman's "flat world" contention. In Article 6, "The World Is Spiky," the author argues that the socioeconomic topography of the world is not flat but, in fact, quite "spiky," and population, urbanization, inventions, science, and the like are not found on a level playing field. How is it possible for two people to look at the same world and see two very different images? Is one right and one wrong? But who is right or wrong is perhaps not the issue at hand. Perhaps it is critical questions we need to address such as, is the socioeconomic topography of access to those "resources" also highly variable, and how may that affect development policy choices? In what ways do the nodes of innovation and technology affect local environments or distant environments? Will achieving environmental sustainability depend on the author's current maps of peaks and valleys changing?

Article 7, "Promises and Poverty," provides another example of our global development policy choices and the impact those choices may have on our environment, but from a different perspective. If one looks hard enough, there frequently seems to be a paradox imbedded in the idea of "eco-friendly" economic production functions. A production function, by its very nature, requires the transformation of some raw material (typically, at some point, that material was a "natural resource") into a finished product that can be "consumed" by the human being. In this article, the product is coffee, and the producer is Starbucks. Although the company has prided itself in its eco-friendly production methods, author Tom Knudson takes a deeper look and finds the paradox. Behind the images of recycled, biodegradable cups, fair trade pricing, happy peasant coffee growers and commitment to sustainably develop lies the commodity production function, with all its negative social–economic–environmental externalities. Exploited farmers, impoverished communities, degraded soils, damaged ecosystems, and deforestation are the prices paid for a global development approach with a cup of coffee as its endgame. Certainly a good cup of coffee can *improve the quality of one's life* any day, but you had better be enjoying a pretty *high standard of living* to pay for that soy java chip Frappuchino.

Renowned economist Jeffrey Sachs describes a twenty-first-century world of converging technology and economics in Article 8, "A User's Guide to the Century." He argues that this could be a good thing, a not-so-good-thing for society and our environment. But the author proposes five guideposts that if followed, may help guide our global development foreign policies in a direction that can help us achieve shared prosperity without degrading the environment or increasing the risk of global conflicts. This article is included at the end of the unit to encourage readers to assess the challenges of development by reflecting on Sachs's interpretation of this converging world and evaluate his guidepost proposals in light of issues addressed in Units 1 (population), 3 (food), 4 (water), and 8 (energy).

It's a Flat World, After All

THOMAS L. FRIEDMAN

In 1492 Christopher Columbus set sail for India, going west. He had the Niña, the Pinta and the Santa Maria. He never did find India, but he called the people he met "Indians" and came home and reported to his king and queen: "The world is round." I set off for India 512 years later. I knew just which direction I was going. I went east. I had Lufthansa business class, and I came home and reported only to my wife and only in a whisper: "The world is flat."

And therein lies a tale of technology and geoeconomics that is fundamentally reshaping our lives—much, much more quickly than many people realize. It all happened while we were sleeping, or rather while we were focused on 9/11, the dot-com bust and Enron—which even prompted some to wonder whether globalization was over. Actually, just the opposite was true, which is why it's time to wake up and prepare ourselves for this flat world, because others already are, and there is no time to waste.

I wish I could say I saw it all coming. Alas, I encountered the flattening of the world quite by accident. It was in late February [2004], and I was visiting the Indian high-tech capital, Bangalore, working on a documentary for the Discovery Times channel about outsourcing. In short order, I interviewed Indian entrepreneurs who wanted to prepare my taxes from Bangalore, read my X-rays from Bangalore, trace my lost luggage from Bangalore and write my new software from Bangalore. The longer I was there, the more upset I became—upset at the realization that while I had been off covering the 9/11 wars, globalization had entered a whole new phase, and I had missed it. I guess the eureka moment came on a visit to the campus of Infosys Technologies, one of the crown jewels of the Indian outsourcing and software industry. Nandan Nilekani, the Infosys C.E.O., was showing me his global video-conference room, pointing with pride to a wall-size flat-screen TV, which he said was the biggest in Asia. Infosys, he explained, could hold a virtual meeting of the key players from its entire global supply chain for any project at any time on that supersize screen. So its American designers could be on the screen speaking with their Indian software writers and their Asian manufacturers all at once. That's what globalization is all about today, Nilekani said. Above the screen there were eight clocks that pretty well summed up the Infosys workday: 24/7/365. The clocks were labeled U.S. West, U.S. East, G.M.T., India, Singapore, Hong Kong, Japan, Australia.

"Outsourcing is just one dimension of a much more fundamental thing happening today in the world," Nilekani explained. "What happened over the last years is that there was a massive investment in technology, especially in the bubble era, when hundreds of millions of dollars were invested in putting broadband connectivity around the world, undersea cables, all those things." At the same time, he added, computers became cheaper and dispersed all over the world, and there was an explosion of e-mail software, search engines like Google and proprietary software that can chop up any piece of work and send one part to Boston, one part to Bangalore and one part to Beijing, making it easy for anyone to do remote development. When all of these things suddenly came together around 2000, Nilekani said, they "created a platform where intellectual work, intellectual capital, could be delivered from anywhere. It could be disaggregated, delivered, distributed, produced and put back together again—and this gave a whole new degree of freedom to the way we do work, especially work of an intellectual nature. And what you are seeing in Bangalore today is really the culmination of all these things coming together."

At one point, summing up the implications of all this, Nilekani uttered a phrase that rang in my ear. He said to me, "Tom, the playing field is being leveled." He meant that countries like India were now able to compete equally for global knowledge work as never before—and that America had better get ready for this. As I left the Infosys campus that evening and bounced along the potholed road back to Bangalore, I kept chewing on that phrase: "The playing field is being leveled."

"What Nandan is saying," I thought, "is that the playing field is being flattened. Flattened? Flattened? My God, he's telling me the world is flat!"

Here I was in Bangalore—more than 500 years after Columbus sailed over the horizon, looking for a shorter route to India using the rudimentary navigational technologies of his day, and returned safely to prove definitively that the world was round—and one of India's smartest engineers, trained at his country's top technical institute and backed by the most modern technologies of his day, was telling me that the world was flat, as flat as that screen on which he can host a meeting of his whole global supply chain. Even more interesting, he was citing this development as a new milestone in human progress and a great opportunity for India and the world—the fact that we had made our world flat!

This has been building for a long time. Globalization 1.0 (1492 to 1800) shrank the world from a size large to a size medium, and the dynamic force in that era was countries globalizing for resources and imperial conquest. Globalization 2.0 (1800 to 2000) shrank the world from a size medium to a size small, and it was spearheaded by companies globalizing for markets and labor. Globalization 3.0 (which started around 2000) is shrinking the world from a size small to a size tiny and flattening the playing field at the same time. And while the dynamic force in Globalization 1.0 was countries globalizing and the dynamic force in Globalization 2.0 was companies globalizing, the dynamic force in Globalization 3.0—the thing that gives it its unique character—is individuals and small groups globalizing. Individuals must, and can, now ask: where do I fit into the global competition and opportunities of the day, and how can I, on my own, collaborate with others globally? But Globalization 3.0 not only differs from the previous eras in how it is shrinking and flattening the world and in how it is empowering individuals. It is also different in that Globalization 1.0 and 2.0 were driven primarily by European and American companies and countries. But going forward, this will be less and less true. Globalization 3.0 is not only going to be driven more by individuals but also by a much more diverse—non-Western, nonwhite—group of individuals. In Globalization 3.0, you are going to see every color of the human rainbow take part.

"Today, the most profound thing to me is the fact that a 14-year-old in Romania or Bangalore or the Soviet Union or Vietnam has all the information, all the tools, all the software easily available to apply knowledge however they want," said Marc Andreessen, a co-founder of Netscape and creator of the first commercial Internet browser. "That is why I am sure the next Napster is going to come out of left field. As bioscience becomes more computational and less about wet labs and as all the genomic data becomes easily available on the Internet, at some point you will be able to design vaccines on your laptop."

Andreessen is touching on the most exciting part of Globalization 3.0 and the flattening of the world: the fact that we are now in the process of connecting all the knowledge pools in the world together. We've tasted some of the downsides of that in the way that Osama bin Laden has connected terrorist knowledge pools together through his Qaeda network, not to mention the work of teenage hackers spinning off more and more lethal computer viruses that affect us all. But the upside is that by connecting all these knowledge pools we are on the cusp of an incredible new era of innovation, an era that will be driven from left field and right field, from West and East and from North and South. Only 30 years ago, if you had a choice of being born a B student in Boston or a genius in Bangalore or Beijing, you probably would have chosen Boston, because a genius in Beijing or Bangalore could not really take advantage of his or her talent. They could not plug and play globally. Not anymore. Not when the world is flat, and anyone with smarts, access to Google and a cheap wireless laptop can join the innovation fray.

When the world is flat, you can innovate without having to emigrate. This is going to get interesting. We are about to see creative destruction on steroids.

How did the world get flattened, and how did it happen so fast?

It was a result of 10 events and forces that all came together during the 1990's and converged right around the year 2000. Let me go through them briefly. The first event was 11/9. That's right—not 9/11, but 11/9. Nov. 9, 1989, is the day the Berlin Wall came down, which was critically important because it allowed us to think of the world as a single space. "The Berlin Wall was not only a symbol of keeping people inside Germany; it was a way of preventing a kind of global view of our future," the Nobel Prize-winning economist Amartya Sen said. And the wall went down just as the windows went up—the breakthrough Microsoft Windows 3.0 operating system, which helped to flatten the playing field even more by creating a global computer interface, shipped six months after the wall fell.

The second key date was 8/9. Aug. 9, 1995, is the day Netscape went public, which did two important things. First, it brought the Internet alive by giving us the browser to display images and data stored on websites. Second, the Netscape stock offering triggered the dot-com boom, which triggered the dot-com bubble, which triggered the massive overinvestment of billions of dollars in fiber-optic telecommunications cable. That overinvestment, by companies like Global Crossing, resulted in the willy-nilly creation of a global undersea-underground fiber network, which in turn drove down the cost of transmitting voices, data and images to practically zero, which in turn accidentally made Boston, Bangalore and Beijing next-door neighbors overnight. In sum, what the Netscape revolution did was bring people-to-people connectivity to a whole new level. Suddenly more people could connect with more other people from more different places in more different ways than ever before.

No country accidentally benefited more from the Netscape moment than India. "India had no resources and no infrastructure," said Dinakar Singh, one of the most respected hedge-fund managers on Wall Street, whose parents earned doctoral degrees in biochemistry from the University of Delhi before emigrating to America. "It produced people with quality and by quantity. But many of them rotted on the docks of India like vegetables. Only a relative few could get on ships and get out. Not anymore, because we built this ocean crosser, called fiber-optic cable. For decades you had to leave India to be a professional. Now you can plug into the world from India. You don't have to go to Yale and go to work for Goldman Sachs." India could never have afforded to pay for the bandwidth to connect brainy India with high-tech America, so American shareholders paid for it. Yes, crazy overinvestment can be good. The overinvestment in railroads turned out to be a great boon for the American economy. "But the railroad overinvestment was confined to your own country and so, too, were the benefits," Singh said. In the case of the digital railroads, "it was the foreigners who benefited." India got a free ride.

The first time this became apparent was when thousands of Indian engineers were enlisted to fix the Y2K—the year 2000—computer bugs for companies from all over the world. (Y2K should be a national holiday in India. Call it "Indian Interdependence Day," says Michael Mandelbaum, a foreign-policy analyst at Johns Hopkins.) The fact that the Y2K work could be outsourced to Indians was made possible by the first two flat-teners, along with a third, which I call "workflow." Workflow is shorthand for all the software applications, standards and electronic transmission pipes, like middleware, that connected all those computers and fiber-optic cable. To put it another way, if the Netscape moment connected people to people like never before, what the workflow revolution did was connect applications to applications so that people all over the world could work together in manipulating and shaping words, data and images on computers like never before.

Indeed, this breakthrough in people-to-people and application-to-application connectivity produced, in short order, six more flatteners—six new ways in which individuals and companies could collaborate on work and share knowledge. One was "outsourcing." When my software applications could connect seamlessly with all of your applications, it meant that all kinds of work—from accounting to software-writing—could be digitized, disaggregated and shifted to any place in the world where it could be done better and cheaper. The second was "offshoring." I send my whole factory from Canton, Ohio, to Canton, China. The third was "open-sourcing." I write the next operating system, Linux, using engineers collaborating together online and working for free. The fourth was "insourcing." I let a company like UPS come inside my company and take over my whole logistics operation—everything from filling my orders online to delivering my goods to repairing them for customers when they break. (People have no idea what UPS really does today. You'd be amazed!). The fifth was "supply-chaining." This is Wal-Mart's specialty. I create a global supply chain down to the last atom of efficiency so that if I sell an item in Arkansas, another is immediately made in China. (If Wal-Mart were a country, it would be China's eighth-largest trading partner.) The last new form of collaboration I call "informing"—this is Google, Yahoo and MSN Search, which now allow anyone to collaborate with, and mine, unlimited data all by themselves.

So the first three flatteners created the new platform for collaboration, and the next six are the new forms of collaboration that flattened the world even more. The 10th flattener I call "the steroids," and these are wireless access and voice over Internet protocol (VoIP). What the steroids do is turbocharge all these new forms of collaboration, so you can now do any one of them, from anywhere, with any device.

The world got flat when all 10 of these flatteners converged around the year 2000. This created a global, Web-enabled playing field that allows for multiple forms of collaboration on research and work in real time, without regard to geography, distance or, in the near future, even language. "It is the creation of this platform, with these unique attributes, that is the truly important sustainable breakthrough that made what you call the flattening of the world possible," said Craig Mundie, the chief technical officer of Microsoft.

No, not everyone has access yet to this platform, but it is open now to more people in more places on more days in more ways than anything like it in history. Wherever you look today—whether it is the world of journalism, with bloggers bringing down Dan Rather; the world of software, with the Linux code writers working in online forums for free to challenge Microsoft; or the world of business, where Indian and Chinese innovators are competing against and working with some of the most advanced Western multinationals—hierarchies are being flattened and value is being created less and less within vertical silos and more and more through horizontal collaboration within companies, between companies and among individuals.

Do you recall "the IT revolution" that the business press has been pushing for the last 20 years? Sorry to tell you this, but that was just the prologue. The last 20 years were about forging, sharpening and distributing all the new tools to collaborate and connect. Now the real information revolution is about to begin as all the complementarities among these collaborative tools start to converge. One of those who first called this moment by its real name was Carly Fiorina, the former Hewlett-Packard C.E.O., who in 2004 began to declare in her public speeches that the dot-com boom and bust were just "the end of the beginning." The last 25 years in technology, Fiorina said, have just been "the warm-up act." Now we are going into the main event, she said, "and by the main event, I mean an era in which technology will truly transform every aspect of business, of government, of society, of life."

As if this flattening wasn't enough, another convergence coincidentally occurred during the 1990's that was equally important. Some three billion people who were out of the game walked, and often ran, onto the playing field. I am talking about the people of China, India, Russia, Eastern Europe, Latin America and Central Asia. Their economies and political systems all opened up during the course of the 1990s so that their people were increasingly free to join the free market. And when did these three billion people converge with the new playing field and the new business processes? Right when it was being flattened, right when millions of them could compete and collaborate more equally, more horizontally and with cheaper and more readily available tools. Indeed, thanks to the flattening of the world, many of these new entrants didn't even have to leave home to participate. Thanks to the 10 flatteners, the playing field came to them!

It is this convergence—of new players, on a new playing field, developing new processes for horizontal collaboration—that I believe is the most important force shaping global economics and politics in the early 21st century. Sure, not all three billion can collaborate and compete. In fact, for most people the world is not yet flat at all. But even if we're talking about only 10 percent, that's 300 million people—about twice the size of the American work force. And be advised: the Indians and Chinese are not racing us to the bottom. They are racing us to the top. What China's leaders really want is that the next generation of underwear and airplane wings not just be "made in China" but also be "designed in China." And that is where things are heading. So in 30 years we will have gone from "sold in China" to "made in China" to "designed in China" to "dreamed up in China"—or from China as collaborator with the worldwide manufacturers on nothing to China as a low-cost, high-quality, hyperefficient collaborator with worldwide manufacturers on everything. Ditto India. Said

Craig Barrett, the C.E.O. of Intel, "You don't bring three billion people into the world economy overnight without huge consequences, especially from three societies"—like India, China and Russia—"with rich educational heritages."

That is why there is nothing that guarantees that Americans or Western Europeans will continue leading the way. These new players are stepping onto the playing field legacy free, meaning that many of them were so far behind that they can leap right into the new technologies without having to worry about all the sunken costs of old systems. It means that they can move very fast to adopt new, state-of-the-art technologies, which is why there are already more cellphones in use in China today than there are people in America.

If you want to appreciate the sort of challenge we are facing, let me share with you two conversations. One was with some of the Microsoft officials who were involved in setting up Microsoft's research center in Beijing, Microsoft Research Asia, which opened in 1998—after Microsoft sent teams to Chinese universities to administer I.Q. tests in order to recruit the best brains from China's 1.3 billion people. Out of the 2,000 top Chinese engineering and science students tested, Microsoft hired 20. They have a saying at Microsoft about their Asia center, which captures the intensity of competition it takes to win a job there and explains why it is already the most productive research team at Microsoft: "Remember, in China, when you are one in a million, there are 1,300 other people just like you."

The other is a conversation I had with Rajesh Rao, a young Indian entrepreneur who started an electronic-game company from Bangalore, which today owns the rights to Charlie Chaplin's image for mobile computer games. "We can't relax," Rao said. "I think in the case of the United States that is what happened a bit. Please look at me: I am from India. We have been at a very different level before in terms of technology and business. But once we saw we had an infrastructure that made the world a small place, we promptly tried to make the best use of it. We saw there were so many things we could do. We went ahead, and today what we are seeing is a result of that. There is no time to rest. That is gone. There are dozens of people who are doing the same thing you are doing, and they are trying to do it better. It is like water in a tray: you shake it, and it will find the path of least resistance. That is what is going to happen to so many jobs—they will go to that corner of the world where there is the least resistance and the most opportunity. If there is a skilled person in Timbuktu, he will get work if he knows how to access the rest of the world, which is quite easy today. You can make a website and have an e-mail address and you are up and running. And if you are able to demonstrate your work, using the same infrastructure, and if people are comfortable giving work to you and if you are diligent and clean in your transactions, then you are in business."

Instead of complaining about outsourcing, Rao said, Americans and Western Europeans would "be better off thinking about how you can raise your bar and raise yourselves into doing something better. Americans have consistently led in innovation over the last century. Americans whining—we have never seen that before."

Rao is right. And it is time we got focused. As a person who grew up during the cold war, I'll always remember driving down the highway and listening to the radio, when suddenly the music would stop and a grim-voiced announcer would come on the air and say: "This is a test. This station is conducting a test of the Emergency Broadcast System." And then there would be a 20-second high-pitched siren sound. Fortunately, we never had to live through a moment in the cold war when the announcer came on and said, "This is a not a test."

That, however, is exactly what I want to say here: "This is not a test."

The long-term opportunities and challenges that the flattening of the world puts before the United States are profound. Therefore, our ability to get by doing things the way we've been doing them—which is to say not always enriching our secret sauce—will not suffice any more. "For a country as wealthy as we are, it is amazing how little we are doing to enhance our natural competitiveness," says Dinakar Singh, the Indian-American hedge-fund manager. "We are in a world that has a system that now allows convergence among many billions of people, and we had better step back and figure out what it means. It would be a nice coincidence if all the things that were true before were still true now, but there are quite a few things you actually need to do differently. You need to have a much more thoughtful national discussion."

If this moment has any parallel in recent American history, it is the height of the cold war, around 1957, when the Soviet Union leapt ahead of America in the space race by putting up the Sputnik satellite. The main challenge then came from those who wanted to put up walls; the main challenge to America today comes from the fact that all the walls are being taken down and many other people can now compete and collaborate with us much more directly. The main challenge in that world was from those practicing extreme Communism, namely Russia, China and North Korea. The main challenge to America today is from those practicing extreme capitalism, namely China, India and South Korea. The main objective in that era was building a strong state, and the main objective in this era is building strong individuals.

Meeting the challenges of flatism requires as comprehensive, energetic and focused a response as did meeting the challenge of Communism. It requires a president who can summon the nation to work harder, get smarter, attract more young women and men to science and engineering and build the broadband infrastructure, portable pensions and health care that will help every American become more employable in an age in which no one can guarantee you lifetime employment.

We have been slow to rise to the challenge of flatism, in contrast to Communism, maybe because flatism doesn't involve ICBM missiles aimed at our cities. Indeed, the hot line, which used to connect the Kremlin with the White House, has been replaced by the help line, which connects everyone in America to call centers in Bangalore. While the other end of the hot line might have had Leonid Brezhnev threatening nuclear war, the other end of the help line just has a soft voice eager to help you sort out your AOL bill or collaborate with you on a new piece of software. No, that voice has none of the menace of Nikita Khrushchev pounding a shoe on the table at the United Nations, and it has none of the sinister snarl of the bad guys in "From Russia with Love." No, that voice on the help line just has a friendly Indian lilt that masks any sense of threat or challenge. It simply says: "Hello, my name is Rajiv. Can I help you?"

No, Rajiv, actually you can't. When it comes to responding to the challenges of the flat world, there is no help line we can call. We have to dig into ourselves. We in America have all the basic economic and educational tools to do that. But we have not been improving those tools as much as we should. That is why we are in what Shirley Ann Jackson, the 2004 president of the American Association for the Advancement of Science and president of Rensselaer Polytechnic Institute, calls a "quiet crisis"—one that is slowly eating away at America's scientific and engineering base.

"If left unchecked," said Jackson, the first African-American woman to earn a Ph.D. in physics from M.I.T., "this could challenge our pre-eminence and capacity to innovate." And it is our ability to constantly innovate new products, services and companies that has been the source of America's horn of plenty and steadily widening middle class for the last two centuries. This quiet crisis is a product of three gaps now plaguing American society. The first is an "ambition gap." Compared with the young, energetic Indians and Chinese, too many Americans have gotten too lazy. As David Rothkopf, a former official in the Clinton Commerce Department, puts it, "The real entitlement we need to get rid of is our sense of entitlement." Second, we have a serious numbers gap building. We are not producing enough engineers and scientists. We used to make up for that by importing them from India and China, but in a flat world, where people can now stay home and compete with us, and in a post-9/11 world, where we are insanely keeping out many of the first-round intellectual draft choices in the world for exaggerated security reasons, we can no longer cover the gap. That's a key reason companies are looking abroad. The numbers are not here. And finally we are developing an education gap. Here is the dirty little secret that no C.E.O. wants to tell you: they are not just outsourcing to save on salary. They are doing it because they can often get better-skilled and more productive people than their American workers.

These are some of the reasons that Bill Gates, the Microsoft chairman, warned the governors' conference in a Feb. 26 speech that American high-school education is "obsolete." As Gates put it: "When I compare our high schools to what I see when I'm traveling abroad, I am terrified for our work force of tomorrow. In math and science, our fourth graders are among the top students in the world. By eighth grade, they're in the middle of the pack. By 12th grade, U.S. students are scoring near the bottom of all industrialized nations. . . . The percentage of a population with a college degree is important, but so are sheer numbers. In 2001, India graduated almost a million

more students from college than the United States did. China graduates twice as many students with bachelor's degrees as the U.S., and they have six times as many graduates majoring in engineering. In the international competition to have the biggest and best supply of knowledge workers, America is falling behind."

We need to get going immediately. It takes 15 years to train a good engineer, because, ladies and gentlemen, this really is rocket science. So parents, throw away the Game Boy, turn off the television and get your kids to work. There is no sugarcoating this: in a flat world, every individual is going to have to run a little faster if he or she wants to advance his or her standard of living. When I was growing up, my parents used to say to me, "Tom, finish your dinner—people in China are starving." But after sailing to the edges of the flat world for a year, I am now telling my own daughters, "Girls, finish your homework—people in China and India are starving for your jobs."

I repeat, this is not a test. This is the beginning of a crisis that won't remain quiet for long. And as the Stanford economist Paul Romer so rightly says, "A crisis is a terrible thing to waste."

Critical Thinking

1. What does the author mean, "the world is flat"?

2. In a world of such environmental, biotic, and resource variability, how can it be "flat"? What are some of the "environmental" assumptions the author is making about the Earth? About its peoples?

3. This article was written in 2005. What were some of the "headline" environmental issues then? What are some of the issues now, in 2012, seven years later? How might our current environmental issues and future ones impact this "flat world" argument?

4. How does "flattening the world" contribute to our goal of "environmental sustainability" or "sustainable development"?

5. What do Articles 5 and 6 have to do with "global development"?

6. Identify in this article three key terms, concepts, or principles that are used in your textbook (environmental science, economics, sociology, history, geography, etc.) or employed in the discipline you are currently studying. (Note: The terms, concepts, or principles may be implicit, explicit, implied, or inferred.)

THOMAS L. FRIEDMAN is the author of "The World Is Flat: A Brief History of the Twenty-First Century," to be published this week by Farrar, Straus & Giroux and from which this article is adapted. His column appears on the Op-Ed page of The Times, and his television documentary "Does Europe Hate Us?" was shown on the Discovery Channel on April 7, 2005.

The World Is Spiky

Globalization has changed the economic playing field, but hasn't leveled it.

RICHARD FLORIDA

The world, according to the title of *The New York Times* columnist Thomas Friedman's book, is flat. Thanks to advances in technology, the global playing field has been leveled, the prizes are there for the taking, and everyone's a player—no matter where on the surface of the earth he or she may reside. "In a flat world," Friedman writes, "you can innovate without having to emigrate."

Friedman is not alone in this belief: for the better part of the past century economists have been writing about the leveling effects of technology. From the invention of the telephone, the automobile, and the airplane to the rise of the personal computer and the Internet, technological progress has steadily eroded the economic importance of geographic place—or so the argument goes.

But in partnership with colleagues at George Mason University and the geographer Tim Gulden, of the Center for International and Security Studies, at the University of Maryland, I've begun to chart a very different economic topography. By almost any measure the international economic landscape is not at all flat. On the contrary, our world is amazingly "spiky." In terms of both sheer economic horsepower and cutting-edge innovation, surprisingly few regions truly matter in today's global economy. What's more, the tallest peaks—the cities and regions that drive the world economy—are growing ever higher, while the valleys mostly languish.

The most obvious challenge to the flat-world hypothesis is the explosive growth of cities worldwide. More and more people are clustering in urban areas—the world's demographic mountain ranges, so to speak. The share of the world's population living in urban areas, just three percent in 1800, was nearly 30 percent by 1950. Today it stands at about 50 percent; in advanced countries three out of four people live in urban areas. Map A shows the uneven distribution of the world's population. Five megacities currently have more than 20 million inhabitants each. Twenty-four cities have more than 10 million inhabitants, sixty more than 5 million, and 150 more than 2.5 million. Population density is of course a crude indicator of human and economic activity. But it does suggest that at least some of the tectonic forces of economics are concentrating people and resources, and pushing up some places more than others.

Still, differences in population density vastly understate the spikiness of the global economy; the continuing dominance of the world's most productive urban areas is astounding. When it comes to actual economic output, the ten largest U.S. metropolitan areas combined are behind only the United States as a whole and Japan. New York's economy alone is about the size of Russia's or Brazil's, and Chicago's is on a par with Sweden's. Together New York, Los Angeles, Chicago, and Boston have a bigger economy than all of China. If U.S. metropolitan areas were countries, they'd make up forty-seven of the biggest 100 economies in the world.

Unfortunately, no single, comprehensive information source exists for the economic production of all the world's cities. A rough proxy is available, though. Map B shows a variation on the widely circulated view of the world at night, with higher concentrations of light— indicating higher energy use and, presumably, stronger economic production—appearing in greater relief. U.S. regions appear almost Himalayan on this map. From their summits one might look out on a smaller mountain range stretching across Europe, some isolated peaks in Asia, and a few scattered hills throughout the rest of the world.

Population and economic activity are both spiky, but it's innovation— the engine of economic growth—that is most concentrated. The World Intellectual Property Organization recorded about 300,000 patents from resident inventors in more than a hundred nations in 2002 (the most recent year for which statistics are available). Nearly two thirds of them went to American and Japanese inventors. Eighty-five percent went to the residents of just five countries (Japan, the United States, South Korea, Germany, and Russia).

Worldwide patent statistics can be somewhat misleading, since different countries follow different standards for granting patents. But patents granted in the United States—which receives patent applications for nearly all major innovations worldwide, and holds them to the same strict standards—tell a similar story. Nearly 90,000 of the 170,000 patents granted in the United States in 2002 went to Americans. Some 35,000 went to Japanese inventors, and 11,000 to Germans. The next ten most innovative countries—including the usual suspects in Europe plus Taiwan, South Korea, Israel, and Canada— produced roughly 25,000 more. The rest of the broad, flat world accounted for just five percent of all innovations patented in the United States. In 2003 India generated 341 U.S. patents and China 297. The University of California alone generated more than either country. IBM accounted for five times as many as the two combined.

This is not to say that Indians and Chinese are not innovative. On the contrary, AnnaLee Saxenian, of the University of California at Berkeley, has shown that Indian and Chinese entrepreneurs founded or co-founded roughly 30 percent of all Silicon Valley startups in the late 1990s. But these fundamentally creative people had to travel to Silicon Valley and be absorbed into its innovative ecosystem before their ideas became economically viable. Such ecosystems matter, and there aren't many of them.

Map C (omitted)—which makes use of data from both the World Intellectual Property Organization and the U.S. Patent and Trademark

1. Peaks, Hills, and Valleys

When looked at through the lens of economic production, many cities with large populations are diminished and some nearly vanish. Three sorts of places make up the modern economic landscape. First are the cities that generate innovations. These are the tallest peaks; they have the capacity to attract global talent and create new products and industries. They are few in number, and difficult to topple. Second are the economic "hills"—places that manufacture the world's established goods, take its calls, and support its innovation engines. These hills can rise and fall quickly; they are prosperous but insecure. Some, like Dublin and Seoul, are growing into innovative, wealthy peaks; others are declining, eroded by high labor costs and a lack of enduring competitive advantage. Finally there are the vast valleys—places with little connection to the global economy and few immediate prospects.

2. The Geography of Innovation

Commercial innovation and scientific advance are both highly concentrated—but not always in the some places. Several cities in East Asia—particularly in Japan—are home to prolific business innovation but still depend disproportionately on scientific breakthroughs made elsewhere. Likewise, some cities excel in scientific research but not in commercial adaptation. The few places that do both well are very strongly positioned in the global economy. These regions have little to fear, and much to gain, from continuing globalization.

A. Population

Urban areas house half of all the worlds people, and continue to grow in both rich and poor countries.

Map reprinted from the Atlantic Monthly, October 2005, "The World Is Spiky," by Richard Florida. Map by Tim Gulden. Data

Source: Center for International Earth Science Information Network, Columbia University; and Centro Internacional de Agricultura Tropical.

B. Light Emissions

Economic activity—roughly estimated here using light-emissions data—is remarkably concentrated. Many cities, despite their large populations, barely register.

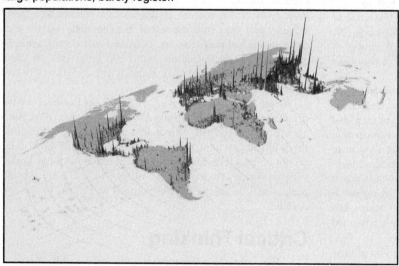

Map reprinted from the Atlantic Monthly, October 2005, "The World Is Spiky," by Richard Florida, Map by Tim Gulden.

Source: U.S. Defense Meteorological Satellite Program.

C. Patents

Just a few places produce most of the world's innovations. Innovation remains difficult without a critical mass of financiers, entrepreneurs, and scientists, often nourished by world-class universities and flexible corporations.

D. Scientific Citations

The worlds most prolific and influential scientific researchers overwhelmingly reside in U.S. and European cities.

Office—shows a world composed of innovation peaks and valleys. Tokyo, Seoul, New York, and San Francisco remain the front-runners in the patenting competition. Boston, Seattle, Austin, Toronto, Vancouver, Berlin, Stockholm, Helsinki, London, Osaka, Taipei, and Sydney also stand out.

Map D (omitted) shows the residence of the 1,200 most heavily cited scientists in leading fields. Scientific advance is even more concentrated than patent production. Most occurs not just in a handful of countries but in a handful of cities—primarily in the United States and Europe. Chinese and Indian cities do not even register. As far as global innovation is concerned, perhaps a few dozen places worldwide really compete at the cutting edge.

Concentrations of creative and talented people are particularly important for innovation, according to the Nobel Prize-winning economist Robert Lucas. Ideas flow more freely, are honed more sharply, and can be put into practice more quickly when large numbers of innovators, implementers, and financial backers are in constant contact with one another, both in and out of the office. Creative people cluster not simply because they like to be around one another or they prefer cosmopolitan centers with lots of amenities, though both those things count. They and their companies also cluster because of the powerful productivity advantages, economies of scale, and knowledge spillovers such density brings.

So although one might not *have* to emigrate to innovate, it certainly appears that innovation, economic growth, and prosperity occur in those places that attract a critical mass of top creative talent. Because globalization has increased the returns to innovation, by allowing innovative products and services to quickly reach consumers worldwide, it has strengthened the lure that innovation centers hold for our planet's best and brightest, reinforcing the spikiness of wealth and economic production.

The main difference between now and even a couple of decades ago is not that the world has become flatter but that the world's peaks have become slightly more dispersed—and that the world's hills, the industrial and service centers that produce mature products and support innovation centers, have proliferated and shifted. For the better part of the twentieth century the United States claimed the lion's share of the global economy's innovation peaks, leaving a few outposts in Europe and Japan. But America has since lost some of those peaks, as such industrial-age powerhouses as Pittsburgh, St. Louis, and Cleveland have eroded. At the same time, a number of regions in Europe, Scandinavia, Canada, and the Pacific Rim have moved up.

The world today looks flat to some because the economic and social distances between peaks worldwide have gotten smaller. Connection between peaks has been strengthened by the easy mobility of the global creative class—about 150 million people worldwide. They participate in a global technology system and a global labor market that allow them to migrate freely among the world's leading cities. In a Brookings Institution study the demographer Robert Lang and the world-cities expert Peter Taylor identify a relatively small group of leading city-regions—London, New York, Paris, Tokyo, Hong Kong, Singapore, Chicago, Los Angeles, and San Francisco among them—that are strongly connected to one another.

But Lang and Taylor also identify a much larger group of city-regions that are far more locally oriented. People in spiky places are often more connected to one another, even from half a world away, than they are to people and places in their veritable back yards.

The flat-world theory is not completely misguided. It is a welcome supplement to the widely accepted view (illustrated by the Live 8 concerts and Bono's forays into Africa, by the writings of Jeffrey Sachs and the UN Millennium project) that the growing divide between rich and poor countries is the fundamental feature of the world economy. Friedman's theory more accurately depicts a developing world with capabilities that translate into economic development. In his view, for example, the emerging economies of India and China combine cost advantages, high-tech skills, and entrepreneurial energy, enabling those countries to compete effectively for industries and jobs. The tensions set in motion as the playing field is leveled affect mainly the advanced countries, which see not only manufacturing work but also higher-end jobs, in fields such as software development and financial services, increasingly threatened by off-shoring.

But the flat-world theory blinds us to far more insidious tensions among the world's growing peaks, sinking valleys, and shifting hills. The innovative, talent-attracting "have" regions seem increasingly remote from the talent-exporting "have-not" regions. Second-tier cities, from Detroit and Wolfsburg to Nagoya and Mexico City, are entering an escalating and potentially devastating competition for jobs, talent, and investment. And inequality is growing across the world and within countries.

This is far more harrowing than the flat world Friedman describes, and a good deal more treacherous than the old rich-poor divide. We see its effects in the political backlash against globalization in the advanced world. The recent rejection of the EU constitution by the French, for example, resulted in large part from high rates of "no" votes in suburban and rural quarters, which understandably fear globalization and integration.

But spiky globalization also wreaks havoc on poorer places. China is seeing enormous concentrations of talent and innovation in centers such as Shanghai, Shenzhen, and Beijing, all of which are a world apart from its vast, impoverished rural areas. According to detailed polling by Richard Burkholder, of Gallup, average household incomes in urban China are now triple those in rural regions, and they've grown more than three times as fast since 1999; perhaps as a result, urban and rural Chinese now have very different, often conflicting political and lifestyle values. India is growing even more divided, as Bangalore, Hyderabad, and parts of New Delhi and Bombay pull away from the rest of that enormous country, creating destabilizing political tensions. Economic and demographic forces are sorting people around the world into geographically clustered "tribes" so different (and often mutually antagonistic) as to create a somewhat Hobbesian vision.

We are thus confronted with a difficult predicament. Economic progress requires that the peaks grow stronger and taller. But such growth will exacerbate economic and social disparities, fomenting political reactions that could threaten further innovation and economic progress. Managing the disparities between peaks and valleys worldwide—raising the valleys without shearing off the peaks—will be among the top political challenges of the coming decades.

Critical Thinking

1. What does the author mean by "the world is spiky"?

2. In a textbook entitled *Environment*, what would an article about a "spiky world" be doing in the text? (Hint: Think natural resources.)

3. List several linkages that the "geography of innovation" would have with a concept like "environmental sustainability."

4. If global climate change is occurring, how might it change the global patterns of the author's spiky world?

5. Identify in this article three key terms, concepts, or principles that are used in your textbook (environmental science, economics, sociology, history, geography, etc.) or employed in the discipline you are currently studying. (Note: The terms, concepts, or principles may be implicit, explicit, implied, or inferred.)

RICHARD FLORIDA, the author of *The Flight of the Creative Class,* was formerly the Hirst Professor of Public Policy at George Mason University.

Promises and Poverty

Starbucks calls its coffee worker-friendly—but in Ethiopia, a day's pay is a dollar.

TOM KNUDSON

Gemadro, Ethiopia—Tucked inside a fancy black box, the $26-a-pound Starbucks Black Apron Exclusives coffee promised to be more than just another bag of beans.

Not only was the premium coffee from a remote plantation in Ethiopia "rare, exotic, cherished," according to Starbucks advertising, it was grown in ways that were good for the environment—and for local people, too.

Companies routinely boast about what they're doing for the planet, in part because guilt-ridden consumers expect as much—and are willing to pay extra for it. But, in this case, Starbucks' eco-friendly sales pitch does not begin to reflect the complex story of coffee in East Africa.

Inside the front flap of Starbucks' box are African arabica beans grown on a plantation in a threatened mountain rain forest. Behind the lofty phrases on the back label are coffee workers who make less than a dollar a day and a dispute between plantation officials and neighboring tribal people, who accuse the plantation of using their ancestral land and jeopardizing their way of life.

"We used to hunt and fish in there, and also we used to have honeybee hives in trees," one tribal member, Mikael Yatola, said through a translator. "But now we can't do that. . . . When we were told to remove our beehives from there, we felt deep sorrow, deep sadness."

25 New U.S. Stores per Week

Few companies have so dramatically conquered the American retail landscape as Starbucks. Last year, the $7.8 billion company opened an average of 25 new stores a week in the United States alone. Nowhere is Starbucks a more common sight than in environmentally conscious California, which has 2,350 outlets, more than New York, Massachusetts, Florida, Oregon and Washington—Starbucks' home state—combined.

No coffee company claims to do more for the environment and Third World farmers than Starbucks either. In full-page ads in *The New York Times,* in brochures and on its Web page, Starbucks says that it pays premium prices for premium beans, protects tropical forests and enhances the lives of farmers by building schools, clinics and other projects.

In places, Starbucks delivers on those promises, certainly more so than other multinational coffee companies. In parts of Latin America, for instance, its work has helped improve water quality, educate children and protect biodiversity.

Inside many Starbucks outlets across America, the African décor is hard to miss. There are photographs and watercolors of quaint coffee-growing scenes from Ethiopia to Tanzania to Zimbabwe. Yet such images clash with the reality of African life.

They don't show the industrial arm of coffee—the large farms and estates that encroach on wild forest regions. They don't reveal that even in the best of times in Ethiopia, the birthplace of wild coffee and the source of some of Starbucks' priciest offerings, there is barely enough for the peasant coffee farmers who still grow most of the nation's beans.

Even where Starbucks has built its bricks-and-mortar projects in Ethiopia, poverty remains a cornerstone of life, visible in the soot-stained cooking pots, spindly legs and ragged T-shirts, in the mad scramble of children for a visitor's cookie or empty water bottle.

"We plant coffee, harvest coffee but we never get anything out of it," said Muel Alema, a rail-thin coffee farmer who lives near a Starbucks-funded footbridge spanning a narrow chasm in Ethiopia's famous Sidamo coffee-growing region.

Alema's tattered shirt looked years old. So did his mud-splattered thongs. The red coffee berries he sold to a local buyer last fall were mixed with mountains of others, stripped of their pulp and sold as beans to distant companies—like other farmers, he did not know which ones—that made millions selling Sidamo coffee. Only $220 dribbled back to him.

This February, after Alema paid workers to pick the beans and bought grain for his family, just $110 remained—not enough, he said, to feed his wife and three children, to buy them clothes until the crop ripens again.

'A Marketing Genius'

Starbucks conveys a different image on the white foil bags of Ethiopia Sidamo whole bean coffee it sells for $10.45 a pound. "Good coffee, doing good," says lettering on the side.

"We believe there's a connection between the farmers who grow our coffees, us and you. That's why we work together with coffee-growing communities—paying prices that help farmers support their families . . . and funding projects like building a bridge in Ethiopia's Sidamo region to help farmers get to market safely. . . . By drinking this coffee, you're helping to make a difference."

And while the Sidamo footbridge does make travel safer, it is but a simple yellow-brown concrete slab, 10 paces long.

Dean Cycon, founder of Dean's Beans, an organic coffee company in Massachusetts, calls Starbucks "a marketing genius."

"They put out cleverly crafted material that makes the consumer feel they are doing everything possible," Cycon said. "But there is no institutional commitment. They do it to capture a market and shut up the activists."

Starbucks officials insist such critics have it wrong. As proof, they point to Latin America, the source of the bulk of the company's beans.

"You go to Nariño, Colombia. We built 1,800 (coffee) washing stations and sanitation facilities and homes," said Dub Hay, Starbucks senior vice president for global coffee procurement. "It's literally changed the face of that whole area."

"The same is true throughout Latin America," Hay added. "They call it the Starbucks effect."

Starbucks' dealings in Latin America have drawn some fire. Near the El Triunfo Biosphere Reserve in Chiapas, Mexico, for instance, farmers cut off relations about three years ago over a dispute about selling to an exporter instead of directly to Starbucks. The new arrangement, farmers said, would drain profits from peasant growers.

Starbucks, the farmers charged in a memo to coffee buyers, was supporting "a pseudo-fair trade system, adapted to their own neo-liberal interests, to dismantle structures and advances that we have made."

In an e-mailed response to The Bee, Starbucks vice president for global communications, Frank Kern, wrote that the Chiapas farmers were ultimately "given the opportunity to ship directly to us as they requested, but they were unable to manage it."

Sharper Focus on Africa

In Ethiopia, Starbucks says, it spent $25,000 on three footbridges in 2004. The company estimates the structures are used by 70,000 farmers and family members—about 1 percent of those who depend upon coffee for income. Some Ethiopian coffee leaders say there is a better way to help.

"If we are paid a (coffee) price which is decent, the people can make the bridge on their own," said Tadesse Meskela, general manager of the Oromia Coffee Farmers' Cooperative Union of 100,000 farmers, which has sold to Starbucks. "We don't have to be always beggars."

Starbucks won't disclose what it pays for Ethiopian coffee. Instead, it lumps its purchases together into a global average, which last year was $1.42 a pound, 16 cents more than the Fair Trade minimum. Much of that money, though, never makes it into the pockets of farmers but instead is siphoned off by buyers, processors and other middlemen.

Starbucks executives say they want to shrink that supply chain. "You end up at least five levels removed from the farmer and that's where the money goes," said Hay. "And that's a shame." Hay said that as the company buys more coffee from Africa—it plans to double its purchases there to 36 million pounds by 2009—the commerce will spur more progress.

"That's our goal," Hay said. "Africa is 6 percent of our purchases. . . . Seventy percent is from Latin America. So that's where our money has gone."

Making an impact in Ethiopia is undeniably a challenge. Good roads, electricity, potable water don't exist in many places. The climate is often hostile. There are ethnic conflicts, border disputes and rebel movements, and a sea of young faces that gather every time a car stops.

Since 1990, Ethiopia's population has jumped from 52 million to about 80 million: two new Los Angeleses. The more people, the less there is to go around. Ethiopia's per capita annual income is only $180, one of the lowest on Earth.

The environment is hurting, too, as coffee and tea plantations—as well as peasant farmers—spread into once wild areas, raising concern about the demise of one of the country's natural treasures: its biologically rich southwestern rain forest.

However, Samuel Assefa, the Ethiopian ambassador to the United States, said human suffering must be taken into consideration, too.

"We have a population of more than 80 million people, many of them living in rural and impoverished circumstances," he wrote in an e-mail. "Returning all cleared land to rain forest might be a victory for some extreme environmentalists, but it would condemn millions of my fellow citizens to starvation."

Atonement in a Cup

Starbucks has bought coffee from Africa for years. But now it is expanding rapidly there because it wants more of the continent's high-quality arabica beans. In Ethiopia alone, Starbucks purchases jumped 400 percent between 2002 and 2006.

While Starbucks is rapidly cloning retail outlets globally, 79 percent of its revenue last year was made in one caffeine-crazed country: the United States. With just 5 percent of the world's people, the United States drinks one-fifth of its coffee, more than any other nation.

Thanks largely to Starbucks, coffee is no longer just coffee. Now it is a vanilla soy latte, a java chip Frappuccino, a grande zebra mocha or—if you're feeling guilty—a Fair Trade-certified, bird-friendly, shade-grown caramel macchiato.

Starbucks did not pioneer the push for more equitable, conservation-based coffee. But it has woven the theme into everything from the earthy feel of its stores to its own certification program—called Coffee and Farmer Equity practices, or C.A.F.E.—that rewards farmers for meeting social and environmental goals.

"Social justice is becoming increasingly important to consumers," said industry analyst Judith Ganes-Chase as she flashed slides across a screen at a Long Beach coffee conference. One slide read: "Fair Trade is absolution in a cup."

Atonement, though, is not as simple as it may seem.

"It's very comfortable to believe Starbucks is doing the right thing—and to some degree, they are," said Eric Perkunder, a Seattle resident who worked as a Starbucks environmental manager in the 1980s and '90s. "They lull us into complacency. The stores are comfortable. You see pictures of people from origin countries. You believe

certain things they are telling you. But there's more to the story."

Dirt Road, Stick Huts

Part of that story lies in the southwestern corner of Ethiopia, in a swath of mountains not far from the Sudan border. There, a dirt road snakes through one of the country's largest coffee plantations—the Ethiopia Gemadro Estate—and comes to a halt in a dense mat of reeds and grasses.

A narrow path winds through the thicket and spills out into a clearing of stick huts. This is the home of an African Sheka tribe that for generations has lived off the land—catching fish, gathering wild honey and trapping animals in the forest. They call themselves the Shabuyye.

"The land over there used to belong to our forefathers," said Yatola, the tribal member in his 20s, as he nodded toward the plantation.

Conflict with local people and tribes is growing across southwest Ethiopia as coffee and tea plantations spread into the region under the government's effort to sow more development. At Gemadro, 2,496 acres of coffee were planted from 1998 to 2001 on land the company obtained from the government in a countylike jurisdiction called the Sheka Zone.

"One of Ethiopia's last remaining forests, Sheka Forest, is under huge pressure. . . . The rate of deforestation is now increasing and threatens the forest biodiversity . . . and the very livelihood" of forest-dwelling tribes, says the 2006 annual report of Melca Mahiber, an environmental group in the nation's capital, Addis Ababa.

However, Haile Michael Shiferaw, the plantation's manager—who attended the Long Beach coffee conference—said the Gemadro Estate had not displaced any tribe members.

"Before our farm was started," he said, "very few people were living in Gemadro."

Last year, Starbucks bought about 75,000 pounds of coffee from the Gemadro plantation and sold it as one of its "Black Apron Exclusives." At the time, the purchase was the 12th in the series of vintage offerings, six of which originated in Africa.

Starbucks packaged the beans in the fancy black box and inserted a flier touting the plantation's environmental and social track record. It also donated $15,000 to the Gemadro Estate for a school and health clinic.

"With its pure water supply, near pristine growing environment and dedication to conservation-based farming methods, this 2,300-hectare (5,700-acre) farm . . . is setting new standards for progressive, sustainable coffee farming," the flier said. "Gemadro workers and their families enjoy access to clean water, health care, housing and schools, all in keeping with the estate's commitment to maintain the highest standards of social and environmental stewardship."

Family's Income: 66 Cents a Day

Hailu is one of those workers. He stood outside his one-room, dirt-floor home, folded his arms across his chest and said that while he was happy to have a job, he was struggling to support his wife and family on just 6 Ethiopian birr per day—66 cents.

"Life is expensive," he said. "We have to go all the way to the town of Tepi (about 35 miles) for supplies." The round-trip bus ticket costs him four days' pay.

Plantation manager Shiferaw said Gemadro Estate wages are higher than the 55 cents a day workers earn at a government plantation near Tepi. Gemadro workers—most of whom are classified as temporary—also subsequently received a raise to between 77 cents and $1.10 a day, he said, adding, "We pay more than the minimum wage of the country."

That's still not a livable wage, according to the U.S. State Department. In a 2006 report on human rights in Ethiopia, the agency said that "there is no national minimum wage" and that public employees earn about $23 a month; private workers, $27. Those wages, it said, do "not provide a decent standard of living."

The Gemadro Estate is owned by Ethiopian-born Saudi Sheik Mohammed Al Amoudi, ranked by Forbes magazine as one of the world's 100 wealthiest individuals, with a net worth of $8 billion.

Al Amoudi owns many businesses in Ethiopia, from the posh Sheraton Hotel in Addis Ababa where rooms start at $270 a night—about a year's wages on the coffee plantation—to Ethio Agri-CEFT Plc, the farm management company that oversees the plantation.

Asked about the contrast between the sheik's wealth and the plantation wages, Assefa Tekle, Ethio Agri-CEFT's commercial manager, said, "There is no additional income that has been given from Sheik Al Amoudi. So what can you do? You have to be profitable to exist."

Ethiopian Ecology Suffers

Plantation wages are only one issue for workers and neighbors, however. The health care partially supported by Starbucks is another.

"When the estate came, they said they were going to give us adequate health service," said Geremew Gelito, an elder in the tiny village of Gemadro, where many estate workers live. He called the clinic bureaucratic and ineffective.

"As far as the promise of adequate health service, we have not received it," Gelito said. Shiferaw—the farm manager—said there are plans to improve the care. "It is not a very big clinic," he said, "but now we are increasing."

In his office above a furniture store in Addis Ababa, 2½ days' drive to the northeast, the agricultural manager for Ethio Agri-CEFT, Biru Abebe, said he was unaware of any complaints.

"Everybody goes and gets treatment," he said.

That even includes the native tribe, said Tekle, the commercial manager, a claim that exceeds that of Starbucks. "They are people of the environment so the farm has to give assistance."

But tribal member Yatola said the Shabuyye who live just downstream cannot get health care. "The company just gives medical attention to people who work for the company," he charged. Sitting in a stick hut, Yatola looked pensive. "We were isolated before. We didn't interact with anyone," he said as women outside pounded grain into mash with heavy wooden sticks. "But since the company has come, the road has acquainted us with the outside world."

Outsiders show up along the river, catching fish that feed the tribe, Yatola said. "There are less fish—and more people fishing," he said. "When we hear that the company, the way they have operated, they have created a nice relationship with the community, we know it's not true."

What's unfolding in the Gemadro region is not unique. An article in the April 2007 Journal of Agrarian Change points out that as coffee growing expands in southwest Ethiopia, the ecology can suffer.

"Environmental degradation is a serious concern with rates of deforestation estimated at 10,000 hectares per year (25,000 acres) in the coffee growing areas," the article said. "High levels of river pollution are also a major problem near coffee pulping and washing stations."

During the fall harvest, Yatola said, coffee processing pulp appears in the river from somewhere upstream. "The river becomes black, almost like oil," he said. "It smells like a dead horse."

Abebe, Ethio Agri-CEFT's agricultural manager, said the plantation has a lagoon to control pollution, that no waste flows from the Gemadro Estate. "The river is clean throughout the year. There is no pollution from our farm," he said. "Zero."

However, plantation manager Shiferaw said that not long ago valuable coffee beans did show up in the estate's wastewater lagoon, where a handful of workers and area farmers had dumped them in a failed theft attempt. "They are in prison now," he said.

In late February, four months after the bustling coffee harvest season, the Gemadro River looked clean. But in one of the ankle-deep side streams flowing into it, a white truck used to haul coffee was parked in the silvery water, being scrubbed of dirt and grease—a different but obvious pollution source.

"Sometimes it happens," acknowledged Tekle, the commercial manager.

"We don't have any control over that," Abebe said. "That road, though we made it, is a public road, so any truck can go and come."

Deforestation Takes a Toll

Although public, the road into the Gemadro Estate felt private. Armed guards checked vehicles entering or leaving to prevent coffee smuggling.

A row of rusty metal shacks looked like tool sheds, but were in fact worker housing. Even Shiferaw, the plantation manager, acknowledged they were not adequate.

"Yes, some is not. OK?" he said. "But we have a program to improve. When you see the standards of the country, it is better than most."

Along the bumpy dirt road, waxy green rows of coffee trees sprouted in rows, shaded by clumps of taller trees that not long ago were part of a denser, more diverse forest.

To Tadesse Gole—an ecologist in Addis Ababa and a native Ethiopian whose doctoral thesis at the University of Bonn, Germany, focused on preservation of wild arabica coffee—that manicured landscape is a biological calamity.

"This is an Afromontane rain forest—an area of high plant diversity, of unique epiphytic plants that grow on the branches of trees," Gole said. "We lose those plant species. And we lose many of the animals, birds and insects dependent on them."

Gole is the author of a recent study about the environmental and cultural impacts of coffee and tea plantations in Ethiopia, including the Gemadro Estate. The estate has spawned a wave of imitators, Gole said: smaller coffee and tea farms that are toppling more trees.

The estate itself is growing, too.

Last year, the Ethiopian Herald cited the plantation's project manager, Asenake Nigatu, telling the Ethiopian News Agency that Gemadro had "developed coffee on 1,000 hectares (about 2,470 acres) of land" it had obtained from the state investment bureau and had begun "activities to develop additional coffee on 1,500 hectares (about 3,700 acres) of land."

Gole tapped the touchpad on his laptop. An image based on satellite photography popped up showing land use changes in the Gemadro region from 1973 to 1987. A small puddle of red blotches appeared, indicating deforested areas. Gole tapped again, bringing up an image through 2001—three years after the plantation started.

The red blotches had spread across the map, like measles. His eyes widened.

"This is quite big," he said.

In his study, prepared for a future book, Gole analyzed coffee planting and forest change across two woredas—local districts—in the Sheka Zone. "The area under forest cover has dropped significantly in all parts," he wrote in the study. One area in particular stood out.

"The highest deforestation rate was observed in Gemadro, with (an) annual deforestation rate of 12.2 percent," he wrote.

But Abebe, the Ethio-Agri CEFT agricultural manager, said Gole's statements are misleading because the region was partially settled and cleared before the plantation came.

Ambassador Assefa agreed. "As I understand it, the land was largely cleared before Gemadro acquired the property," he said.

In addition, Abebe said, Gemadro is helping the land recover by incorporating conservation principles into its practices. To bolster his point, he pointed to an award the estate recently received from the Southern Nations, Nationalities and Peoples' Regional State for, he said, "being a model coffee farm."

Those model practices include planting grasses and reeds to slow erosion, planting shade trees for coffee and leaving 3,200 acres untouched for wildlife, Abebe said, adding that they "have even planted indigenous trees in the appropriate areas."

Gole, however, said such practices come up short. Many trees are non-native, he said, changing the composition of the forests. The plantation's Web site says cover crops planted there include some from South America, Mexico and India.

Struggle over Trademarks

The Ethiopian government's advocacy for its coffee industry—as well as for tea and other rural developments—has drawn international concern about the fate of its highland rain forests. A 2007 article by two German scientists blamed a lack of consistent forest policies.

"Ethiopia's montane rain forests are declining at an alarming rate," scientists Carmen Richerzhagen and Detlef Virchow wrote in the *International Journal of Biotechnology*. "The absence of a land use policy in Ethiopia creates spontaneous decisions on land allocations in a disorganized manner—therefore the forest is always the one to suffer."

But Ethiopia's push to grow more coffee drew plenty of encouragement at a conference in Addis Ababa in

February attended by representatives of the world's leading coffee companies, including Starbucks.

At the time, Starbucks and Ethiopia were locked in a struggle over the government's effort to trademark its famous coffee names—including Muel Alema's Sidamo—to create a distinctive brand that could funnel more profits back to the countryside.

Starbucks fought the effort, saying geographical certification programs, such as those in place for Colombian coffees, keep prices higher by guaranteeing that coffee from Colombia really is from Colombia.

"A trademark does not do that," said Hay, the Starbucks vice president, in a February interview in Addis Ababa. "It could be Sidamo and toilet paper. It doesn't mean anything about the region or the quality."

At times, the dispute turned bitter.

"What I don't understand is why Starbucks is resisting this," said Getachew Mengistie, director general of Ethiopia's Intellectual Property Office. "They are for improving the lives of the farmers. We are for improving the lives of the farmers. Where is the problem?"

In May, the issue was resolved in the government's favor, a positive step, said the ambassador.

"Starbucks is an important supporter of Ethiopia's efforts to control our specialty coffee brands, and it is critical that our relationship isn't about charity but about sound business," Ambassador Assefa said. "While the company only buys a small fraction of Ethiopia's coffee, our agreement will encourage market forces to allow Ethiopian farmers to capture a greater share of retail prices. This broad effort already has benefited thousands of poor farmers and could potentially benefit millions more."

Who Checked the Plantation?

Improving the lives of farmers and the environment is the goal of many coffee certification systems, such as Fair Trade, Rain Forest Alliance and Smithsonian Bird-Friendly. The Gemadro Estate has been approved by the European-based Utz Certified, an organization started by a Dutch coffee roaster and Guatemalan growers.

"It's good," said Yehasab Aschale, Utz's field representative in Ethiopia who said he toured the Gemadro plantation last year. "They are environmentally friendly. They are planting shade trees of the indigenous types. And they are improving the working conditions of the workers."

Starbucks also gave the estate's beans its own C.A.F.E. practices approval last year, signifying that the plantation protected the environment, paid workers fairly and provided them with decent housing.

Yet no one from Starbucks ever inspected the Gemadro plantation for C.A.F.E. certification. Dub Hay—the

Starbucks global purchasing executive—said he knew little about the plantation because he hadn't been there. Starbucks bought the coffee after tasting it in Europe, Hay said, adding that a coffee buyer from Switzerland visited at some point.

No one from the company Starbucks pays to oversee C.A.F.E. verification, Scientific Certification Systems of Emeryville, inspected Gemadro either. Instead, the plantation hired and paid an Africa-based company to do the job—a common industry practice.

Then, something out of the ordinary happened: The African company's inspector was fired for doing a poor job, a fact that emerged only after The Bee asked about the verification process.

"Clearly, the inspector didn't do as good, or as thorough, a job as is to be expected in C.A.F.E. practices," said Ted Howes, vice president of corporate social responsibility for Scientific Certification Systems.

Howes declined to release a copy of the inspection report, but he said his company would visit the plantation this year. "There are issues we want to look at more closely," he said.

Dennis Macray, Starbucks' director of corporate responsibility, added that such problems "can happen in any kind of a system. . . . You can have something go wrong."

If the C.A.F.E. certification process was flawed, why did Starbucks certify the beans?

"I can't tell you," Macray said at the Long Beach conference. With that, the trail went cold. Macray did not return follow-up calls. Starbucks spokeswoman Stacey Krum said details are confidential.

Krum, however, defended C.A.F.E. practices in general.

"There isn't a standard code for the coffee industry; this is something we are learning as we go along," she said. "And we are proud of it and confident it is achieving results."

Critical Thinking

1. Sketch out a "concept map" of the linkages between this article about Starbucks coffees, and Articles 5 and 6.

2. How does/do the evidence/facts regarding "development" presented in this article compare to what was presented in Article 3?

3. What "environmental economics" are illustrated in this article?

4. How might issues like population growth, demographic trends, consumerism, climate change, or habitat destruction affect the future of Starbucks? Of African coffee production? (Pick two issues and discuss.)

5. Make the argument that "eco-friendly" does not necessarily mean the same thing as "sustainable."

6. Identify in this article three key terms, concepts, or principles that are used in your textbook

(environmental science, economics, sociology, history, geography, etc.) or employed in the discipline you are currently studying. (Note: The terms, concepts, or principles may be implicit, explicit, implied, or inferred.)

Bee reporter **TOM KNUDSON** spent three weeks in Ethiopia during his four months of reporting this story, including journeys to the Gemadro Estate, the Sidamo and Yirgacheffe regions, and an international coffee conference in Addis Ababa. He interviewed coffee farmers, plantation workers, tribal people, scientists and coffee industry leaders in Africa and the United States. And he reviewed dozens of studies, reports, books and scientific journal articles about coffee growing and marketing. Travel and research were underwritten by a grant from the Alicia Patterson Foundation in Washington, D.C.

A User's Guide to the Century

JEFFREY D. SACHS

The "new world order" of the twenty-first century holds the promise of shared prosperity . . . and also the risk of global conflict. This is the paradox of our time. The scale of human society—in population, level of economic production and resource use, and global reach of production networks—gives rise to enormous hopes and equally momentous challenges. Old models of statecraft and economics won't suffice. Solutions to our generation's challenges will require an unprecedented degree of global cooperation, though the need for such cooperation is still poorly perceived and highly contested by political elites and intellectuals in the United States and elsewhere.

Our world is characterized by three dominant patterns: rapid technological diffusion, which creates strong tendencies toward technological and economic convergence among major regions of the world; extensive environmental threats resulting from the unprecedented scale of global economic activity and population; and vast current inequalities of income and power, both between and within countries, resulting from highly diverse patterns of demography, regional endowments of natural resources, and vulnerabilities to natural and societal disruptions. These characteristics hold the possibilities of rapid and equalizing economic growth, but also of regional and global instability and conflict.

The era of modern economic growth is two centuries old. For the first one hundred years, this was a strong *divergence* in economic growth, meaning a widening gap in production and income between the richest regions and the rest of the world. The dramatic divergence of per capita output, industrial production and living standards during the nineteenth century between the North Atlantic (that is, Western Europe and the United States) and the rest of the world was accentuated by several factors. The combination of first-mover industrialization, access to extensive coal deposits, early development of market-based institutions, military dominance resulting from vast industrial power, and then colonial dominance over Africa and Asia all contributed to a century of *economic divergence,* in which the North Atlantic greatly expanded its technological lead (and also military advantage) vis-à-vis the rest of the world. The apogee of "Western" relative dominance was roughly the year 1910. Until the start of World War I, this economic and technological dominance was nearly overpowering.

The period 1910–1950 marked a transition from global economic divergence to economic convergence. Most importantly, of course, was Europe's self-inflicted disaster of two world wars and an intervening Great Depression, which dramatically weakened Europe and proved to be the downfall of the continent's vast overseas empires. Below the surface, longer-term forces of convergence were also stirring. These deeper forces included the global spread of literacy, Western science, the modern technologies of transport and communications, and the political ideas of self-determination and economic development as core national objectives.

Since 1950, we have entered into an era of global convergence, in which much of the non-Western world is gradually catching up, technologically, economically, geopolitically and militarily. The North Atlantic is losing its uniquely dominant position in the world economy. The technological and economic catching-up, most notable of course in Asia, is facilitated by several factors—the spread of national sovereignty following European colonialism; vastly improved transport and communications technologies; the spread of infectious-disease control, mass literacy and public education; the dissemination of global scientific and engineering knowledge; and the broad adoption of a valid "catch-up model" of economic development based on technology imports within a mixed public-private system. The system was modeled heavily on the state-led market development of Japan, the only non-Western country to succeed in achieving modern industrialization during the nineteenth century. Japan's economic development following the Meiji Restoration in 1868 can indeed be viewed as the invention of "catch-up growth."

The modern age of convergence, begun with Japan's rapid rebuilding after World War II, was extended in the 1950s and 1960s by the rise of Korea, Taiwan, Hong Kong and Singapore, all built on an export-led growth model using U.S. and Japanese technologies and institutions. Convergent economic growth then spread through Southeast Asia (notably Indonesia, Malaysia and Thailand) in the 1970s and 1980s, again supported by Japanese and U.S. technologies, and Japanese aid and development concepts. The convergence patterns were greatly expanded with the initiation of rapid market-based growth in China after 1978 (which imitated strategies in East and Southeast Asia) and then India in the 1980s (and especially after market-based reforms initiated in 1991). In the early twenty-first century, both Brazil and Mexico are similarly experiencing rapid technological catch-up.

In economic terms, the share of global income in the North Atlantic is now declining quickly as the emerging economies of Asia, the Middle East and Latin America grow rapidly. This is, of course, especially true when output and income are measured in purchasing-power-adjusted terms, thereby adding weight to the share of the emerging economies. By 2050, Asia will be home to more than half of global production, up from around 20 percent as of 1970. In geopolitical terms, the unipolar world of the North Atlantic is over. China, India, Brazil and other regional powers now fundamentally constrain the actions of the United States and Western Europe. This shift to multipolarity in geopolitics is bound to accelerate in the coming decades.

Modern economic growth did not end humans' dependence on their physical environment, contrary to the false impressions sometimes given by modern urban life. Our food still comes from farms, not from supermarkets and bakeries. Our crops still demand land and water, not simply microwaves and gas grills. Our industrial prowess has been built mainly on fossil fuels (first coal, then oil and natural gas), not merely on cleverness and efficiency. Our food production demands enormous inputs of energy and water, not only high-yield seeds. The bottom line is that the growth of the world economy has meant a roughly commensurate growth in human impacts on the physical world, not an escape from such impacts. These anthropogenic impacts are now so significant, and indeed threatening to the sustainable well-being of humans and other species, that Nobel Laureate Paul Crutzen (a codiscoverer of the human-induced loss of stratospheric ozone) has termed our age the Anthropocene, meaning the geological epoch when human activity dominates or deranges the earth's major biogeophysical fluxes (including the carbon, nitrogen and water cycles, among others).

The world economy is now characterized by 6.7 billion people—roughly ten times more than in 1750—producing output at a rate of roughly $10,000 per person per year in purchasing-power-adjusted prices. The resulting $67 trillion annual output (in approximate terms, as precision here should not be pretended) is at least one hundredfold larger than at the start of the industrial era. The human extent of natural-resource use is unprecedented—indeed utterly unrecognizable—in historical perspective, and is now dangerous to long-term well-being. While the typical economist's lighthearted gloss is that Malthusian resource pessimism was utterly and fully debunked generations ago—overcome by human ingenuity and technical know-how—it is more correct to say that the unprecedented level of global human output has been achieved not by overcoming resource constraints, but by an unprecedented appropriation of the earth's natural resources.

In fact, the current rate of resource use, if technologies remain constant, is literally unsustainable. Current fossil-fuel use would lead to the imminent peak of oil and gas production within years or decades, and of conventional coal deposits within decades or a century or two. We would see dangerous human impacts on the global climate system, and hence regional climates in all parts of the world, through greenhouse-gas emissions. The appropriation of up to half of the earth's photosynthetic potential, at the cost of other species, would occur. There would be massive deforestation and land degradation as a result of the increasing spatial range and the intensification of farming and pasture use; massive appropriation of freshwater resources, through depletion of fossil aquifers, diversion of rivers, melting of glaciers, drainage of wetlands, destruction of mangroves and estuaries, and other processes. And, an introduction of invasive species, pests and pathogens through a variety of human-induced changes.

The mistaken belief that we've overcome "similar" resource constraints in the past is no proof that global society will do it again, or at least do it successfully without massive economic and social upheavals, especially in view of the fact that our earlier "solutions" were rarely based on resource-saving technologies. Indeed, most earlier "solutions" to resource constraints typically involved new ways to "mine" the natural environment, not to conserve it. This time around, human societies will have to shift from resource-using technologies to resource-saving technologies. Some of the needed technologies are already known but often not widely used, while others will still have to be developed, demonstrated and diffused on a global scale.

Human pressures on the earth's ecological systems are bound to increase markedly in the years ahead. The global economy has been growing between 3 and 5 percent per year, meaning the economy will take fourteen to twenty-three years to double. Thus, the intense environmental and resource pressures now occurring will increase markedly and in short order. The catch-up growth of the largest emerging markets—Brazil, China and India, with around 40 percent of the world's population—is based squarely on the adoption and diffusion of resource-intensive technologies, such as coal-fired power plants and standard internal-combustion-engine vehicles.

The age of convergence offers the realistic possibility of ending extreme poverty and narrowing the vast inequalities within and between countries. The catching-up of China and India, for example, is rapidly reducing the national poverty rates in both countries. Other regions will also experience rapid declines in poverty rates. Yet the actual record of poverty reduction and trends in inequality leave major gaps in success. There are many parts of the planet where the numbers, and sometimes even proportions, of people in extreme poverty are rising rather than falling. Even more generally, the gaps between the rich and poor within nations seem to be widening markedly in most parts of the world.

Significant regions of the world—including sub-Saharan Africa, Central Asia, and parts of the Andean and Central American highlands—have experienced increasing poverty during the past generation. These places left behind by global economic growth tend to display some common infirmities. For example: long distances from major global trade routes, landlocked populations, heavy burdens of tropical diseases, great vulnerabilities to natural hazards (such as earthquakes, tropical storms and the like), lack of nonbiomass energy resources, lack of low-cost access to irrigation, difficult topography (e.g., high elevations and steep slopes), widespread illiteracy and a rapid growth of population due to consistently high fertility rates.

These conditions tend to perpetuate extreme poverty, and often lead to a vicious circle in which poverty contributes to further environmental degradation, persistence of high fertility rates, and social conflicts and violence, which in turn perpetuate or intensify the extreme poverty. These vicious circles (or "poverty traps") can be broken, but to succeed often requires external financial and technological assistance. Assistance like building infrastructure raises productivity and thereby controls the interlocking problems of transport costs, disease, illiteracy, vulnerability to hazards and high fertility. Without the external assistance, a continuing downward spiral becomes much more likely. The adverse consequences can then include war, the spread of epidemic diseases, displaced popula-

tions and mass illegal migration. On top of this can be the spread of illicit activities (drug trafficking, smuggling, kidnapping and piracy) and continued serious environmental degradation with large-scale poaching, land degradation and rampant deforestation, to name a few.

The global forces of demographic change, economic convergence and global production systems are also apparently contributing to rising inequalities within societies. Technological advances favor educated workers and leave uneducated workers behind. The entry of China and India into the global trading and production system, similarly, has pushed down the relative wages of unskilled workers in all parts of the world. Geography has played a key role, favoring those regions and parts of countries which are most easily incorporated into global production systems and which are well endowed with energy, fertile land, water and climate conducive to food production. Rapid population growth in rural and poverty-stricken regions (sub-Saharan Africa) has dramatically lowered well-being in these places. In general, urban dwellers have done better than rural dwellers in the past twenty years in almost all parts of the world.

Even relatively homogenous societies are facing major challenges of social stability as a result of massive changes in demographic patterns and economic trends across ethnic, linguistic and religious communities. By 2050, roughly half of the U.S. population will be "white, non-Hispanic," down from around 80 percent as of 1950. This trend reflects both the differential fertility rates across different subpopulations as well as the continued rapid in-migration of Hispanics into the United States. Such large demographic changes can potentially create major fissures in society, especially when there have been long histories of intercommunal strife and suspicion.

The new world order is therefore crisis prone. The existence of rapidly emerging regional powers, including Brazil, China and India, can potentially give rise to conflicts with the United States and Europe.

The combination of rapid technological diffusion and therefore convergent economic growth, coupled with the natural-resource constraints of the Anthropocene, could trigger regional-scale or global-scale tensions and conflicts. China's rapid economic growth could turn into a strenuous, even hot, competition with the United States over increasingly scarce hydrocarbons in the Middle East, Africa and Central Asia. Conflicts over water flow in major and already-contested watersheds (among India, Bangladesh and Pakistan; China and Southeast Asia; Turkey, Israel, Iraq and Jordan; the countries of the Nile basin; and many others) could erupt into regional conflicts. Disagreements over management of the global commons—including ocean fisheries, greenhouse gases, the Arctic's newly accessible resources, species extinctions and much more—could also be grounds for conflict.

The continuation of extreme poverty, and the adverse spillovers from laggard regions, could trigger mass violence. Local conflicts can draw in major powers, which then threaten expanded wars—as in Afghanistan, Somalia and Sudan. When poverty is combined with rapid population growth and major environmental shocks (such as prolonged droughts in the Sahel and the Horn of Africa) there is a distinct likelihood of mass population movements, such as large-scale illegal migrations of populations escaping hunger and destitution. Such movements in the past have contributed to local violence, as in South Africa of late, and even to war, as in Darfur.

These intersecting challenges of our crowded world, multi-polarity, unprecedented demographic and environmental stresses, and the growing inequalities both within and between countries, can trigger spirals of conflict and instability—disease, migration, state failure and more— and yet are generally overlooked by the broad public and even by many, if not most, foreign-policy analysts. The instability of the Horn of Africa, the Middle East and Central Asia has been viewed wrongly by many in the U.S. public and foreign-policy community mainly as the battleground over Islamic extremism and fundamentalism, with little reflection on the fact that the extremism and fundamentalism is often secondary to illiteracy, youth unemployment, poverty, indignation, economic hopelessness and hunger, rather than religion per se. The swath of "Islamic" extremist violence across the African Sahel, Horn of Africa, and into the Middle East and Central Asia lies in the world's major dryland region, characterized by massive demographic, environmental and economic crises.

The security institutions—such as ministries of defense—of the major powers are trained to see these crises through a military lens, and to look for military responses, rather than see the underlying demographic, environmental and economic drivers—and the corresponding developmental options to address them. Genuine global security in the next quarter century will depend on the ability of governments to understand the true interconnected nature of these crises, and to master the scientific and technological knowledge needed to find solutions.

In the United States, I propose a new Department for International Sustainable Development, which would oversee U.S. foreign assistance and initiatives related to sustainable development in low-income countries, including water, food production, disease control and climate-change adaptation and mitigation.

I propose five major guideposts for a more-functional foreign policy in the coming years. First, we will need, on a global scale, to develop and diffuse new sustainable technologies so that the global economy can continue to support broad-based economic growth. If we remain stuck with our current technologies, the world will face a zero-sum struggle for increasingly scarce resources across competing regions. The new sustainable technologies will not arise from market forces alone. All major technological advances, such as the introduction of large-scale solar or nuclear power, will require massive public-sector investments (in basic science, demonstration projects, diffusion of proven technologies and regulatory framework) alongside the R&D of private markets. These public investments will be global-scale, internationally cooperative efforts.

Free-market ideologues who are convinced that technologies emerge from market forces alone should think again. They might compare the successful government-led promotion of nuclear power in France with the failure of the private-sector-led nuclear-power industry in the United States, which failed because of a collapse in U.S. public confidence in the safety of the technology. Similarly, they can examine the highly successful public-private partnerships linking the public-sector National Institutes of Health with the private-sector pharmaceutical industry, or the public-sector investments that underpinned the start-up of computer and Internet technologies.

Second, we will need to address the still-rapid rise of the world's population, heavily centered in the world's poorest countries. Sub-Saharan Africa is on a trajectory that will expand its population from around 800 million to 1.8 billion by 2050, according to the medium-fertility forecast of the United Nations Population Division. Yet that extent of population increase, an added 1 billion people, resulting from Africa's very high fertility rates, would actually be a grave threat to Africa's economy, political stability and environment, and would inevitably spill over adversely into the rest of the world. Rapid and voluntary fertility reduction in Africa is possible, if girls can be encouraged to stay in school through the secondary level; if family planning and contraception are made widely available; if child mortality is reduced (giving confidence to parents to reduce fertility rates); and if women are economically empowered.

Third, the world will need to address critical failings in the management of the global commons, most importantly, by restricting greenhouse-gas emissions, protecting the oceans and biodiversity, and managing transnational water resources sustainably at the regional level. Of course several global treaties have committed the world's nations to do just this, but these treaties have yet to be implemented. Three treaties of overriding importance are the UN Framework Convention on Climate

Change, the UN Convention on Biological Diversity and the UN Convention to Combat Desertification. If these treaties are honored, the global commons can be sustainably managed.

Fourth, we will need to take seriously the risks of impoverished "failed states," to themselves, to their neighborhoods and to the world. The poorest and least-stable countries are rife with risks to peace and avoidable human tragedies like the 10 million children each year who die tragically and unnecessarily before their fifth birthday, largely the result of extreme poverty. Darfur, the Horn of Africa, Yemen, Afghanistan, Pakistan, Sri Lanka and elsewhere are places trapped in vicious cycles of extreme violence and poverty. These poverty-conflict traps can be broken, most importantly if the donors of the G-8, the oil-rich states in the Middle East, and the new donors in Latin America and Asia will pool their efforts to ensure the success of the Millennium Development Goals in today's impoverished and fragile regions.

Fifth, and finally, we require a new analytical framework for addressing our generation's challenges, and a new governmental machinery to apply that framework. Traditional problems of statecraft—the balance of power, alliances, arms control and credible deterrence—certainly will continue to play a role, but we need to move beyond these traditional concepts to face the challenges of sustainable development ahead. Will our era be a time of wondrous advances, based on our unprecedented scientific and technological know-how, or will we succumb to a nightmare of spreading violence and conflict? We face world-shaping choices. Our global challenges are unique to our generation, in scale and character. Vision, leadership and global cooperation will be our most important resources for ensuring our future well-being.

Critical Thinking

1. What are the three dominant patterns the author describes as characterizing our world? Does this "characterize" the whole world, or just certain regions? Explain.

2. What facts does the author use to support his argument that "modern economic growth did not end humans' dependence on their physical environment? Explain how the articles about "ecosystems services" in this text support his argument.

3. What does the "age of convergence" mean? What impacts may the population growth and demographic issues discussed in Unit 1 have on this "age of convergence" in the near future?

4. According to the article, why is the current rate of resource use unsustainable? What regions of the earth populations are contributing most to this unsustainably?

5. The author proposes five major guideposts for a more functional foreign policy in the future. What are those guideposts? Describe the linkages that *nonresolution* of environmental issues like population growth, food, water, climate change, energy, and consumption will have with *achieving* of those guideposts.

6. Identify in this article three key terms, concepts, or principles that are used in your textbook (environmental science, economics, sociology, history, geography, etc.) or employed in the discipline you are currently studying. (Note: The terms, concepts, or principles may be implicit, explicit, implied, or inferred.)

JEFFREY D. SACHS is the director of the Earth Institute at Columbia University and author of *Common Wealth: Economics for a Crowded Planet* (Penguin, 2008).

UNIT 3
Feeding Humanity

Unit Selections

Learning Outcomes

After reading this Unit, you should be able to:

- Identify the root causes of global hunger and offer insight on the controversy regarding how best to meet the challenge of environmental issue #4: feeding humanity.

- Describe the argument for radically rethinking agricultural production patterns.

- Discuss the linkages between modern agricultural technologies and improving global security.

- Assess the benefits and risks of agricultural genetic engineering.

- Offer insight regarding the politically challenging steps to feeding the world.

- Discuss some of the potential negative consequences of expanding commercial agriculture at the expense of reducing peasant agriculture.

- Outline the connections between food production, economics, and geography.

- Explain how nations may better manage their natural resources necessary for growing food.

- Recognize the connections between land productivity, water resources, grain production consumption, and our future ability to feed 8 billion people.

- Compare/contrast arguments regarding the need for increased application of agricultural technologies/commercialization versus agricultural localizations and changing grain consumption patterns.

Student Website

www.mhhe.com/cls

Internet References

The Hunger Project
    www.thp.org
National Geographic Society
    www.nationalgeographic.com
Penn Library: Resources by Subject
    www.library.upenn.edu/cgi-bin/res/sr.cgi
Sustainable Development.Org
    www.sustainabledevelopment.org

Give a man a fish, and he can eat for a day. But teach a man how to fish, and he'll be dead of mercury poisoning inside of three years.

—Charles Haas

Humans consume the Earth to survive. We need food, we need water. Some clothes would be nice, as well as a roof over our heads. Nice, but not essential. Simple food. Freshwater. It's not rocket science. At least it wasn't for most of the generations of people before us. However, feeding our near future 8, 9, or 10 billion people may take some serious science according to one camp of scholars. Others think not. A little common sense and a bit of reprioritizing traditional food consumption patterns of industrialized nations may be what it really takes to feed humanity now and into the future—sustainably.

Like addressing climate change, resolving *Top Ten Environmental Issue 3,* Feeding Humanity, will require not only an interdisciplinary approach, but an integrated one as well. The challenge cannot be met via applications of agricultural science alone. Sociologists, economists, political scientists, foreign policy analysts, anthropologists, biologists, chemists, physicists, mathematicians, geographers, environmental scientists, civic leaders, teachers, world leaders—the contributions will be endless. The *fundamental* science of growing food, like "global warming," we understand. Reverse climate change? The consensus seems to be that we need to stop filling our atmosphere with life-threatening chemical emissions. There is pretty widespread agreement among the world's scientific community that approach would work. But now comes the political part. The implementing part. The people part. The place part. And what a messy part that is. World hunger? More and more humans, especially the hungry ones—the poor, the working class, the middle class—are asking why we don't simply take the food that is grown now and feed people, not livestock, provide affordable and fair access to basic foodstuffs, give farmers (especially, smallholder and intensive subsistence) the status they deserve, re-localize food production, grow food simply and sustainably (as it has been done by many peoples in many places for any generations). But then comes the political part . . .

The intent of **Unit 3** is to provide food for critical thought. The four articles provide examples of some of the common themes fueling the debate that whirls around the environmental issue of feeding people. Matters of science, politics, economics, foreign affairs, environmental management, and human behavior are just of few of the topics that come into play when trying to resolve hunger. Article 9, "Radically Rethinking Agriculture for the 21st Century," and Article 12, "How to Feed 8 Billion People," can be viewed perhaps as anchoring the two ends of a feeding-fix spectrum, a sort of "how-to." On the one end, Article 9 focuses on encouraging and implementing more "hard science" to meet our agricultural needs. The authors argue that success in feeding the world's billions will depend on acceptance and use of our agricultural science knowledge: genetic engineering, pesticides, herbicides, irrigation techniques. They believe we need to scale up and build on existing technologies and innovations, and do

so immediately. And, at the same time, we need to reevaluate existing agricultural regulatory frameworks that are based on scientific data and not fear.

On the other end of the food-fix spectrum, there is a call for encouraging resource management and changing human consumption attitudes/behaviors to meet our food needs. Although Article 8 employs the phrase "radically rethinking," Lester Brown's Article 12 is perhaps no less characterized by "radically thinking." The author believes the world is seeing diminishing returns on the agricultural science of the Green Revolution. That combined with the increasing global food demand—for grain in particular—is leading us to an impending food crisis. Using grain, says Brown, to grow grain-based animal protein and to fuel cars is the crux of our food problem. This agricultural-orientation choice is leading to other collateral environmental damage such as soil erosion, freshwater shortages, and decreasing productivity. According to the article, what we need to do in order to feed the world is to better manage our limited agriculture resources and the factors (e.g., climate change, population growth, urbanization, water use) that affect those resources. To do this, Lester Brown says we need to institute (radical) changes in conservation policies, food production systems, and the food consumption behaviors of the industrialized world.

In between these food-fix philosophies, Articles 10 and 11 provide two examples more reflecting of what is happening now: the politics, illusions, and greed currently fanning the fire of the food crisis, and the simple, everyday face of hunger and the anger it fuels. Paul Collier, in Article 10, "The Politics of Hunger," argues that the science of food production is well understood and that the challenge will be political, not scientific. However, unlike Lester Brown's approach in Article 12, Collier is a champion of bigger is better (more large-scale commercial agriculture) and more science (the ban needs to be lifted on genetically modified crop production). They both, nevertheless, see the counterproductiveness of using agriculture land to produce biofuels.

Article 11, "Across Globe, Empty Bellies Bring Rising Anger," provides a succinct example of what happens when people are hungry. And although there appears to be a strong relationship between poverty and hunger, the author argues that even working-class and middle-class income groups are beginning to feel the pinch of higher food prices triggered by rising demand, high oil prices, and the growing of fuel instead of food. The reader is asked to reflect on what may be the global geographical pattern of eating mud pies in the future.

As an environmental issue of impending magnitude, humans will have to believe intimately and whole-heartedly that food is fundamental to life, and not a luxury. Science can help, but it will ultimately come down to the food consumption choices (needs versus wants) we make and the kind of sustainable agricultural relationship we and future generations develop with our environment.

Radically Rethinking Agriculture for the 21st Century

N. V. FEDOROFF ET AL.

Population experts anticipate the addition of another roughly 3 billion people to the planet's population by the mid-21st century. However, the amount of arable land has not changed appreciably in more than half a century. It is unlikely to increase much in the future because we are losing it to urbanization, salinization, and desertification as fast as or faster than we are adding it.[1] Water scarcity is already a critical concern in parts of the world.[2]

Climate change also has important implications for agriculture. The European heat wave of 2003 killed some 30,000 to 50,000 people.[3] The average temperature that summer was only about 3.5°C above the average for the last century. The 20 to 36% decrease in the yields of grains and fruits that summer drew little attention. But if the climate scientists are right, summers will be that hot on average by midcentury, and by 2090 much of the world will be experiencing summers hotter than the hottest summer now on record.

The yields of our most important food, feed, and fiber crops decline precipitously at temperatures much above 30°C.[4] Among other reasons, this is because photosynthesis has a temperature optimum in the range of 20° to 25°C for our major temperate crops, and plants develop faster as temperature increases, leaving less time to accumulate the carbohydrates, fats, and proteins that constitute the bulk of fruits and grains.[5] Widespread adoption of more effective and sustainable agronomic practices can help buffer crops against warmer and drier environments,[6] but it will be increasingly difficult to maintain, much less increase, yields of our current major crops as temperatures rise and drylands expand.[7]

Climate change will further affect agriculture as the sea level rises, submerging low-lying cropland, and as glaciers melt, causing river systems to experience shorter and more intense seasonal flows, as well as more flooding.[7]

Recent reports on food security emphasize the gains that can be made by bringing existing agronomic and food science technology and know-how to people who do not yet have it,[8,9] as well as by exploring the genetic variability in our existing food crops and developing more ecologically sound farming practices.[10] This requires building local educational, technical, and research capacity, food processing capability, storage capacity, and other aspects of agribusiness, as well as rural transportation and water and communications infrastructure. It also necessitates addressing the many trade, subsidy, intellectual property, and regulatory issues that interfere with trade and inhibit the use of technology.

What people are talking about today, both in the private and public research sectors, is the use and improvement of conventional and molecular breeding, as well as molecular genetic modification (GM), to adapt our existing food crops to increasing temperatures, decreased water availability in some places and flooding in others, rising salinity,[8,9] and changing pathogen and insect threats.[11] Another important goal of such research is increasing crops' nitrogen uptake and use efficiency, because nitrogenous compounds in fertilizers are major contributors to waterway eutrophication and greenhouse gas emissions.

There is a critical need to get beyond popular biases against the use of agricultural biotechnology and develop forward-looking regulatory frameworks based on scientific evidence. In 2008, the most recent year for which statistics are available, GM crops were grown on almost 300 million acres in 25 countries, of which 15 were developing countries.[12] The world has consumed GM crops for 13 years without incident. The first few GM crops that have been grown very widely, including insect-resistant and herbicide-tolerant corn, cotton, canola, and soybeans, have increased agricultural productivity and farmers' incomes. They have also had environmental and health benefits, such as decreased use of pesticides and herbicides and increased use of no-till farming.[13]

Despite the excellent safety and efficacy record of GM crops, regulatory policies remain almost as restrictive as they were when GM crops were first introduced. In the United States, case-by-case review by at least two and sometimes three regulatory agencies (USDA, EPA, and FDA) is still commonly the rule rather than the exception. Perhaps the most detrimental effect of this complex, costly, and time-intensive regulatory apparatus is the virtual exclusion of public-sector researchers from the use of molecular methods to improve crops for farmers. As a result, there are still only a few GM crops, primarily those for which there is a large seed market,[12] and the benefits of biotechnology have not been realized for the vast majority of food crops.

What is needed is a serious reevaluation of the existing regulatory framework in the light of accumulated evidence and experience. An authoritative assessment of existing data on GM crop safety is timely and should encompass protein safety, gene stability, acute toxicity, composition, nutritional value, allergenicity, gene flow, and effects on nontarget organisms. This would establish a foundation for reducing the complexity of the regulatory process without affecting the integrity of the safety assessment. Such an evolution of the regulatory process in the United States would be a welcome precedent globally.

It is also critically important to develop a public facility within the USDA with the mission of conducting the requisite safety testing of GM crops developed in the public sector. This would make it possible for university and other public-sector researchers to use contemporary molecular knowledge and techniques to improve local crops for farmers.

However, it is not at all a foregone conclusion that our current crops can be pushed to perform as well as they do now at much higher temperatures and with much less water and other agricultural inputs. It will take new approaches, new methods, new technology—indeed, perhaps even new crops and new agricultural systems.

Aquaculture is part of the answer. A kilogram of fish can be produced in as little as 50 liters of water,[14] although the total water requirements depend on the feed source. Feed is now commonly derived from wild-caught fish, increasing pressure on marine fisheries. As well, much of the growing aquaculture industry is a source of nutrient pollution of coastal waters, but self-contained and isolated systems are increasingly used to buffer aquaculture from pathogens and minimize its impact on the environment.[15]

Another part of the answer is in the scale-up of dryland and saline agriculture.[16] Among the research leaders are several centers of the Consultative Group on International Agricultural Research, the International Center for Biosaline Agriculture, and the Jacob Blaustein Institutes for Desert Research of the Ben-Gurion University of the Negev.

Systems that integrate agriculture and aquaculture are rapidly developing in scope and sophistication. A 2001 United Nations Food and Agriculture Organization report[17] describes the development of such systems in many Asian countries. Today, such systems increasingly integrate organisms from multiple trophic levels[18]. An approach particularly well suited for coastal deserts includes inland seawater ponds that support aquaculture, the nutrient efflux from which fertilizes the growth of halophytes, seaweed, salt-tolerant grasses, and mangroves useful for animal feed, human food, and biofuels, and as carbon sinks.[19] Such integrated systems can eliminate today's flow of agricultural nutrients from land to sea. If done on a sufficient scale, inland seawater systems could also compensate for rising sea levels.

The heart of new agricultural paradigms for a hotter and more populous world must be systems that close the loop of nutrient flows from microorganisms and plants to animals and back, powered and irrigated as much as possible by sunlight and seawater. This has the potential to decrease the land, energy, and freshwater demands of agriculture, while at the same time ameliorating the pollution currently associated with agricultural chemicals and animal waste. The design and large-scale implementation of farms based on nontraditional species in arid places will undoubtedly pose new research, engineering, monitoring, and regulatory challenges, with respect to food safety and ecological impacts as well as control of pests and pathogens. But if we are to resume progress toward eliminating hunger, we must scale up and further build on the innovative approaches already under development, and we must do so immediately.

References and Notes

1. *The Land Commodities Global Agriculture & Farmland Investment Report 2009* (Land Commodities Asset Management AG, Baar, Switzerland, 2009; www.landcommodities.com).
2. *Water for Food, Water for Life: A Comprehensive Assessment of Water Management* (International Water Management Institute, Colombo, Sri Lanka, 2007).
3. D. S. Battisti, R. L. Naylor, Historical warnings of future food insecurity with unprecedented seasonal heat. *Science* **323,** 240 (2009). [Abstract/Free Full Text].
4. W. Schlenker, M. J. Roberts, Nonlinear temperature effects indicate severe damages to U.S. crop yields under climate change. *Proc. Natl. Acad. Sci. U.S.A.* **106,** 15594 (2009). [Abstract/Free Full Text].
5. M. M. Qaderi, D. M. Reid, in *Climate Change and Crops,* S. N. Singh, Ed. (Springer-Verlag, Berlin, 2009), pp. 1–9.
6. J. I. L. Morison, N. R. Baker, P. M. Mullineaux, W. J. Davies, Improving water use in crop production. *Philos. Trans. R. Soc. London Ser. B* **363,** 639 (2008). [Abstract/Free Full Text].
7. Intergovernmental Panel on Climate Change, *Climate Change 2007: Impacts, Adaptation and Vulnerability* (Cambridge Univ. Press, Cambridge, 2007; www.ipcc.ch/publications_and_data/publications_ipcc_fourth_assessment_report_wg2_report_impacts_adaptation_and_vulnerability.htm).
8. *Agriculture for Development* (World Bank, Washington, DC, 2008; http://siteresources.worldbank.org/INTWDR2008/Resources/WDR_00_book.pdf).
9. *Reaping the Benefits: Science and the Sustainable Intensification of Global Agriculture* (Royal Society, London, 2009; http://royalsociety.org/Reapingthebenefits).
10. *The Conservation of Global Crop Genetic Resources in the Face of Climate Change* (Summary Statement from a Bellagio Meeting, 2007; http://iis-db.stanford.edu/pubs/22065/Bellagio_final1.pdf).
11. P. J. Gregory, S. N. Johnson, A. C. Newton, J. S. I. Ingram, Integrating pests and pathogens into the climate change/food security debate. *J. Exp. Bot.* **60,** 2827 (2009). [Abstract/Free Full Text].
12. C. James, *Global Status of Commercialized Biotech/GM Crops: 2008* (International Service for the Acquisition of Agri-biotech Applications, Ithaca, NY, 2008).
13. G. Brookes, P. Barfoot, *AgBioForum* **11,** 21 (2008).
14. S. Rothbard, Y. Peretz, in *Tilapia Farming in the 21st Century,* R. D. Guerrero III, R. Guerrero-del Castillo, Eds. (Philippines Fisheries Associations, Los Baños, Philippines, 2002), pp. 60–65.
15. *The State of World Fisheries and Aquaculture 2008* (United Nations Food and Agriculture Organization, Rome, 2009; www.fao.org/docrep/011/i0250e/i0250e00.HTM).
16. M. A. Lantican, P. L. Pingali, S. Rajaram, Is research on marginal lands catching up? The case of unfavourable wheat growing environments. *Agric. Econ.* **29,** 353 (2003). [CrossRef].

17. *Integrated Agriculture-Aquaculture* (United Nations Food and Agriculture Organization, Rome, 2001; www.fao.org/DOCREP/005/Y1187E/y1187e00.htm).

18. T. Chopin *et al.,* in *Encyclopedia of Ecology,* S. E. Jorgensen, B. Fath, Eds. (Elsevier, Amsterdam, 2008), pp. 2463–2475.

19. The Seawater Foundation, www.seawaterfoundation.org.

20. The authors were speakers in a workshop titled "Adapting Agriculture to Climate Change: What Will It Take?" held 14 September 2009 under the auspices of the Office of the Science and Technology Adviser to the Secretary of State. The views expressed here should not be construed as representing those of the U.S. government. N.V.F. is on leave from Pennsylvania State University. C.N.H. is co-chair of Global Seawater, which promotes creation of Integrated Seawater Farms.

Critical Thinking

1. What does it mean to "radically re-examine" agriculture?

2. Who is doing the "re-examining"; who will "implement" the "conclusions" of the re-examination; who will "pay" for the "application" of the conclusions; who will benefit?

3. Compare the agriculture ideas presented in Article 9 to the ideas presented in Article 12. Where do they differ?

4. Locate on the map provided in this text the following: regions of food production; regions of consumption; regions where risks of global food security are highest.

5. Referring to question 4 above, list some general characteristics like socioeconomic, cultural, and environmental of the places you have identified. Describe what you see.

6. Identify in this article three key terms, concepts, or principles that are used in your textbook (environmental science, economics, sociology, history, geography, etc.) or employed in the discipline you are currently studying. (Note: The terms, concepts, or principles may be implicit, explicit, implied, or inferred.)

From *Science Magazine*, February 12, 2010, pp. 833–844. Copyright © 2010 by American Association for the Advancement of Science. Reprinted by permission.

The Politics of Hunger

How Illusion and Greed Fan the Food Crisis

PAUL COLLIER

After many years of stability, world food prices have jumped 83 percent since 2005—prompting warnings of a food crisis throughout much of the world earlier this year. In the United States and Europe, the increase in food prices is already yesterday's news; consumers in the developed world now have more pressing concerns, such as the rising price of energy and the falling price of houses. But in the developing world, a food shock of this magnitude is a major political event. To the typical household in poor countries, food is the equivalent of energy in the United States, and people expect their government to do something when prices rise. Already, there have been food riots in some 30 countries; in Haiti, they brought down the prime minister. And for some consumers in the world's poorest countries, the true anguish of high food prices is only just beginning. If global food prices remain high, the consequences will be grim both ethically and politically.

Politicians and policymakers do, in fact, have it in their power to bring food prices down. But so far, their responses have been less than encouraging: beggar-thy-neighbor restrictions, pressure for yet larger farm subsidies, and a retreat into romanticism. In the first case, neighbors have been beggared by the imposition of export restrictions by the governments of food-exporting countries. This has had the immaculately dysfunctional consequence of further elevating world prices while reducing the incentives for the key producers to invest in the agricultural sector. In the second case, the subsidy hunters have, unsurprisingly, turned the crisis into an opportunity; for example, Michel Barnier, the French agricultural minister, took it as a chance to urge the European Commission to reverse its incipient subsidy-slashing reforms of the Common Agricultural Policy. And finally, the romantics have portrayed the food crisis as demonstrating the failure of scientific commercial agriculture, which they have long found distasteful. In its place they advocate the return to organic small-scale farming—counting on abandoned technologies to feed a prospective world population of nine billion.

The real challenge is not the technical difficulty of returning the world to cheap food but the political difficulty of confronting the lobbying interests and illusions on which current policies rest. Feeding the world will involve three politically challenging steps. First, contrary to the romantics, the world needs more commercial agriculture, not less. The Brazilian model of high-productivity large farms could readily be extended to areas where land is underused. Second, and again contrary to the romantics, the world needs more science: the European ban and the consequential African ban on genetically modified (GM) crops are slowing the pace of agricultural productivity growth in the face of accelerating growth in demand. Ending such restrictions could be part of a deal, a mutual de-escalation of folly, that would achieve the third step: in return for Europe's lifting its self-damaging ban on GM products, the United States should lift its self-damaging subsidies supporting domestic biofuel.

Supply-Side Solutions

Typically, in trying to find a solution to a problem, people look to its causes—or, yet more fatuously, to its "root" cause. But there need be no logical connection between the cause of a problem and appropriate or even just feasible solutions to it. Such is the case with the food crisis. The root cause of high food prices is the spectacular economic growth of Asia. Asia accounts for half the world's population, and because its people are still poor, they devote much of their budgets to food. As Asian incomes rise, the world demand for food increases. And not only are Asians eating more, but they are also eating better: carbohydrates are being replaced by protein. And because it takes six kilograms of grain to produce one kilogram of beef, the switch to a protein-heavy diet further drives up demand for grain.

The two key parameters in shaping demand are income elasticity and price elasticity. The income elasticity of demand for food is generally around 0.5, meaning that if income rises by, say, 20 percent, the demand for food rises by 10 percent. (The price elasticity of demand for food is only around 0.1: that is, people simply have to eat, and they do not eat much less in response to higher prices.) Thus, if the supply of food were fixed, in order to choke off an increase in demand of 10 percent after a 20 percent rise in income, the price of food would need to double. In other words, modest increases in global income will drive prices up alarmingly unless matched by increases in supply.

In recent years, the increase in demand resulting from gradually increasing incomes in Asia has instead been matched with several supply shocks, such as the prolonged drought in Australia. These shocks will only become more common with the climatic volatility that accompanies climate change. Accordingly, against a backdrop of relentlessly rising demand, supply will fluctuate more sharply as well.

Because food looms so large in the budgets of the poor, high world food prices have a severely regressive effect in their toll. Still, by no means are all of the world's poor adversely affected by expensive food. Most poor people who are farmers are largely self-sufficient. They may buy and sell food, but the rural markets in which they trade are often not well integrated into global markets and so are largely detached from the surge in prices. Where poor farmers are integrated into global markets, they are likely to benefit. But even the good news for farmers needs to be qualified. Although most poor farmers will gain most of the time, they will lose precisely when they are hardest hit: when their crops fail. The World Food Program is designed to act as the supplier of last resort to such localities. Yet its budget, set in dollars rather than bushels, buys much less when food prices surge. Paradoxically, then, the world's insurance program against localized famine is itself acutely vulnerable to global food shortages. Thus, high global food prices are good news for farmers but only in good times.

The unambiguous losers when it comes to high food prices are the urban poor. Most of the developing world's large cities are ports, and, barring government controls, the price of their food is set on the global market. Crowded in slums, the urban poor cannot grow their own food; they have no choice but to buy it. Being poor, they would inevitably be squeezed by an increase in prices, but by a cruel implication of the laws of necessity, poor people spend a far larger proportion of their budgets on food, typically around a half, in contrast to only around a tenth for high-income groups. (Hungry slum dwellers are unlikely to accept their fate quietly. For centuries, sudden hunger in slums has provoked the same response: riots. This is the classic political base for populist politics, such as Peronism in Argentina, and the food crisis may provoke its ugly resurgence.)

At the end of the food chain comes the real crunch: among the urban poor, those most likely to go hungry are children. If young children remain malnourished for more than two years, the consequence is stunted growth—and stunted growth is not merely a physical condition. Stunted people are not just shorter than they would have been; their mental potential is impaired as well. Stunted growth is irreversible. It lasts a lifetime, and indeed, some studies find that it is passed down through the generations. And so although high food prices are yesterday's news in most of the developed world, if they remain high for the next few years, their consequences will be tomorrow's nightmare for the developing world.

In short, global food prices must be brought down, and they must be brought down fast, because their adverse consequences are so persistent. The question is how. There is nothing to be done about the root cause of the crisis—the increasing demand for food. The solution must come from dramatically increasing world food supply. That supply has been growing for decades, more than keeping up with population growth, but it now must be accelerated, with production increasing much more rapidly than it has in recent decades. This must happen in the short term, to bring prices down from today's levels, and in the medium and long terms, since any immediate increase in supply will soon be overtaken by increased demand.

Fortunately, policymakers have the power to do all of this: by changing regulation, they can quickly generate an increase in supply; by encouraging organizational changes, they can raise the growth of production in the medium term; and by encouraging innovations in technology, they can sustain this higher growth indefinitely. But currently, each of these steps is blocked by a giant of romantic populism: all three must be confronted and slain.

The First Giant of Romantic Populism

The first giant that must be slain is the middle- and upper-class love affair with peasant agriculture. With the near-total urbanization of these classes in both the United States and Europe, rural simplicity has acquired a strange allure. Peasant life is prized as organic in both its literal and its metaphoric sense. (Prince Charles is one of its leading apostles.) In its literal sense, organic agricultural production is now a premium product, a luxury brand. (Indeed, Prince Charles has his own such brand, Duchy Originals.) In its metaphoric sense, it represents the antithesis of the large, hierarchical, pressured organizations in which the middle classes now work. (Prince Charles has built a model peasant village, in traditional architectural style.) Peasants, like pandas, are to be preserved.

Peasants, like pandas, show little inclination to reproduce themselves.

But distressingly, peasants, like pandas, show little inclination to reproduce themselves. Given the chance, peasants seek local wage jobs, and their offspring head to the cities. This is because at low-income levels, rural bliss is precarious, isolated, and tedious. The peasant life forces millions of ordinary people into the role of entrepreneur, a role for which most are ill suited. In successful economies, entrepreneurship is a minority pursuit; most people opt for wage employment so that others can have the worry and grind of running a business. And reluctant peasants are right: their mode of production is ill suited to modern agricultural production, in which scale is helpful. In modern agriculture, technology is fast-evolving, investment is lumpy, the private provision of transportation infrastructure is necessary to counter the lack of its public provision, consumer food fashions are fast-changing and best met by integrated marketing chains, and regulatory standards are rising toward the holy grail of the traceability of produce back to its source. Far from being the answer to global poverty, organic self-sufficiency is a luxury lifestyle. It is appropriate for burnt-out investment bankers, not for hungry families.

Large organizations are better suited to cope with investment, marketing chains, and regulation. Yet for years, global development agencies have been leery of commercial agriculture, basing their agricultural strategies instead on raising peasant production. This neglect is all the more striking given the standard account of how economic development started in Europe: the English enclosure movement, which was enabled by legislative changes, is commonly supposed to have launched development by permitting large farms that could achieve higher productivity. Although current research qualifies the conventional account, reducing the estimates of productivity gains to the range of 10–20 percent, to ignore commercial agriculture as a force for rural development and enhanced food supply is surely ideological.

Innovation, especially, is hard to generate through peasant farming. Innovators create benefits for the local economy, and to the extent that these benefits are not fully captured by the innovators, innovation will be too slow. Large organizations can internalize the effects that in peasant agriculture are localized externalities—that is, benefits of actions that are not reflected in costs or profits—and so not adequately taken into account in decision-making. In the European agricultural revolution, innovations occurred on small farms as well as large, and today many peasant farmers, especially those who are better off and better educated, are keen to innovate. But agricultural innovation is highly sensitive to local conditions, especially in Africa, where the soils are complex and variable. One solution is to have an extensive network of publicly funded research stations with advisers who reach out to small farmers. But in Africa, this model has largely broken down, an instance of more widespread malfunctioning of the public sector. In eighteenth-century Great Britain, the innovations in small-holder agriculture were often led by networks among the gentry, who corresponded with one another on the consequences of agricultural experimentation. But such processes are far from automatic (they did not occur, for example, in continental Europe). Commercial agriculture is the best way of making innovation quicker and easier.

Over time, African peasant agriculture has fallen further and further behind the advancing commercial productivity frontier, and based on present trends, the region's food imports are projected to double over the next quarter century. Indeed, even with prices as high as they currently are, the United Nations Food and Agriculture Organization is worried that African peasants are likely to reduce production because they cannot afford the increased cost of fertilizer inputs. There are partial solutions to such problems through subsidies and credit schemes, but it should be noted that large-scale commercial agriculture simply does not face this particular problem: if output prices rise by more than input prices, production will be expanded.

A model of successful commercial agriculture is, indeed, staring the world in the face. In Brazil, large, technologically sophisticated agricultural companies have demonstrated how successfully food can be mass-produced. To give one remarkable example, the time between harvesting one crop and planting the next—the downtime for land—has been reduced to an astounding 30 minutes. Some have criticized the Brazilian model for displacing peoples and destroying rain forest, which has indeed happened in places where commercialism has gone unregulated. But in much of the poor world, the land is not primal forest; it is just badly farmed. Another benefit of the Brazilian model is that it can bring innovation to small farmers as well. In the "out-growing," or "contract farming," model, small farmers supply a central business. Depending on the details of crop production, sometimes this can be more efficient than wage employment.

There are many areas of the world that have good land that could be used far more productively if properly managed by large companies. Indeed, large companies, some of them Brazilian, are queuing up to manage those lands. Yet over the past 40 years, African governments have worked to scale back large commercial agriculture. At the heart of the matter is a reluctance to let land rights be marketable, and the source of this reluctance is probably the lack of economic dynamism in Africa's cities. As a result, land is still the all-important asset (there has been little investment in others). In more successful economies, land has become a minor asset, and thus the rights of ownership, although initially assigned based on political considerations, are simply extensions of the rights over other assets; as a result, they can be acquired commercially. A further consequence of a lack of urban dynamism is that jobs are scarce, and so the prospect of mass landlessness evokes political fears: the poor are safer on the land, where they are less able to cause trouble.

Commercial agriculture is not perfect. Global agribusiness is probably overly concentrated, and a sudden switch to an unregulated land market would probably have ugly consequences. But allowing commercial organizations to replace peasant agriculture gradually would raise global food supply in the medium term.

The War on Science

The second giant of romantic populism is the European fear of scientific agriculture. This has been manipulated by the agricultural lobby in Europe into yet another form of protectionism: the ban on GM crops. GM crops were introduced globally in 1996 and already are grown on around ten percent of the world's crop area, some 300 million acres. But due to the ban, virtually none of this is in Europe or Africa.

Robert Paarlberg, of Wellesley College, brilliantly anatomizes the politics of the ban in his new book, *Starved for Science*. After their creation, GM foods, already so disastrously named, were described as "Frankenfoods"—sounding like a scientific experiment on consumers. Just as problematic was the fact that genetic modification had grown out of research conducted by American corporations and so provoked predictable and deep-seated hostility from the European left. Although Monsanto, the main innovator in GM-seed technology, has undertaken never to market a seed that is incapable of reproducing itself, skeptics propagated a widespread belief that farmers will be trapped into annual purchases of "terminator" seeds from a monopoly supplier. Thus were laid the political foundations for a winning coalition: onto the base of national agricultural protectionism was added the anti-Americanism of the left and the paranoia of health-conscious consumers who, in the wake of the mad cow disease outbreak in the United Kingdom in the 1990s, no longer trusted their governments' assurances. In the 12 years since the ban was introduced, in 1996, the scientific case for lifting it has become progressively more robust, but the political coalition against GM foods has only expanded.

The GM-crop ban has had three adverse effects. Most obviously, it has retarded productivity growth in European agriculture. Prior to 1996, grain yields in Europe tracked those in the United States. Since 1996, they have fallen behind by 1–2 percent a year. European grain production could be increased by around 15 percent were the ban lifted. Europe is a major cereal producer, so this is a large loss. More subtly, because Europe is out of the market for GM-crop technology, the pace of research has slowed. GM-crop research takes a very long time to come to fruition, and its core benefit, the permanent reduction in food prices, cannot fully be captured through patents. Hence, there is a strong case for supplementing private research with public money. European governments should be funding this research, but instead research is entirely reliant on the private sector. And since private money for research depends on the prospect of sales, the European ban has also reduced private research.

However, the worst consequence of the European GM-crop ban is that it has terrified African governments into themselves banning GM crops, the only exception being South Africa. They fear that if they chose to grow GM crops, they would be permanently shut out of European markets. Now, because most of Africa has banned GM crops, there has been no market for discoveries pertinent to the crops that Africa grows, and so little research—which in turn has led to the critique that GM crops are irrelevant for Africa.

Africa cannot afford this self-denial; it needs all the help it can possibly get from genetic modification. For the past four decades, African agricultural productivity per acre has stagnated; raising production has depended on expanding the area under cultivation. But with Africa's population still growing rapidly, this option is running out, especially in light of global warming. Climate forecasts suggest that in the coming years, most of Africa will get hotter, the semiarid parts will get drier, and rainfall variability on the continent will increase, leading to more droughts. It seems likely that in southern Africa, the staple food, maize, will at some point become nonviable.

Whereas for other regions the challenge of climate change is primarily about mitigating carbon emissions, in Africa it is primarily about agricultural adaptation.

It has become commonplace to say that Africa needs a green revolution. Unfortunately, the reality is that the green revolution in the twentieth century was based on chemical fertilizers, and even when fertilizer was cheap, Africa did not adopt it. With the rise in fertilizer costs, as a byproduct of high-energy prices, any African green revolution will perforce not be chemical. To counter the effects of Africa's rising population and deteriorating climate, African agriculture needs a biological revolution. This is what GM crops offer, if only sufficient money is put into research. There has as yet been little work on the crops of key importance to the region, such as cassava and yams. GM-crop research is still in its infancy, still on the first generation: single-gene transfer. A gene that gives one crop an advantage is identified, isolated, and added to another crop. But even this stage offers the credible prospect of vital gains. In a new scientific review, Jennifer Thomson, of the Department of Molecular and Cell Biology at the University of Cape Town, considers the potential of GM technology for Africa. Maize, she reports, can be made more drought-resistant, buying Africa time in the struggle against climatic deterioration. Grain can be made radically more resistant to fungi, reducing the need for chemicals and cutting losses due to storage. For example, stem borer beetles cause storage losses in the range of 15–40 percent of the African maize crop; a new GM variety is resistant.

It is important to recognize that genetic modification, like commercialization, is not a magic fix for African agriculture: there is no such fix. But without it, the task of keeping Africa's food production abreast of its population growth looks daunting. Although Africa's coastal cities can be fed from global supplies, the vast African interior cannot be fed in this way other than in emergencies. Lifting the ban on GM crops, both in Africa and in Europe, is the policy that could hold down global food prices in the long term.

The final giant of romantic populism is the American fantasy that the United States can escape dependence on Arab oil by growing its own fuel—making ethanol or other biofuels, largely from corn. There is a good case for growing fuel. But there is not a good case for generating it from American grain: the conversion of grain into ethanol uses almost as much energy as it produces. This has not stopped the American agricultural lobby from gouging out grotesquely inefficient subsidies from the government; as a result, around a third of American grain has rapidly been diverted into energy. This switch demonstrates both the superb responsiveness of the market to price signals and the shameful power of subsidy-hunting lobbying groups. If the United States wants to run off of agrofuel instead of oil, then Brazilian sugar cane is the answer; it is a far more efficient source of energy than American grain. The killer evidence of political capture is the response of the U.S. government to this potential lifeline: it has actually restricted imports of Brazilian ethanol to protect American production. The sane goal of reducing dependence on Arab oil has been sacrificed to the self-serving goal of pumping yet more tax dollars into American agriculture.

Inevitably, the huge loss of grain for food caused by its diversion into ethanol has had an impact on world grain prices. Just how large an impact is controversial. An initial claim by the Bush administration was that it had raised prices by only three percent, but a study by the World Bank suggests that the effect has been much larger. If the subsidy were lifted, there would probably be a swift impact on prices: not only would the supply of grain for food increase, but the change would shift speculative expectations. This is the policy that could bring prices down in the short term.

Striking a Deal

The three policies—expanding large commercial farms, ending the GM-crop ban, and doing away with the U.S. subsidies on ethanol—fit together both economically and politically. Lifting the ethanol subsidies would probably puncture the present ballooning of prices. The expansion of commercial farms could, over the next decade, raise world output by a further few percentage points. Both measures would buy the time needed for GM crops to deliver on their potential (the time between starting research and the mass application of its results is around 15 years). Moreover, the expansion of commercial farming in Africa would encourage global GM-crop research on Africa-suited crops, and innovations would find a ready market not so sensitive to political interference. It would also facilitate the localized adaptation of new varieties. It is not by chance that the only African country in which GM crops have not been banned is South Africa, where the organization of agriculture is predominantly commercial.

Politically, the three policies are also complementary. Homegrown energy, keeping out "Frankenfoods," and preserving the peasant way of life are all classic populist programs: they sound instantly appealing but actually do harm. They must be countered by messages of equal potency.

One such message concerns the scope for international reciprocity. Although Americans are attracted to homegrown fuel, they are infuriated by the European ban on GM crops. They see the ban for what it is: a standard piece of anti-American protectionism. Europeans, for their part, cling to the illusory comfort of the ban on high-tech crops, but they are infuriated by the American subsidies on ethanol. They see the subsidies for what they are: a greedy deflection from the core task of reducing U.S. energy profligacy. Over the past half century, the United States and Europe have learned how to cooperate. The General Agreement on Tariffs and Trade was fundamentally a deal between the United States and Europe that virtually eliminated tariffs on manufactured goods. NATO is a partnership in security. The Organization for Economic Cooperation and Development is a partnership in economic governance. Compared to the difficulties of reaching agreement in these areas, the difficulties of reaching a deal on the mutual de-escalation of recent environmental follies is scarcely daunting:

the United States would agree to scrap its ethanol subsidies in return for Europe's lifting the ban on GM crops. Each side can find this deal infuriating and yet attractive. It should be politically feasible to present this to voters as better than the status quo.

The romantic hostility to scientific and commercial agriculture must be countered.

How might the romantic hostility toward commercial and scientific agriculture be countered politically? The answer is to educate the vast community of concern for the poorest countries on the bitter realities of the food crisis. In both the United States and Europe, millions of decent citizens are appalled by global hunger. Each time a famine makes it to television screens, the popular response is overwhelming, and there is a large overlap between the constituency that responds to such crises and the constituency attracted by the idea of preserving organic peasant lifestyles. The cohabitation of these concerns needs to be challenged. Many people will need to agonize over their priorities. Some will decide that the vision articulated by Prince Charles is the more important one: a historical lifestyle must be preserved regardless of the consequences. But however attractive that vision, these people must come face-to-face with the prospect of mass malnutrition and stunted children and realize that the vital matter for public policy is to increase food supplies. Commercial agriculture may be irredeemably unromantic, but if it fills the stomachs of the poor, then it should be encouraged.

American environmentalists will also need to do some painful rethinking. The people most attracted to achieving energy self-sufficiency through the production of ethanol are potentially the constituency that could save the United States from its ruinous energy policies. The United States indeed needs to reduce its dependence on imported oil, but growing corn for biofuel is not the answer. Americans are quite simply too profligate when it comes to their use of energy; Europeans, themselves pretty profligate, use only half the energy per capita and yet sustain a high-income lifestyle. The U.S. tax system needs to be shifted from burdening work to discouraging energy consumption.

The mark of a good politician is the ability to guide citizens away from populism.

The mark of a good politician is the ability to guide citizens away from populism. Unless countered, populism will block the policies needed to address the food crisis. For the citizens of the United States and Europe, the continuation of high food prices will be an inconvenience, but not sufficiently so to slay the three giants on which the current strain of romantic populism rests. Properly informed, many citizens will rethink their priorities, but politicians will need to deliver these messages and forge new alliances. If food prices are not brought down fast and then kept down, slum children will go hungry, and their future lives will be impaired. Shattering a few romantic illusions is a small price to pay.

Critical Thinking

1. If, as the author argues, the world needs more commercial agriculture and more science, who will pay for (and own) this commercialization? Who will pay for and own the science?

2. How, precisely, will food become affordable to all the world's poor by taking the steps outlined in this article?

3. Locate on your map where this commercialization of agriculture will occur. Locate on your map also areas of the world that are predicted to be at high risk for climate change-induced impacts. Offer insights regarding what connections you see in global food production and climate change.

4. Given the dynamics of climate induced conflict, the current uncertainties of the global economy, and "peak oil," is the kind of agricultural production described in this article "sustainable"? Explain why or why not.

5. In the year 2050, will the poor be feeding themselves, or will "we" be feeding the poor? Discuss the social, economic, and environmental implications of each "feeding arrangement."

6. Identify in this article three key terms, concepts, or principles that are used in your textbook (environmental science, economics, sociology, history, geography, etc.) or employed in the discipline you are currently studying. (Note: The terms, concepts, or principles may be implicit, explicit, implied, or inferred.)

PAUL COLLIER is Professor of Economics and Director of the Center for the Study of African Economies at Oxford University and the author of *The Bottom Billion: Why the Poorest Countries Are Failing and What Can Be Done About It.*

From *Foreign Affairs*, November/December 2008, pp. 67–79. Copyright © 2008 by Council on Foreign Relations, Inc. Reprinted by permission of Foreign Affairs. www.ForeignAffairs.com.

Across the Globe, Empty Bellies Bring Rising Anger

MARC LACEY

Hunger bashed in the front gate of Haiti's presidential palace. Hunger poured onto the streets, burning tires and taking on soldiers and the police. Hunger sent the country's prime minister packing.

Haiti's hunger, that burn in the belly that so many here feel, has become fiercer than ever in recent days as global food prices spiral out of reach, spiking as much as 45 percent since the end of 2006 and turning Haitian staples like beans, corn and rice into closely guarded treasures.

Saint Louis Meriska's children ate two spoonfuls of rice apiece as their only meal recently and then went without any food the following day. His eyes downcast, his own stomach empty, the unemployed father said forlornly, "They look at me and say, 'Papa, I'm hungry,' and I have to look away. It's humiliating and it makes you angry."

That anger is palpable across the globe. The food crisis is not only being felt among the poor but is also eroding the gains of the working and middle classes, sowing volatile levels of discontent and putting new pressures on fragile governments.

In Cairo, the military is being put to work baking bread as rising food prices threaten to become the spark that ignites wider anger at a repressive government. In Burkina Faso and other parts of sub-Saharan Africa, food riots are breaking out as never before. In reasonably prosperous Malaysia, the ruling coalition was nearly ousted by voters who cited food and fuel price increases as their main concerns.

"It's the worst crisis of its kind in more than 30 years," said Jeffrey D. Sachs, the economist and special adviser to the United Nations secretary general, Ban Ki-moon. "It's a big deal and it's obviously threatening a lot of governments. There are a number of governments on the ropes, and I think there's more political fallout to come."

Indeed, as it roils developing nations, the spike in commodity prices—the biggest since the Nixon administration—has pitted the globe's poorer south against the relatively wealthy north, adding to demands for reform of rich nations' farm and environmental policies. But experts say there are few quick fixes to a crisis tied to so many factors, from strong demand for food from emerging economies like China's to rising oil prices to the diversion of food resources to make biofuels.

There are no scripts on how to handle the crisis, either. In Asia, governments are putting in place measures to limit hoarding of rice after some shoppers panicked at price increases and bought up everything they could.

Even in Thailand, which produces 10 million more tons of rice than it consumes and is the world's largest rice exporter, supermarkets have placed signs limiting the amount of rice shoppers are allowed to purchase.

But there is also plenty of nervousness and confusion about how best to proceed and just how bad the impact may ultimately be, particularly as already strapped governments struggle to keep up their food subsidies.

'Scandalous Storm'

"This is a perfect storm," President Elías Antonio Saca of El Salvador said Wednesday at the World Economic Forum on Latin America in Cancún, Mexico. "How long can we withstand the situation? We have to feed our people, and commodities are becoming scarce. This scandalous storm might become a hurricane that could upset not only our economies but also the stability of our countries."

In Asia, if Prime Minister Abdullah Ahmad Badawi of Malaysia steps down, which is looking increasingly likely amid postelection turmoil within his party, he may be that region's first high-profile political casualty of fuel and food price inflation.

In Indonesia, fearing protests, the government recently revised its 2008 budget, increasing the amount it will spend on food subsidies by about $280 million.

"The biggest concern is food riots," said H.S. Dillon, a former adviser to Indonesia's Ministry of Agriculture. Referring to small but widespread protests touched off by a rise in soybean prices in January, he said, "It has happened in the past and can happen again."

Last month in Senegal, one of Africa's oldest and most stable democracies, police in riot gear beat and used tear gas against people protesting high food prices and later raided a television station that broadcast images of the event. Many Senegalese have expressed anger at President Abdoulaye Wade for spending lavishly on roads and five-star hotels for an Islamic summit meeting last month while many people are unable to afford rice or fish.

"Why are these riots happening?" asked Arif Husain, senior food security analyst at the World Food Program, which has issued urgent appeals for donations. "The human instinct is to survive, and people are going to do no matter what to survive. And if you're hungry you get angry quicker."

Leaders who ignore the rage do so at their own risk. President René Préval of Haiti appeared to taunt the populace as the chorus of complaints about la vie chère—the expensive life—grew. He said if Haitians could afford cellphones, which many do carry, they should be able to feed their families. "If there is a protest against the rising prices," he said, "come get me at the palace and I will demonstrate with you."

When they came, filled with rage and by the thousands, he huddled inside and his presidential guards, with United Nations peacekeeping troops, rebuffed them. Within days, opposition lawmakers had voted out Mr. Préval's prime minister, Jacques-Édouard Alexis, forcing him to reconstitute his government. Fragile in even the best of times, Haiti's population and politics are now both simmering.

"Why were we surprised?" asked Patrick Élie, a Haitian political activist who followed the food riots in Africa earlier in the year and feared they might come to Haiti. "When something is coming your way all the way from Burkina Faso you should see it coming. What we had was like a can of gasoline that the government left for someone to light a match to it."

Dwindling Menus

The rising prices are altering menus, and not for the better. In India, people are scrimping on milk for their children. Daily bowls of dal are getting thinner, as a bag of lentils is stretched across a few more meals.

Maninder Chand, an auto-rickshaw driver in New Delhi, said his family had given up eating meat altogether for the last several weeks.

Another rickshaw driver, Ravinder Kumar Gupta, said his wife had stopped seasoning their daily lentils, their chief source of protein, with the usual onion and spices because the price of cooking oil was now out of reach. These days, they eat bowls of watery, tasteless dal, seasoned only with salt.

Down Cairo's Hafziyah Street, peddlers selling food from behind wood carts bark out their prices. But few customers can afford their fish or chicken, which bake in the hot sun. Food prices have doubled in two months.

Ahmed Abul Gheit, 25, sat on a cheap, stained wooden chair by his own pile of rotting tomatoes. "We can't even find food," he said, looking over at his friend Sobhy Abdullah, 50. Then raising his hands toward the sky, as if in prayer, he said, "May God take the guy I have in mind."

Mr. Abdullah nodded, knowing full well that the "guy" was President Hosni Mubarak.

The government's ability to address the crisis is limited, however. It already spends more on subsidies, including gasoline and bread, than on education and health combined.

"If all the people rise, then the government will resolve this," said Raisa Fikry, 50, whose husband receives a pension equal to about $83 a month, as she shopped for vegetables. "But everyone has to rise together. People get scared. But we will all have to rise together."

It is the kind of talk that has prompted the government to treat its economic woes as a security threat, dispatching riot forces with a strict warning that anyone who takes to the streets will be dealt with harshly.

Niger does not need to be reminded that hungry citizens overthrow governments. The country's first postcolonial president, Hamani Diori, was toppled amid allegations of rampant corruption in 1974 as millions starved during a drought.

More recently, in 2005, it was mass protests in Niamey, the Nigerien capital, that made the government sit up and take notice of that year's food crisis, which was caused by a complex mix of poor rains, locust infestation and market manipulation by traders.

"As a result of that experience the government created a cabinet-level ministry to deal with the high cost of living," said Moustapha Kadi, an activist who helped organize marches in 2005. "So when prices went up this year the government acted quickly to remove tariffs on rice, which everyone eats. That quick action has kept people from taking to the streets."

The Poor Eat Mud

In Haiti, where three-quarters of the population earns less than $2 a day and one in five children is chronically malnourished, the one business booming amid all the gloom is the selling of patties made of mud, oil and sugar, typically consumed only by the most destitute.

"It's salty and it has butter and you don't know you're eating dirt," said Olwich Louis Jeune, 24, who has taken to eating them more often in recent months. "It makes your stomach quiet down."

But the grumbling in Haiti these days is no longer confined to the stomach. It is now spray-painted on walls of the capital and shouted by demonstrators.

In recent days, Mr. Préval has patched together a response, using international aid money and price reductions by importers to cut the price of a sack of rice by about 15 percent. He has also trimmed the salaries of some top officials. But those are considered temporary measures.

Real solutions will take years. Haiti, its agriculture industry in shambles, needs to better feed itself. Outside investment is the key, although that requires stability, not the sort of widespread looting and violence that the Haitian food riots have fostered.

Meanwhile, most of the poorest of the poor suffer silently, too weak for activism or too busy raising the next generation of hungry. In the sprawling slum of Haiti's Cité Soleil, Placide Simone, 29, offered one of her five offspring to a stranger. "Take one," she said, cradling a listless baby and motioning toward four rail-thin toddlers, none of whom had eaten that day. "You pick. Just feed them."

Critical Thinking

1. List the places mentioned in this article which are used to illustrate the thesis. Are all the people in those countries you listed hungry and angry? Explain.

2. If Thailand produces so much rice, why are its people limited on the amount they can buy? Who is consuming the rice they grow? Explain.

3. Do you know anyone personally who makes less than $2/day? Have you ever had to eat a "mud pie"? How do you think such a personal experience would affect you?

4. How can so many people be hungry when the world is also struggling with an obesity epidemic?

5. Identify in this article three key terms, concepts, or principles that are used in your textbook (environmental science, economics, sociology, history, geography, etc.) or employed in the discipline you are currently studying. (Note: The terms, concepts, or principles may be implicit, explicit, implied, or inferred).

Reporting was contributed by Lydia Polgreen from Niamey, Niger, Michael Slackman from Cairo, Somini Sengupta from New Delhi, Thomas Fuller from Bangkok and Peter Gelling from Jakarta, Indonesia.

How to Feed 8 Billion People

Record grain shortages are threatening global food security in the immediate future. A noted environmental analyst shows how nations can better manage their limited resources.

LESTER R. BROWN

The world is entering a new food era. It will be marked by higher food prices, rapidly growing numbers of hungry people, and an intensifying competition for land and water resources that crosses national boundaries when food-importing countries buy or lease vast tracts of land in other countries. Because some of the countries where land is being acquired do not have enough land to adequately feed their own people, the stage is being set for future conflicts.

The sharp rise of grain prices in recent years underlines the gravity of the situation. From mid-2006 to mid-2008, world prices of wheat, rice, corn, and soybeans roughly tripled, reaching historic highs. It was not until the global economic crisis beginning in 2008 that grain prices began to level off and recede slightly.

The world has experienced several grain price surges over the last half century, but none like this. Earlier surges were event-driven, weather-related, and temporary—caused by monsoons, droughts, heat waves, etc.

The recent record surge in grain prices has been trend-driven. Working our way out of this tightening food situation means reversing the trends that are causing it, such as soil erosion, falling water tables, and rising carbon emissions.

As a result of persistently high food prices, hunger is spreading. In the mid-1990s, the number of hungry people had fallen to 825 million. But instead of continuing to decline, the number of people facing chronic food insecurity and undernourishment started to edge upward, jumping to more than 1 billion in 2009.

Rising food prices and the swelling ranks of the hungry are among the early signs of a tightening world food situation. More and more, food is looking like the weak link in our civilization, much as it was for the earlier ones whose archaeological sites we now study.

Food: The Weak Link

As the world struggles to feed all its people, farmers are facing some worrying trends. On the demand side of the food equation are three consumption-boosting trends: population growth, the growing consumption of grain-based animal protein, and, most recently, the massive use of grain to fuel cars.

Each year there are 79 million more people at the dinner table, and the overwhelming majority of these individuals are being added in countries where soils are eroding, water tables are falling, and irrigation wells are going dry.

Even as our numbers are multiplying, some 3 billion people are trying to add to their diets, consuming more meat and dairy products. At the top of the food-consumption ranking are the United States and Canada, where people consume on average 800 kilograms of grain per year, most of it indirectly as beef, pork, poultry, milk, and eggs. Near the bottom of this ranking is India, where people have less than 200 kilograms of grain each, and thus must consume nearly all of it directly, leaving little for conversion into animal protein.

The orgy of investment in ethanol fuel distilleries that followed the 2005 surge in U.S. gas prices doubled grain consumption to 40 million tons by 2008.

On the supply side, ongoing environmental trends are making it very difficult to expand food production fast enough. These include soil erosion, aquifer depletion, crop-shrinking heat waves, melting ice sheets and rising sea levels, and the melting of the mountain glaciers that feed major rivers and irrigation systems. In addition, three resource trends are affecting our food supply: the loss of cropland to non-farm uses, the diversion of irrigation water to cities, and the coming reduction in oil supplies.

Soil erosion is currently lowering the inherent productivity of some 30% of the world's cropland. In some countries, it has reduced grain production by half or more over the last three decades. Vast dust storms coming out of sub-Saharan Africa, northern China, western Mongolia, and Central Asia remind us that the loss of topsoil is not only continuing but expanding. Advancing deserts in China—the result of overgrazing, overplowing, and deforestation—have forced the complete or partial abandonment of some 24,000 villages and the cropland surrounding them.

The loss of topsoil began with the first wheat and barley plantings, but falling water tables are historically quite recent, simply because the pumping capacity to deplete aquifers has evolved only in recent decades. Water tables are now falling in countries that together contain half the world's people. An estimated 400 million people (including 175 million in India and 130 million in China) are being fed by overpumping, a process that is by definition short term. Saudi Arabia has announced that, because its major aquifer, a nonreplenishable fossil aquifer, is largely depleted, it will be phasing out wheat production entirely by 2016.

An estimated 400 million people are being fed by overpumping water, a process that is by definition short term.

Climate change also threatens food security. For each 1°C rise in temperature above the norm during the growing season, farmers can expect a 10% decline in wheat, rice, and corn yields. Since 1970, the earth's average surface temperature has increased by 0.6°C. And the Intergovernmental Panel on Climate Change projects that the temperature will rise by up to 6°C during this century.

As the earth's temperature continues to rise, mountain glaciers are melting throughout the world. The projected melting of the glaciers on which China and India depend presents the most massive threat to food security that humanity has ever faced. China and India are the world's leading wheat producers and also dominate the world rice harvest. Whatever happens to the wheat and rice harvests in these two population giants will affect food prices everywhere.

The accelerating melting of the Greenland and West Antarctic ice sheets combined with thermal expansion of the oceans could raise sea level by up to six feet during this century. Every rice-growing river delta in Asia is threatened by the melting of these ice sheets. Even a three-foot rise would devastate the rice harvest in the Mekong Delta, which produces more than half the rice in Vietnam, the world's number-two rice exporter.

Three-fourths of oceanic fisheries are now being fished at or beyond capacity or are recovering from over-exploitation. If we continue with business as usual, many of these fisheries will collapse. We are taking fish from the oceans faster than they can reproduce.

With additional water no longer available in many countries, growing urban thirst can be satisfied only by taking irrigation water from farmers. Thousands of farmers in California find it more profitable to sell their irrigation water to Los Angeles and San Diego and leave their land idle. China's farmers are also losing irrigation water to the country's fast-growing cities.

If we paid the full cost of producing it—including the true cost of the oil used in producing it, the future costs of overpumping aquifers, the destruction of land through erosion, and the carbon-dioxide emissions from land clearing—food would cost far more than we now pay for it in the supermarket.

The question—at least for now—is: Will the world grain harvest expand fast enough to keep pace with steadily growing demand? Food security will deteriorate further unless leading countries collectively mobilize to stabilize population, stabilize climate, stabilize aquifers, conserve soils, protect cropland, and restrict the use of grain to produce fuel for cars.

The Emerging Geopolitics of Food Scarcity

As world food security deteriorates, individual countries, acting in their narrowly defined self-interest, are banning or limiting grain exports.

In response, other countries have been trying to nail down long-term bilateral trade agreements that would lock up future grain supplies. Several have succeeded. Egypt, for example, has reached a long-term agreement with Russia for more than 3 million tons of wheat each year.

The more affluent food-importing countries have sought to buy or lease for the long term large blocks of land to farm in other countries. Libya, which imports 90% of its grain and has been worried about

access to supplies, was one of the first to look abroad for land. After more than a year of negotiations, it reached an agreement to farm 100,000 hectares (250,000 acres) of land in Ukraine to grow wheat for its own people.

Countries selling or leasing their land are often low-income countries and, more often than not, those where chronic hunger and malnutrition are commonplace. A major acquisition site for Saudi Arabia and several other countries is Sudan—the site of the World Food Programme's largest famine relief effort.

The growing competition for land across national boundaries is also an indirect competition for water. In effect, land acquisitions are also water acquisitions. Land acquisitions in Sudan that tap water from the Nile, which is already fully utilized, may mean that Egypt will get less water from the river—making it even more dependent on imported grain.

Such bilateral land acquisitions raise many questions. To begin with, negotiations and the agreements they lead to tend to lack transparency. Typically, only a few high-ranking officials are involved, and the terms are confidential. Not only are many stakeholders such as farmers not at the table when the agreements are negotiated, they do not even learn about the deals until after they have been signed. And since there is rarely idle productive land in the countries where the land is being purchased or leased, the agreements suggest that many local farmers will be displaced. Their land may be confiscated or bought from them at a price over which they have little say.

This helps explain the public hostility that often arises within host countries. China, for example, signed an agreement with the Philippine government to lease more than a million hectares of land on which to produce crops that would be shipped home. Once word leaked out, the public outcry—much of it from Filipino farmers—forced the government to suspend the agreement. A similar situation developed in Madagascar, where South Korea's Daewoo Logistics had pursued rights to an area half the size of Belgium. The political furor led to a change in government and cancellation of the agreement.

Raising Land Productivity

There are many things that can be done in agriculture to raise land and water productivity. The challenge is for each country to fashion agricultural and economic policies that enable it to realize its unique potential.

Prior to 1950, expansion of the food supply came almost entirely from expanding cropland area. Then as frontiers disappeared and population growth accelerated after World War II, the world quickly shifted to raising land productivity. After several decades of rapid rise, however, it is now becoming more difficult to continue increasing productivity.

Gains in land productivity have come primarily from three sources: the growing use of fertilizer, the spread of irrigation, and the development of higher-yielding varieties of wheat, rice, and corn.

Among the three grains, corn is the only one where the yield is continuing to rise in high-yield countries. Even though fertilizer use has not increased since 1980, corn yields continue to edge upward as seed companies invest huge sums in corn breeding.

Despite dramatic past leaps in grain yields, it is becoming more difficult to expand world food output. There is little productive new land to bring under the plow. Expanding the irrigated area is difficult. Returns on the use of additional fertilizer are mostly diminishing. In the more arid countries of Africa, there is not enough rainfall to raise yields dramatically.

One way is to breed crops that are more tolerant of drought and cold. Another way to raise land productivity, where soil moisture permits, is to expand the area of land that produces more than one crop per year. Indeed, the tripling in the world grain harvest from 1950 to 2000 was due in part to widespread increases in multiple cropping in Asia. The spread of double cropping of winter wheat and corn on the North China Plain helped boost China's grain production to where it now rivals that of the United States.

A concerted U.S. effort to both breed earlier-maturing varieties and develop cultural practices that would facilitate multiple cropping could boost crop output. If China's farmers can extensively double crop wheat and corn, then U.S. farmers—at a similar latitude and with similar climate patterns—could do more if agricultural research and farm policy were reoriented to support it. Western Europe, with its mild winters and high-yielding winter wheat, might also be able to double crop more with a summer grain, such as corn, or an oilseed crop. Brazil and Argentina, which have extensive frost-free growing seasons, commonly multicrop wheat or corn with soybeans.

One encouraging effort to raise cropland productivity in Africa is the simultaneous planting of grain and leguminous trees. At first the trees grow slowly, permitting the grain crop to mature and be harvested; then the saplings grow quickly to several feet in height, dropping leaves that provide nitrogen and organic matter, both sorely needed in African soils. The wood is then cut and used for fuel. This simple, locally adapted technology, developed by scientists at the International Centre for Research in Agroforestry in Nairobi, has enabled farmers to double their grain yields within a matter of years as soil fertility builds.

Raising Water Productivity

Since it takes 1,000 tons of water to produce one ton of grain, it is not surprising that 70% of world water use is devoted to irrigation. Thus, raising irrigation efficiency is central to raising water productivity overall.

Data on the efficiency of surface water irrigation projects—that is, dams that deliver water to farmers through a network of canals—show that crop usage of irrigation water never reaches 100% because some irrigation water evaporates, some percolates downward, and some runs off. Water policy analysts have found that surface water irrigation efficiency is well below 50% in a number of countries, including India and Thailand.

Irrigation water efficiency is affected not only by the type and condition of irrigation systems but also by soil type, temperature, and humidity. In hot, arid regions, the evaporation of irrigation water is far higher than in cooler, humid regions. In a May 2004 meeting, China's Minister of Water Resources Wang Shucheng outlined for me in some detail the plans to raise China's irrigation efficiency from 43% in 2000 to 51% in 2010 and 55% in 2030. The steps he described included raising the price of water, providing incentives for adopting more irrigation-efficient technologies, and developing the local institutions to manage this process. Reaching these goals, he believes, would assure China's future food security.

Raising irrigation efficiency typically means shifting to overhead sprinklers or drip irrigation, the gold standard of irrigation efficiency. Switching to low-pressure sprinkler systems reduces water use by an estimated 30%, while switching to drip irrigation typically cuts water use in half.

A drip system also raises yields because it provides a steady supply of water with minimal losses to evaporation. Since drip systems are both labor-intensive and water-efficient, they are well suited to countries with a surplus of labor and a shortage of water. Israel (where the

method was pioneered) and neighboring Jordan both rely heavily on drip irrigation. In contrast, among the big three agricultural producers, this more-efficient technology is used on roughly 3% of irrigated land in India and China and on roughly 4% in the United States.

In recent years, small-scale drip-irrigation systems—literally a bucket or drum with flexible plastic tubing to distribute the water—have been developed to irrigate small vegetable gardens. The containers are elevated slightly so that gravity distributes the water. Large-scale drip systems using plastic lines that can be moved easily are also becoming popular. These simple systems can pay for themselves in one year. By simultaneously reducing water costs and raising yields, they can dramatically raise incomes.

Shifting to more water-efficient crops wherever possible also boosts water productivity. Rice production is being phased out around Beijing because rice is such a thirsty crop. Similarly, Egypt restricts rice production in favor of wheat.

Strategic Reductions in the Demand for Grain

Although we seldom consider the climate effect of various dietary options, they are substantial, to say the least. A plant-based diet requires roughly one-fourth as much energy as a diet rich in red meat. Shifting to a vegetarian diet cuts greenhouse gas emissions almost as much as shifting from an SUV to a hybrid vehicle does. Shifting to less grain-intensive forms of animal protein such as poultry or certain types of fish can also reduce pressure on the earth's land and water resources.

Shifting to a vegetarian diet cuts greenhouse gas emissions almost as much as shifting from an SUV to a hybrid vehicle does.

When considering how much animal protein to consume, it is useful to distinguish between grass-fed and grain-fed products. For example, most of the world's beef is produced with grass. Even in the United States, with an abundance of feedlots, over half of all beef cattle weight gain comes from grass rather than grain. Grasslands are usually too steeply sloping or too arid to plow, and can contribute to the food supply only if used for grazing.

Beyond the role of grass in providing high-quality protein in our diets, it is sometimes assumed that we can increase the efficiency of land and water use by shifting from animal protein to high-quality plant protein, such as that from soybeans. It turns out, however, that since corn yields in the U.S. Midwest are three to four times those of soybeans, it may be more resource-efficient to produce corn and convert it into poultry or catfish at a ratio of two to one than to have everyone heavily reliant on soy.

The massive conversion of grain into biofuel began just a few years ago. If we are to reverse the spread of hunger, we will almost certainly have to cut back on ethanol production. Removing the incentives for converting food into fuel will help ensure that everyone has enough to eat. It will also lessen the pressures that lead to overpumping of groundwater and the clearing of tropical rain forests. If the U.S. government were to abolish the subsidies and mandates that are driving the conversion of grain into fuel, it would help stabilize grain prices and set the stage for relaxing the political tensions that have emerged within importing countries.

The Localization of Agriculture

In the United States, there has been a surge of interest in eating fresh local foods, corresponding with mounting concerns about the climate effects of consuming food from distant places. This is reflected in the rise in urban gardening, school gardening, and farmers' markets.

Food from more distant locations boosts carbon emissions while losing flavor and nutrition. A localized food economy reduces fossil fuel usage. Supermarkets are increasingly contracting with local farmers, and upscale restaurants are emphasizing locally grown food on their menus.

In school gardens, children learn how food is produced, a skill often lacking in urban settings, and they may get their first taste of freshly picked peas or vine-ripened tomatoes. School gardens also provide fresh produce for school lunches. California, a leader in this area, has 6,000 school gardens.

Many universities are now making a point of buying local food as well. Some universities compost kitchen and cafeteria food waste and make the compost available to the farmers who supply them with fresh produce.

Community gardens can be used by those who would otherwise not have access to land for gardening. Providing space for community gardens is seen by many local governments as an essential service.

Many market outlets are opening up for local produce. Perhaps the best known of these are the farmers' markets where local farmers bring their produce for sale. Many farmers' markets also now take food stamps, giving low-income consumers access to fresh produce that they might not otherwise be able to afford.

A survey of food consumed in Iowa showed conventional produce traveled on average 1,500 miles, not including food imported from other countries. In contrast, locally grown produce traveled on average 56 miles—a huge difference in fuel investment.

Concerns about the climate effects of transporting food long distances has led Tesco, the leading U.K. supermarket chain, to begin labeling products with their carbon footprint, indicating the greenhouse gas contribution of food items from the farm to supermarket shelf.

The shift from factory farm production of milk, meat, and eggs to mixed crop—livestock operations also facilitates nutrient recycling as local farmers return livestock manure to the land. The combination of high prices of natural gas, which is used to make nitrogen fertilizer, and of phosphate, as reserves are depleted, suggests a much greater future emphasis on nutrient recycling—an area where small farmers producing for local markets have a distinct advantage over massive feeding operations.

Costs and Solutions

If we cannot quickly cut carbon emissions, the world will face crop-shrinking heat waves that can massively and unpredictably reduce harvests. A hotter world will mean melting ice sheets, rising sea levels, and the inundation of the highly productive rice-growing river deltas of Asia. The loss of glaciers in the Himalayas and on the Tibetan Plateau will shrink wheat and rice harvests in both India and China, the world's most populous countries. Both are already facing water shortages driven by aquifer depletion and melting glaciers.

Since hunger is almost always the result of poverty, eradicating hunger depends on eradicating poverty. And where people are outrunning their land and water resources, this means stabilizing population.

Given that a handful of the more affluent grain-importing countries are reportedly investing some $20–30 billion in land acquisition, there is no shortage of capital to invest in agricultural development. Why not invest it across the board in helping low-income countries develop their unrealized potential for expanding food production, enabling them to export more grain?

We have a role to play as individuals. Whether we bike, bus, or drive to work will affect carbon emissions, climate change, and food security. The size of the car we drive to the supermarket and its effect on climate may indirectly affect the size of the bill at the supermarket checkout counter. If we are living high on the food-consumption chain, we can move down, improving our health while helping to stabilize climate. Food security is something in which we all have a stake—and a responsibility.

Critical Thinking

1. Sketch out a concept map illustrating the linkages between the three consumption-boosting trends the author suggests. Now describe the geographic location of the (1) population growth, (2) grain-based animal protein consumption, and (3) automobile driving. What are your observations?

2. The author states that food security will deteriorate further unless leading countries collectively mobilize to do several things. What are those actions, and how do these recommendations differ from those argued in Article 9?

3. Describe what is meant by the "localization of agriculture."

4. How might the promotion of "agricultural localization" have more global benefit and better environmental resiliency than large, commercial agricultural production operations?

5. Identify in this article three key terms, concepts, or principles that are used in your textbook (environmental science, economics, sociology, history, geography, etc.) or employed in the discipline you are currently studying. (Note: The terms, concepts, or principles may be implicit, explicit, implied, or inferred.)

LESTER R. BROWN is the founder and president of the Washington, D.C.-based nonprofit Earth Policy Institute. This article draws from his most recent book, *Plan B 4.0: Mobilizing to Save Civilization* (W.W. Norton and Co., 2009). For additional information, visit www.earth-policy.org.

UNIT 4
A Thirsty Planet

Unit Selections

Learning Outcomes

After reading this Unit, you should be able to:

- Articulate the current state of environmental issue #5: freshwater, in terms of global availability, future supply shifts, geopolitical access, and sustainability challenges.

- Explain how the concept of environmental scarcity is linked to political unrest and possibly future conflicts.

- Outline the reasons why water may become a key environmental variable associated with potential future conflict in the Middle East and North Africa.

- Describe the environmental linkages between global water supply, access, consumption, and regional population growth.

- Explain the relationship of glaciers, agricultural production, urbanization, and Asian political geography.

- Outline linkages between a region's natural resources and its social structures.

- Offer insights into a potential world water crisis per its (1) environmental dimensions and (2) geopolitical dimensions.

- Outline a list of changes that will need to be addressed to meet future water demand.

- Expand on the social and environmental consequences of continued unsustainable water consumption patterns.

Student Website

www.mhhe.com/cls

Internet References

Freshwater Society
   www.freshwater.org
EnviroLink
   www.envirolink.org
National Geographic Society
   www.nationalgeographic.com
The World's Water
   www.worldwater.org
United Nations Environment Program (UNEP)
   www.unep.org

The whole moon and the entire sky are reflected in one dewdrop on the grass.

—*Zen Master Basho*

A dewdrop is freshwater. And within that drop of water is contained the planet Earth. And currently, our planet Earth is getting a little dirtier with each passing year. And, to make matters worse, there is much speculation that many of our dewdrops in the future may be turning to dust. Food. Water. Essentials of life. Unit 3 contends we have a food issue. **Unit 4** contends that *Top Ten Environmental Issue 4:* A Thirsty Planet, may become a crisis of such magnitude that unless we begin to address it now, the food debate may become a moot point, and attending to global conflict resulting from freshwater scarcity and changing geographic patterns of freshwater access will be the order of the day. Go several days without food, and you can lose some weight. Go several days without water, and you will most probably lose your life. Unlike food, freshwater provides for myriad Human–Earth ecosystems needs. If water were simply called on to slake our thirst (thus meeting our requisite biological functions) as food is essentially called on to fill our bellies and provide for cellular respiration (give us the energy needed to consume the Earth), our freshwater issue may be simpler.

But water is also used to grow our food, prevent dehydration of our livestock, keep plants alive (and atmospheric oxygen available), create our soils, help regulate and maintain Earth's organic material cycles, flush our toilets, wash our clothes, bathe our bodies, brush our teeth, maintain terrestrial biota and ecosystems, cook our food, enhance our economies (snow skiing, freshwater sports/recreation, water parks), green our lawns, wash our cars, and make our Starbucks cup of coffee. However, that is not the miracle of it. The fresh, liquid surface water that terrestrial life and human activity have easy access to and depend on makes up less than 0.02% of all water on the planet! And to make matters more interesting, freshwater supplies are characterized by an alarming uneven global distribution. Add climate change's potential to repattern current freshwater sources and access, increasing domestic, industrial, and agricultural demand, and growing regional water stresses, and we have a sure recipe for future resource-scarcity-induced conflicts. Finally, water scarcity and food scarcity reveal strong associated global patterns and cannot be addressed as though they were independent socioenvironmental variables.

One of the primary reasons for including "older" articles in this text is so the reader can compare earlier scholarly predictions regarding our environmental issues and assess the accuracy of those predictions based on the data then

© Design Pics / Kelly Redinger

and the data now. Only by such assessments of our science past and present can we hope to improve the accuracy, and thus planning, for our future. And so it is with inclusion of Article 13, "Where Oil and Water Do Mix: Environmental Scarcity and Future Conflict in the Middle East and North Africa," published in 2004. The authors explain the relationship of environmental scarcity to political unrest. They then continue to argue that the environmental resource water may become a key variable in future regional conflicts, especially in regions like the Middle East and North Africa. Although overt conflict over water specifically has yet to emerge since the article was written, there is no shortage of subliminal warnings of the potential remaining very real. Fortunately, global food aid remains available, and oil prices have allowed the idea of "virtual water" sources/supplies to step in (if only for now) and ward off the regional impacts of water scarcities.

Article 14, "The Big Melt," provides an illustration of the complex "environmental science" linkages between the earth's hydrological cycle, climate change, freshwater availability, agriculture, politics, and conflict. Glacial shrinkage may become a very serious geopolitical issue. Although places like the United States, Europe, and Latin America may not depend currently on glaciers for their water, other places like India, China, and Pakistan do. And to make matters worse, these regions also have population growth stresses, which mean higher demands for food, which means increasing demand for water. Locate this situation in an already contentious sociopolitical arena, and the majesty of glaciers takes on a whole new meaning.

Finally, Article 15, "The World's Water Challenge," provides a summary view of the world's water challenge for now and into the future. Again the themes of the issue remain consistent: availability, access, use and resource competition. However, the authors argue that despite recognition of this issue and its themes, focusing on managing for the sustainability of the resource may be advanced by establishing a value for the resource. Data is available that suggests water can be ascribed a market value against which more sustainable consumption decisions and policies can be made. For example, the article points out the World Health Organization estimates that the global return on every dollar invested in water and sanitation programs is $4 to $9, respectively.

However, on the other hand, without better understanding and appreciation of the linkages between access to and availability of freshwater, such a valuation can lead to inappropriate pricing and denied access. We need simply look at many of today's controversies surrounding the privatization of water supply at not only the municipal level but national level as well. Ultimately we will have to decide if the Earth's fresh water resources are a public or private good, if access to freshwater is a right or a privilege.

Where Oil and Water Do Mix

Environmental Scarcity and Future Conflict in the Middle East and North Africa

Jason J. Morrissette and Douglas A. Borer

"Many of the wars of the 20th century were about oil, but wars of the 21st century will be over water."

—Isamil Serageldin
World Bank Vice President

In the eyes of a future observer, what will characterize the political landscape of the Middle East and North Africa? Will the future mirror the past or, as suggested by the quote above, are significant changes on the horizon? In the past, struggles over territory, ideology, colonialism, nationalism, religion, and oil have defined the region. While it is clear that many of those sources of conflict remain salient today, future war in the Middle East and North Africa also will be increasingly influenced by economic and demographic trends that do not bode well for the region. By 2025, world population is projected to reach eight billion.[1] As a global figure, this number is troubling enough; however, over 90 percent of the projected growth will take place in developing countries in which the vast majority of the population is dependent on local renewable resources. For instance, World Bank estimates place the present annual growth rate in the Middle East and North Africa at 1.9 percent versus a worldwide average of 1.4 percent.[2] In most of these countries, these precious renewable resources are controlled by small segments of the domestic political elite, leaving less and less to the majority of the population. As a result, if present population and economic trends continue, we project that many future conflicts throughout the region will be directly linked to what academic researchers term "environmental scarcity"[3]— the scarcity of renewable resources such as arable land, forests, and fresh water.

The purpose of this article is twofold. In the first section, we conceptualize how environmental scarcity is linked to domestic political unrest and the subsequent crisis of domestic political legitimacy that may ultimately result in conflict. We review the academic literature which suggests that competition over water is the key environmental variable that will play an increasing role in future domestic challenges to governments throughout the region. We then describe how these crises of domestic political legitimacy may result in both intrastate and interstate conflict. Even though the Middle East can generally be characterized as an arid climate, two great river systems, the Nile and the Tigris/Euphrates, serve to anchor the major population centers in the region. Conflict over the water of the Nile may someday come to pass between Egypt, Sudan, and Ethiopia; while Turkey, Syria, and Iraq all are located along the Tigris/Euphrates watershed and compete for its resources. Further conflict over water may embroil Israel, Syria, and the Palestinians.

Despite many existing predictions of war over water, we investigate the intriguing question: How have governments in the Middle East thus far avoided conflict over dwindling water supplies? In the second section of the article, we discuss the concept of "virtual water" and use this concept to illustrate the important linkages between water usage and the global economy, showing how existing tangible water shortages have been ameliorated by a combination of economic factors, which may or may not be sustainable into the future.

Environmental Scarcity and Conflict: An Overview

Mostafa Dolatyar and Tim Gray identify water resources as "the principal challenge for humanity from the early days of civilization."[4] The 1998 United Nations Development Report estimates that almost a third of the 4.4 billion people currently living in the developing world have no access to clean water. The report goes on to note that the world's population tripled in the 20th century, resulting in a corresponding sixfold increase in the use of water resources. Moreover, infrastructure problems related to water supply abound in much of the developing world; the United Nations estimates that between 30 and 50 percent of the water presently diverted for irrigation purposes is lost through leaking pipes alone. In turn, roughly 20 countries in the developing world presently suffer from water stress (defined as having less than 1,000 cubic meters of available freshwater per capita), and 25 more are expected to join that list by 2050.[5] In response to these trends, the United Nations resolved in 2002 to reduce by half the proportion of people in the developing world who are unable to reach—or afford—safe drinking water.

In turn, numerous scholars in recent years have conceptualized water in security terms as a key strategic resource in many regions of the world. Thomas Naff maintains that water scarcity holds significant potential for conflict in large part because it is fundamentally essential to life. Naff identifies six basic characteristics that distinguish water as a vital and potentially contentious resource. (1) Water is necessary for sustaining life and has no substitute for human or animal use. (2) Both in terms of domestic and international policy, water issues are typically addressed by policymakers in a piecemeal fashion rather than comprehensively. (3) Since countries typically feel compelled by security concerns to control the ground on or under which water flows, by its nature, water is also a terrain security issue. (4) Water issues are frequently perceived as zero-sum, as actors compete for the same limited water resources. (5) As a result of the competition for these limited resources, water presents a constant potential for conflict. (6) International law concerning water resources remains relatively "rudimentary" and "ineffectual."[6] As these factors suggest, water is a particularly volatile strategic issue, especially when it is in severe shortage.

Arguing that environmental concerns have gained prominence in the post-Cold War era, Alwyn R. Rouyer establishes a basic paradigm of contemporary environmental conflict. Rouyer argues that "rapid population growth, particularly in the developing world, is putting severe stress on the earth's physical environment and thus creating a growing scarcity of renewable resources, including water, which in turn is precipitating violent civil and international conflict that will escalate in severity as scarcity increases."[7] Rouyer goes on to assert that this potential conflict over scarce resources will likely be most disruptive in states with rapidly expanding populations in which policymakers lack the political and economic capability to minimize environmental damage.

Almost a third of the 4.4 billion people currently living in the developing world have no access to clean water.

Security concerns linked fundamentally to environmental scarcity are far from a contrivance of the post-Cold War era, however. Ulrich Küffner asserts that conflicts over water "have occurred between many countries in all climatic regions, but between countries in arid regions they appear to be unavoidable. Claims over water have led to serious tensions, to threats and counter threats, to hostilities, border clashes, and invasions."[8] Moreover, as Miriam Lowi notes, "Well before the emergence of the nation-state, the arbitrary political division of a unitary river basin . . . led to problems regarding the interests of the states and/or communities located within the basin and the manner in which conflicting interests should be resolved."[9] Lowi fundamentally frames the issue of water scarcity in terms of a dilemma of collective action and failed cooperation—the archetypal "Tragedy of the Common"—in which communal resources are abused by the greediness of individuals. In many regions of the world, the international agreements and coordinating institutions necessary to lower the likelihood of conflict over water are either inadequate or altogether nonexistent.[10]

Thomas Homer-Dixon argues that the environmental resource scarcity that potentially results in conflict, including water scarcity, fundamentally derives from one of three sources. The first, *supply-induced scarcity,* is caused when a resource is either degraded (for example, when cropland becomes unproductive due to overuse) or depleted (for example, when cropland is converted into suburban housing). Throughout most of the Middle East and North Africa countries, both environmental and resource degradation and depletion are of relevant concern. For instance, many of these countries face significant decreases in the agricultural productivity of their arable soil as a result of ongoing trends of desertification, soil erosion, and pollution. This problem is coupled with the continued loss of croplands to urbanization, as rural dwellers move to cities in search of employment and opportunity. The second source of environmental scarcity, *demand-induced scarcity,* is caused by either an increase in per-capita consumption or by simple population growth. If the supply remains constant, and demand increases by existing users consuming more, or more users each consuming the same amount, eventually scarcity will result as demand overtakes supply. The third type of environmental scarcity is known as *structural scarcity,* a phenomenon that results when resource supplies are unequally distributed. In this case the "haves" in any given society generally control and consume an inordinate amount of the existing supply, which results in the more numerous "have-nots" experiencing the scarcity.[11]

These three sources of scarcity routinely overlap and interact in two common patterns: "resource capture" and ecological marginalization. Resource capture occurs when both demand-induced and supply-induced scarcities interact to produce structural scarcity. In this pattern, powerful groups within society foresee future shortages and act to ensure the protection of their vested interests by using their control of state structures to capture control of a valuable resource. An example of this pattern occurred in Mauritania (one of Algeria's neighbors) in the 1970s and 1980s when the countries bordering the Senegal River built a series of dams to boost agricultural production. As a result of the new dams, the value of land adjacent to the river rapidly increased—an economic development that motivated Mauritanian Moors to abandon their traditional vocation as cattle grazers located in the arid land in the north and, instead, to migrate south onto lands next to the river. However, black Mauritanians already occupied the land on the river's edge. As a result, the Moorish political elite that controlled the Mauritanian government rewrote the legislation on citizenship and land rights to effectively block black Mauritanians from land ownership. By declaring blacks as non-citizens, the Islamic Moors managed to capture the land through nominally legal (structural) means. As a result, high levels of violence later arose between Mauritania and Senegal, where hundreds of thousands of the black Mauritanians had become refugees after being driven from their land.[12]

The second pattern, ecological marginalization, occurs when demand-induced and structural scarcities interact in a way that results in supply-induced scarcity. An example of this pattern comes from the Philippines, a country whose agricultural lands traditionally have been controlled by a small group of dominant landowners who, prior to the election of former President Estrada, have controlled Filipino politics since colonial times. Population growth in the 1960s and 1970s forced many poor peasants to settle in the marginal soils of the upland interior. This more mountainous land could not sustain the lowland slash-and-burn farming practices that they brought with them. As a result, the Philippines suffered serious ecological damage in the form of water pollution, soil erosion, landslides, and changes in the hydrological cycle that led to further hardship for the peasantry as the land's capacity shrunk. As a result of their economic marginalization, many upland peasants became increasingly susceptible to the revolutionary rhetoric promoted by the communist-led New People's Army, or they supported the "People Power" movement that ousted US-backed Ferdinand Marcos from power in 1986.[13]

Thus, as shown in the Philippines, social pressures created by environmental scarcity can have a direct influence on the ruling legitimacy of the state, and may cause state power to crumble. Indeed, reductions in agricultural and economic production can produce objective socio-economic hardship; however, deprivation does not necessarily produce grievances against the government that result in serious domestic unrest or rebellion. One can look at the relative stability in famine-stricken North Korea as a poignant example of a polity whose citizens have suffered widespread physical deprivation under policies of the existing regime, but who are unwilling or unable to risk their lives to challenge the state.

This phenomenon is partly explained by conflict theorists who argue that individuals and groups have feelings of "relative deprivation" when they perceive a gap between what they believe they deserve and what in reality they actually have achieved.[14] In other words, can a government meet the expectations of the masses enough to avoid conflict? For example, in North Korea—a regime that tightly controls the information that its people receive—many people understand that they are suffering, but they may not know precisely how much they are suffering relative to others, such as their brethren in the South. The North Korean government indoctrinates its people to expect little other than hardship, which in turn it blames on outside enemies of the state. Thus, the people of North Korea have very low expectations, which their government has been able to meet. More important, then, is the question of whom do the people perceive as being responsible for their plight? If the answer is the people's own government—whether as a result of supply-induced, demand-induced, or structural resource scarcity—then social discord and rebellion are more likely to result in intrastate conflict, as citizens challenge the ruling legitimacy of the state itself. If the answer is someone else's government, then interstate conflict may result.

On numerous occasions, history has shown that governments whose people are suffering can remain in power for long periods of time by pointing to external sources for the people's hardship.[15] As noted above regarding political legitimacy, perception is politically more important than any standard of objective truth.[16] When faced with a crisis of legitimacy derived from environmental resource scarcity, any political regime essentially has a choice of two options in dealing with the situation. The regime may choose temporarily not to respond to looming challenges to its authority because water-induced stress may in fact pass when sufficient heavy rainfall occurs. However, most regimes in the Middle East and North Africa have sought more proactive ways to ensure their survival. Indeed, a people might forgive its government for one drought, but if governmental action is not taken, a subsequent drought-induced crisis of legitimacy could result in significant social upheaval by an unforgiving public. Furthermore, if the government itself is perceived to be the direct source of the scarcity—through structural arrangements, resource capture, or other means—these trends of social unrest are likely to be exacerbated. Thus, in order to survive, most states have developed policies to increase their water supplies and to address issues of environmental scarcity. The problem with doing so throughout most of the Middle East and North Africa, however, is that increasing supply in one state often creates environmental scarcity problems in another. If Turkey builds dams, Iraq and Syria are vulnerable; if Ethiopia or the Sudan builds dams, Egypt feels threatened. Thus far, interstate water problems leading to war have been avoided due to the economic interplay between oil wealth and the importation of "virtual water," which will be discussed at greater length below.

As noted above, resource scarcity issues centered on water are particularly prominent in the Middle East and North Africa. Ewan Anderson notes that resource geopolitics in the Middle East "has long been dominated by one liquid—oil. However, another liquid, water, is now recognized as the fundamental political weapon in the region."[17] Ecologically speaking, water scarcity in the Middle East and North Africa results from four primary causes: fundamentally dry climatic conditions, drought, desiccation (the degradation of land due to the drying up of the soil), and water stress (the low availability of water resulting from a growing population).[18] These resource scarcity problems are exacerbated in the Middle East by such factors as poor water quality and inadequate—and, at times, purposefully discriminatory—resource planning. As a result of these ecological and political trends, Nurit Kliot states, "water, not oil, threatens the renewal of military conflicts and social and economic disruptions" in the Middle East.[19] In the case of the Arab-Israeli conflict, Alwyn Rouyer suggests that "water has become inseparable from land, ideology, and religious prophecy."[20] Martin Sherman echoes these sentiments in the following passage, describing specifically the Arab-Israeli conflict:

> In recent years, particularly since the late 1980s, water has become increasingly dominant as a bone of contention between the two sides. More than one Arab leader, including those considered to be among the most moderate,

such as King Hussein of Jordan and former UN Secretary General, Boutros Boutros-Ghali of Egypt, have warned explicitly that water is the issue most likely to become the cause of a future Israeli-Arab war.[21]

Water is a particularly volatile strategic issue, especially when it is in severe shortage.

While Jochen Renger contends that a conflict waged explicitly over water may not lie on the immediate horizon, he notes that "it is likely that water might be used as leverage during a conflict."[22] As a result of such geopolitical trends, managing these water resources in the Middle East and North Africa—and, in turn, managing the conflict over these resources—should be considered a primary concern of both scholars and policymakers.

Keeping the Peace: The Importance of Virtual Water

The warning signals that war over water may replace war over oil and other traditional sources of conflict are very real in recent history. Yet, for more than 25 years, despite increasing demand, water has not been the primary cause of war in the Middle East and North Africa. The scenarios outlined in the preceding section have yet to fully address the fundamental questions of

why and how governments in the region have thus far avoided major interstate conflict over water. In order to understand the likelihood of war, we must address the foundation of the past peace, testing whether or not this foundation remains strong for the foreseeable future. How have the governments of the region been able to avoid the apparently inevitable consequences of conflict that derive from the interlinked problems of water deficits, population growth, and weak economic performance? In this section of the article, we turn our attention to the important linkages between water usage and the global economy, showing how existing water shortages have been ameliorated by a combination of economic factors.

To understand the politics of water in the Middle East and North Africa, one must first look at the region's most fungible resource: oil. For much of the post-World War II era, the growing need for oil to fuel economic growth has served as the dominant motivating factor in US security policy in the Middle East. Conventional wisdom in the United States holds that US dependency on Middle Eastern oil is a strategic weakness. Indeed, the specter of a regional hegemonic power that controls the oil and that is also hostile to the United States strikes fear into the hearts of policymakers in Washington. Thus, for roughly the past 50 years, the United States has sought to prop up "moderate" (meaning pro-US) regimes while denying hegemony to "radical" (meaning anti-US) regimes.[23] However, we contend that both policymakers and the public at large in the United States generally misunderstand the politics of oil as they relate to water in the Middle East.

Country	Total Water Resources per Capita (cubic meters) in 2000	Percent of Population with Access to Adequate, Improved Water Source, 2000	GDP per Capita, 2000 Estimate
Algeria	477	94%	$5,500
Egypt	930	95%	$3,600
Iran	2,040	95%	$6,300
Iraq	1,544	85%	$2,500
Israel	180	99%	$18,900
Jordan	148	96%	$3,500
Kuwait	—	100%	$15,000
Lebanon	1,124	100%	$5,000
Libya	148	72%	$8,900
Morocco	1,062	82%	$3,500
Oman	426	39%	$7,700
Qatar	—	100%	$20,300
Saudi Arabia	119	95%	$10,500
Sudan	5,312	—	$1,000
Syria	2,845	80%	$3,100
Tunisia	434	—	$6,500
Turkey	3,162	83%	$6,800
Yemen	241	69%	$820

Figure 1 Water Resources and Economics in the Middle East and North Africa.

Sources: World Bank Development Indicators, Country-at-a-Glance Tables, Freshwater Resources, and *CIA World Factbook*, at www.worldbank.org and www.cia.gov.

In absolute terms, problems arising from US vulnerability to foreign oil are basically true—it would be better to be free of dependency on oil from any foreign source than to be dependent. However, the other side of the equation is often forgotten: oil-producing states are dependent on the United States and other major oil importers for their economic livelihood. More bluntly put, oil-exporting states are dependent on the influx of dollars, euros, and yen to purchase goods, services, and commodities that they lack. Thus, oil-producing countries in the Middle East and North Africa, few of whom have managed to successfully diversify their economies beyond the petroleum sector, exist in an interdependent world economy. The world depends on their oil, and they depend on the world's goods and services—including that most valuable life-sustaining resource, water.

On the surface, this perhaps seems to be a contentious claim. Outgoing oil tankers do not return with freshwater used to grow crops, and Middle East countries do not rely on the importation of bottled water for their daily consumption needs. However, according to hydrologists, each individual needs approximately 100 cubic meters of water each year for personal needs, and an additional 1,000 cubic meters are required to grow the food that person consumes. Thus, every person alive requires approximately 1,100 cubic meters of water every year. In 1970, the water needs of most Middle Eastern and North African countries could be met from sources within the region. During the colonial and early post-colonial eras, regional governments and their engineers had effectively managed supply to deliver new water to meet the requirements of the growing urban populations, industrial requirements, agricultural needs, and other demand-induced factors. What is clear is that in the past 30 years, the status of the region's water resources has significantly worsened as populations have increased (an example of demand-induced scarcity). Since the mid-1970s, most countries have been able to supply daily consumption and industrial needs; however, as indicated in Figure 1, the approximate 1,000 cubic meters of water per capita that is required for self-sufficient agricultural production represents a seemingly impossible challenge for some Middle Eastern and North African economies.

Simply put, many countries of the region cannot presently meet the irrigation requirements needed to feed their own growing populations.[24] Furthermore, for those countries that have sufficient resources to meet this need in aggregate (such as Syria), resource capture and structural distribution problems keep water out of the hands of many citizens. If this situation has been deteriorating for nearly three decades, the question remains: Why has there been no war over water? The answer, according to Tony Allen, lies in an extremely important hidden source of water, which he describes as "virtual water."[25] Virtual water is the water contained in the food that the region imports—from the United States, Australia, Argentina, New Zealand, the countries of the European Union, and other major food-exporting countries. If each person of the world consumes food that requires 1,000 cubic meters of water to grow, plus 100 additional cubic meters for drinking, hygiene, and industrial production, it is still possible that any country that cannot supply the water to produce food may have sufficient water to

meet its needs—if it has the economic capacity to buy, or the political capacity to beg, the remaining virtual water in the form of imported food.

Many countries of the region cannot presently meet the irrigation requirements needed to feed their own growing populations.

According to Allen, more water flows into the countries of the Middle East and North Africa as virtual water each year than flows down the Nile for Egypt's agriculture. Virtual water obtained in the food available on the global market has enabled the governments of the region's countries to augment their inadequate and declining water resources. For instance, despite its meager freshwater resources of 180 cubic meters per capita, Israel—otherwise self-sufficient in terms of food production—manages its problems of water scarcity in part by importing large supplies of grain each year. As noted in Figure 1, this pattern is replicated by eight other countries in the region that have less than 1,100 cubic meters of water per person. Thus, the global cereal grain commodity markets have proven to be a very accessible and effective system for importing virtual water needs. In the Middle East and North Africa, politicians and resource managers have thus far found this option a better choice than resorting to war over water with their neighbors. As a result, the strategic imperative for maintaining peace has been met through access to virtual water in the form of food imports from the global market.[26]

The global trade in food commodities has been increasingly accessible, even to poor economies, for the past 50 years. During the Cold War, food that could not be purchased was often provided in the form of grants by either the United States or the Soviet Union, and in times of famine, international relief efforts in various parts of the globe have fed the starving. Over time, competition by the generators of the global grain surplus—the United States, Australia, Argentina, and the European Community—brought down the global price of grain. As a result, the past quarter-century, the period during which water conflicts in the Middle East and North Africa have been most insistently predicted, was also a period of global commodity markets awash with surplus grain. This situation allowed the region's states to replace domestic water supply shortages with subsidized virtual water in the form of purchases from the global commodities market. For example, during the 1980s, grain was being traded at about $100 (US) a ton, despite costing about $200 a ton to produce.[27] Thus, US and European taxpayers were largely responsible for funding the cost of virtual water (in the form of significant agricultural subsidies they paid their own farmers) which significantly benefited the countries of the Middle East and North Africa.

For the most part we concur with Allen's evaluation that countries have not gone to war primarily over water, and that they have not done so because they have been able to purchase virtual water on the international market. However, the key question for the future is, Will this situation continue? If the answer is yes, and grain will remain affordable to the countries of the region, then it is relatively safe to conclude that conflict derived from environmental scarcity (in the form of water deficits) will not be a significant problem in the foreseeable future. However, if the answer is no, and grain will not be as affordable as it has been in the past, then future conflict scenarios based on environmental scarcity must be seriously considered.

Global Economic Restructuring: The World Trade Organization's Impact on Subsidies

Regrettably, a trend toward the answer "no" appears to be gaining some momentum due to ongoing structural changes in the global economy. The year 1995 witnessed a dramatic change in the world grain market, when wheat prices rose rapidly, eventually reaching $250 a ton by the spring of 1996. With the laws of supply and demand kicking in, this increased price resulted in greater production; by 1998, world wheat prices had fallen back to $140 a ton, but had risen again to over $270 by June 2001.[28] These rapid wheat price fluctuations reemphasize the strategic importance and volatility of virtual water. If the global price of food staples remains affordable, many countries in the Middle East and North Africa may struggle to meet the demand-induced scarcity resulting from their growing populations, but they most likely will succeed. However, if basic food staple prices rise significantly in the coming decades and the existing economic growth patterns that have characterized the region's economies over the past 30 years remain constant, an outbreak of war is more likely.

It is clear that recent structural changes in the world economy do not favor the continuation of affordable food prices for the region's countries in the future. As noted above, wheat that costs $200 a ton to produce has often been sold for $100 a ton on world markets. This situation is possible only when the supplier is compensated for the lost $100 per ton in the form of a subsidy. Historically, these subsidies have been paid by the governments of major cereal grain-producing countries, primarily the United States and members of the European Union. Indeed, for the last 100 years, farm subsidies have been a bedrock public policy throughout the food-exporting countries of the first world. However, with the steady embrace of global free-trade economics and the establishment of the World Trade Organization (WTO), agricultural subsidies have come under pressure in most major grain-producing countries. According to a recent US Department of Agriculture (USDA) study, "The elimination of agriculture trade and domestic policy distortions could raise world agriculture prices about 12 percent.[29]

Thus, as the WTO gains systematic credibility over the coming decades, its free-trade policies will further erode the practice of farm price supports, and it is highly unlikely that the aggregate farm subsidies of the past will continue at historic levels in the future. Under the new WTO regime, global food production will be increasingly based on the real cost of production plus whatever profit is required to keep farmers in business. Therefore, as global food prices rise in the future, and American and European governments are restricted by the new global trading regime from subsidizing their farmers, the price of virtual water in the Middle East and North Africa and throughout the food-importing world will also rise. According to the USDA report mentioned above, both developed and developing countries will gain from WTO liberalization. Developed countries that are major food exporters will gain immediately from the projected $31 billion in increased global food prices, of which they will share $28.5 billion ($13.3 billion to the United States), with $2.6 billion going to food exporters in the developing world. However, the report also claims food-importing countries will gain because global food price increases will spur more efficient production in their own economies, thus enabling them a "potential benefit" of $21 billion.[30] Even if accepted at face value, it is clear that such benefits will occur mostly in those developing countries with an abundance of water resources. Indeed, developing countries that produce fruits, vegetables, and other high-value crops for export to first-world markets may indeed benefit from the reduction of farm subsidies, which today undercut their competitive advantage. But when it comes to basic foodstuffs—wheat, corn, and rice—the cereal grains that sustain life for most people, the developing world cannot compete with the highly efficient mechanized corporate farms of the first world.

In future research, basic intelligence is needed on two fronts. First, we must obtain a clearer understanding of the capacity of global commodity markets to meet future virtual water needs in the form of food. Second, we must identify which Middle East and North Africa governments will most likely have the economic capacity to meet their virtual water needs though food purchases—or, perhaps more important, which ones will not. In short, is there food available in the global market, and can countries afford to buy it? Countries that cannot afford virtual water may choose instead to pursue war as a means of achieving their national interest goals. Clearly the strongest countries, or those least susceptible to intrastate or interstate conflict arising from environmental scarcity, are those that have significant water resources or the economic capacity to purchase virtual water. However, it is also clear that the relative condition of peace that has existed in the Middle East and North Africa has been maintained historically through deeply buried linkages between American and European taxpayers, their massive farm subsidies programs, and world food prices. In the future, it appears that these hidden links may be radically altered if not broken by the World Trade Organization, and, as a result, the likelihood of conflict will increase.

Conclusion: Why War Will Come

Having moved away from the conventional understanding of water strictly as a zero-sum environmental resource by reconceptualizing it in more fungible economic terms, we nevertheless believe two incompatible social trends will collide to make war in the Middle East and North Africa virtually inevitable in the future. The first trend is economic globalization. As capitalism becomes ever more embraced as the global economic philosophy, and the world increasingly embraces free-trade economics, economic growth is both required and is inevitable. The WTO will facilitate this aggregate global growth, which, on the plus side, will undoubtedly increase the basic standard of living for the average world citizen. However, the global economy will be required to meet the needs of an estimated eight billion citizens in the year 2025. Achieving growth will demand an ever-greater share of the world's existing natural resources, including water. Thus, if present regional economic and demographic trends continue, resource shortfalls will occur, with water being the most highly stressed resource in the Middle East and North Africa.

Globalization is both a cause and a consequence of the rapid spread of information technology. Thus, in the globalized world, the figurative distance between cultures, philosophies of rule, and, perhaps more important, a basic understanding of what is possible in life, becomes much shorter. Personal computers, the internet, cellular phones, fax machines, and satellite television are all working in partnership to rewire the psychological infrastructure of the citizens of the Middle East and North Africa, and the world at large. As a result, by making visible what is possible in the outside world, this cognitive liberation will bring heightened material expectations of a better life, both economically and politically. Consequently, citizens will demand more from their governments. This emerging reality will collide head-on with the second trend—political authoritarianism—that characterizes most Middle East and North Africa governments.

Throughout the region there are few governments that allow for public expression of dissent. Although Turkey, Algeria, Tunisia, and Egypt are democracies in name, these states have exhibited a propensity to revert to authoritarian tactics when deemed necessary to limit political activity among their respective populaces.[31] Likewise, while Israel is institutionally a democracy, ethnic minorities are all but excluded from the democratic process. The remainder of the Middle East and North Africa states can be described only as authoritarian regimes. In retrospect, the most fundamental common denominator of all authoritarian regimes throughout history is their fierce resistance to change. Change is seen as a threat to the regime because most authoritarian regimes base their right to rule in some form of infallibility: the infallibility of the sultan, the king, or the ruling party and its ideology. Any admission that change is needed strikes at the foundation of this inflexible infallibility. Historically, most change has occurred in the Middle East and North Africa during times of intrastate unrest and interstate war. In the coming decades, globalization will bring change that will be resisted by governments of the region. As a result, to the distant observer the future will resemble the past: periods of wholesale peace will be a rare occurrence, intense competition and low-intensity conflict will be the norm, and major wars will occur at sporadic intervals.

The wild card in this equation may be post-2004 Iraq. Operation Iraqi Freedom and the ouster of Saddam Hussein have altered the strategic political landscape. If a sustainable democracy indeed emerges in Iraq, the country may turn away from future conflicts with its neighbors. Potential conflict between Turkey and Iraq over water may now be averted due to the fact that both countries may choose nonviolent solutions to their disputes. If

President Bush's vision of a democratic Middle East comes to fruition, war may be averted. After all, there is a rich body of scholarly research regarding the "democratic peace" that suggests liberal democracies are significantly less likely to resort to war to resolve interstate disputes, and post-Saddam Iraq could serve as a key litmus test for the future of democratic reform in the region. However, it is also highly unlikely that regime change will come quickly to the moderate authoritarian states of the region that are also US allies. Decisionmakers in Washington may be able to dictate the political future of Iraq, but even America's mighty arsenal of political, economic, and military power cannot alter the basic demographic and environmental trends in the region.

Notes

1. Alex Marshall, ed., *The State of World Population 1997* (New York: United Nations Population Fund, 1997), p. 70.

2. "The World Bank: Middle East and North Africa Data Profile," *The World Bank Group Country Data* (2000), http://worldbank.org/data/countrydata/countrydata.html.

3. The leading scholar in this area is Thomas Homer-Dixon. For example, see his recent book (coedited with Jessica Blitt), *Ecoviolence: Links Among Environment, Population, and Security* (New York: Rowman & Littlefield, 1998), which focuses on Chiapas, Gaza, South Africa, Pakistan, and Rwanda.

4. Mostafa Dolatyar and Tim S. Gray, *Water Politics in the Middle East: A Context for Conflict or Co-operation?* (New York: St. Martin's Press, 2000), p. 6.

5. *Human Development Report: Consumption for Human Development* (New York: United Nations Development Programme, Oxford Univ. Press, 1998), p. 55; "Water Woes Around the World," MSNBC, 9 September 2002, http://msnbc.com/news/802693.asp

6. Thomas Naff, "Conflict and Water Use in the Middle East," in *Water in the Arab World: Perspectives and Prognoses,* ed. Peter Rogers and Peter Lydon (Cambridge, Mass.: Harvard Univ. Press, 1994), p. 273.

7. Alwyn R. Rouyer, *Turning Water into Politics: The Water Issue in the Palestinian-Israeli Conflict* (New York: St. Martin's Press, 2000), p. 7.

8. Ulrich Küffner, "Contested Waters: Dividing or Sharing?" in *Water in the Middle East: Potential for Conflicts and Prospects for Cooperation,* ed. Waltina Scheumann and Manuel Schiffler (New York: Springer, 1998), p. 71.

9. Miriam R. Lowi, *Water and Power: The Politics of a Scarce Resource in the Jordan River Basin* (Cambridge, Eng.: Cambridge Univ. Press, 1993), p. 1.

10. Ibid., pp. 2ff.

11. Thomas Homer-Dixon and Jessica Blitt, "Introduction: A Theoretical Overview," in *Ecoviolence: Links Among Environment, Population, and Scarcity,* ed. Thomas Homer-Dixon and Jessica Blitt (New York: Rowman & Littlefield, 1998), p. 6.

12. Thomas Homer-Dixon and Valerie Percival, "The Case of Senegal-Mauritania," in *Environmental Scarcity and Violent Conflict: Briefing Book* (Washington: American Association for the Advancement of Science and the University of Toronto, 1996), pp. 35–38.

13. Douglas Borer witnessed this agricultural problem while visiting rural areas on the Bataan peninsula in late 1985 and

early 1986. The members of the New People's Army which he met were uninterested in Marxism, but they were very interested in ridding themselves of the Marcos regime. See Thomas Homer-Dixon and Valerie Percival, "The Case of the Philippines," ibid., p. 49.

14. Ted Gurr, *Why Men Rebel* (Princeton, N.J.: Princeton Univ. Press, 1970).

15. One need only look 90 miles southward from the Florida coast to find proof of this reality in Castro's Cuba.

16. Thus, Saddam Hussein was able to remain in power in Iraq until 2003 due to two essential factors. First, as noted in a recent article by James Quinlivan, Saddam had created "groups with special loyalties to the regime and the creation of parallel military organizations and multiple internal security agencies," that made Iraq essentially a "coup-proof" regime. (See James T. Quinlivan, "Coup-Proofing: Its Practice and Consequences in the Middle East," *International Security,* 24 [Fall 1999], 131–65.) Second, Saddam had convinced a significant portion of his people that the United States (and Britain) were responsible for their suffering. Thus, as long as these perceptions held and Saddam was able to command loyalty of the inner regime, his ouster from power by domestic sources remained unlikely.

17. Ewan W. Anderson, "Water: The Next Strategic Resource," in *The Politics of Scarcity: Water in the Middle East,"* ed. Joyce R. Starr and Daniel C. Stoll (Boulder, Colo.: Westview Press, 1988), p. 1.

18. Hussein A. Amery and Aaron T. Wolf, "Water, Geography, and Peace in the Middle East," in *Water in the Middle East: A Geography of Peace,* ed. Hussein A. Amery and Aaron T. Wolf (Austin: Univ. of Texas Press, 2000).

19. Nubit Kliot, *Water Resources and Conflict in the Middle East* (London and New York: Routledge, 1994), p. v, as quoted in Dolatyar and Gray, p. 9.

20. Rouyer, p. 9.

21. Martin Sherman, *The Politics of Water in the Middle East: An Israeli Perspective on the Hydro-Political Aspects of the Conflict* (New York: St. Martin's Press, 1999), p. xi.

22. Jochen Renger, "The Middle East Peace Process: Obstacles to Cooperation Over Shared Waters," in *Water in the Middle East: Potential for Conflict and Prospects for Cooperation,* ed. Waltina Scheumann and Manuel Schiffler (New York: Springer, 1998), p. 50.

23. Thus, even though the Saudi government is much more Islamized in religious terms than that of the Iraqis or Syrians, as long as the Saudi government is pro-US and serves US interests in supplying cheap oil, it receives the benevolent "moderate" label, while the more secularized Iraqis and Syrians have been labeled with the prerogative labels "radical" or "rogue-states."

24. Tony Allan, "Watersheds and Problemsheds: Explaining the Absence of Armed Conflict over Water in the Middle East," in *MERNIA: Middle East Review of International Affairs Journal,* 2 (March 1998), http://biu.ac.il/SOC/besa/meria/journal/1998/issue1/jv2n1a7.html.

25. Ibid.

26. Ibid.

27. Ibid.

28. Prices from 26 June 2001 quoted at http://usafutures.com/commodityprices.htm.

29. "Agricultural Policy Reform in the WTO—The Road Ahead," in *ERS Agricultural Economics Report,* No. 802, ed. Mary E. Burfisher (Washington: US Department of Agriculture, May 2001), p. iii.

30. Ibid., p. 6.

31. For instance, as of 2004, Freedom House (http://freedomhouse.org) classifies Algeria, Egypt, and Tunisia as "not free," and Turkey as only "partly free."

Critical Thinking

1. Explain the idea of "environmental scarcity" and its relationship to domestic political unrest.

2. The author argues that consumption over water may become the key environmental variable in future conflicts. Locate on your map where these conflicts may occur. What scarcity patterns do you see?

3. What is the idea of "virtual water"? Why do some nations have access to virtual water and others do not?

4. Sketch a concept map illustrating linkages between climate change, water resources, virtual water, agricultural production, agricultural consumption, oil production, and oil consumption. What do you see?

5. The article was published seven years ago. What is your assessment of the article's thesis in regard to today, 2012?

6. Identify in this article three key terms, concepts, or principles that are used in your textbook (environmental science, economics, sociology, history, geography, etc.) or employed in the discipline you are currently studying. (Note: The terms, concepts, or principles may be implicit, explicit, implied, or inferred.)

JASON J. MORRISSETTE is a doctoral candidate and instructor of record in the School of Public and International Affairs at the University of Georgia. He is currently writing his dissertation on the political economy of water scarcity and conflict. **DOUGLAS A. BORER** (PhD, Boston University, 1993) is an Associate Professor in the Department of Defense Analysis at the Naval Postgraduate School. He recently served as Visiting Professor of Political Science at the US Army War College. Previously he was Director of International Studies at Virginia Tech, and he has taught overseas in Fiji and Australia. Dr. Borer is a former Fulbright Scholar at the University of Kebangsaan Malaysia, and has published widely in the areas of security, strategy, and foreign policy.

The Big Melt

BROOK LARMER

G laciers in the high heart of Asia feed its greatest rivers, lifelines for two billion people. Now the ice and snow are diminishing.

The gods must be furious.

It's the only explanation that makes sense to Jia Son, a Tibetan farmer surveying the catastrophe unfolding above his village in China's mountainous Yunnan Province. "We've upset the natural order," the devout, 52-year-old Buddhist says. "And now the gods are punishing us."

On a warm summer afternoon, Jia Son has hiked a mile and a half up the gorge that Ming-yong Glacier has carved into sacred Mount Kawagebo, looming 22,113 feet high in the clouds above. There's no sign of ice, just a river roiling with silt-laden melt. For more than a century, ever since its tongue lapped at the edge of Mingyong village, the glacier has retreated like a dying serpent recoiling into its lair. Its pace has accelerated over the past decade, to more than a football field every year— a distinctly unglacial rate for an ancient ice mass.

"This all used to be ice ten years ago," Jia Son says, as he scrambles across the scree and brush. He points out a yak trail etched into the slope some 200 feet above the valley bottom. "The glacier sometimes used to cover that trail, so we had to lead our animals over the ice to get to the upper meadows."

Around a bend in the river, the glacier's snout finally comes into view: It's a deathly shade of black, permeated with pulverized rock and dirt. The water from this ice, once so pure it served in rituals as a symbol of Buddha himself, is now too loaded with sediment for the villagers to drink. For nearly a mile the glacier's once smooth surface is ragged and cratered like the skin of a leper. There are glimpses of blue-green ice within the fissures, but the cracks themselves signal trouble. "The beast is sick and wasting away," Jia Son says. "If our sacred glacier cannot survive, how can we?"

It is a question that echoes around the globe, but nowhere more urgently than across the vast swath of Asia that draws its water from the "roof of the world." This geologic colossus—the highest and largest plateau on the planet, ringed by its tallest mountains—covers an area greater than western Europe, at an average altitude of more than two miles. With nearly 37,000 glaciers on the Chinese side alone, the Tibetan Plateau and its surrounding arc of mountains contain the largest volume of ice outside the polar regions. This ice gives birth to Asia's largest

and most legendary rivers, from the Yangtze and the Yellow to the Mekong and the Ganges—rivers that over the course of history have nurtured civilizations, inspired religions, and sustained ecosystems. Today they are lifelines for some of Asia's most densely settled areas, from the arid plains of Pakistan to the thirsty metropolises of northern China 3,000 miles away. All told, some two billion people in more than a dozen countries— nearly a third of the world's population—depend on rivers fed by the snow and ice of the plateau region.

But a crisis is brewing on the roof of the world, and it rests on a curious paradox: For all its seeming might and immutability, this geologic expanse is more vulnerable to climate change than almost anywhere else on Earth. The Tibetan Plateau as a whole is heating up twice as fast as the global average of 1.3°F over the past century—and in some places even faster. These warming rates, unprecedented for at least two millennia, are merciless on the glaciers, whose rare confluence of high altitudes and low latitudes make them especially sensitive to shifts in climate.

For thousands of years the glaciers have formed what Lonnie Thompson, a glaciologist at Ohio State University, calls "Asia's freshwater bank account"—an immense storehouse whose buildup of new ice and snow (deposits) has historically offset its annual runoff (withdrawals). Glacial melt plays its most vital role before and after the rainy season, when it supplies a greater portion of the flow in every river from the Yangtze (which irrigates more than half of China's rice) to the Ganges and the Indus (key to the agricultural heartlands of India and Pakistan). But over the past half century, the balance has been lost, perhaps irrevocably. Of the 680 glaciers Chinese scientists monitor closely on the Tibetan Plateau, 95 percent are shedding more ice than they're adding, with the heaviest losses on its southern and eastern edges. "These glaciers are not simply retreating," Thompson says. "They're losing mass from the surface down."

The ice cover in this portion of the plateau has shrunk more than 6 percent since the 1970s—and the damage is still greater in Tajikistan and northern India, with 35 percent and 20 percent declines respectively over the past five decades. The rate of melting is not uniform, and a number of glaciers in the Karakoram Range on the western edge of the plateau are actually advancing. This anomaly may result from increases in snowfall in the higher latitude—and therefore colder—Karakorams, where snow and ice are less vulnerable to small temperature increases. The gaps in scientific knowledge are still great, and in the Tibetan Plateau they are deepened by the region's remoteness and political sensitivity—as well as by the inherent complexities of climate science.

Though scientists argue about the rate and cause of glacial retreat, most don't deny that it's happening. And they believe the worst may be yet to come. The more dark areas that are exposed by melting, the more sunlight is absorbed than reflected, causing temperatures to rise faster. (Some climatologists believe this warming feedback loop could intensify the Asian monsoon, triggering more violent storms and flooding in places such as Bangladesh and Myanmar.) If current trends hold, Chinese scientists believe that 40 percent of the plateau's glaciers could disappear by 2050. "Full-scale glacier shrinkage is inevitable," says Yao Tandong, a glaciologist at China's Institute of Tibetan Plateau Research. "And it will lead to ecological catastrophe."

The potential impacts extend far beyond the glaciers. On the Tibetan Plateau, especially its dry northern flank, people are already affected by a warmer climate. The grasslands and wetlands are deteriorating, and the permafrost that feeds them with spring and summer melt is retreating to higher elevations. Thousands of lakes have dried up. Desert now covers about one-sixth of the plateau, and in places sand dunes lap across the highlands like waves in a yellow sea. The herders who once thrived here are running out of options.

Along the plateau's southern edge, by contrast, many communities are coping with too much water. In alpine villages like Mingyong, the glacial melt has swelled rivers, with welcome side effects: expanded croplands and longer growing seasons. But such benefits often hide deeper costs. In Mingyong, surging meltwater has carried away topsoil; elsewhere, excess runoff has been blamed for more frequent flooding and landslides. In the mountains from Pakistan to Bhutan, thousands of glacial lakes have formed, many potentially unstable. Among the more dangerous is Imja Tsho, at 16,400 feet on the trail to Nepal's Island Peak. Fifty years ago the lake didn't exist; today, swollen by melt, it is a mile long and 300 feet deep. If it ever burst through its loose wall of moraine, it would drown the Sherpa villages in the valley below.

This situation—too much water, too little water—captures, in miniature, the trajectory of the overall crisis. Even if melting glaciers provide an abundance of water in the short run, they portend a frightening endgame: the eventual depletion of Asia's greatest rivers. Nobody can predict exactly when the glacier retreat will translate into a sharp drop in runoff. Whether it happens in 10, 30, or 50 years depends on local conditions, but the collateral damage across the region could be devastating. Along with acute water and electricity shortages, experts predict a plunge in food production, widespread migration in the face of ecological changes, even conflicts between Asian powers.

The nomads' tent is a pinprick of white against a canvas of green and brown. There is no other sign of human existence on the 14,000-foot-high prairie that seems to extend to the end of the world. As a vehicle rattles toward the tent, two young men emerge, their long black hair horizontal in the wind. Ba O

and his brother Tsering are part of an unbroken line of Tibetan nomads who for at least a thousand years have led their herds to summer grazing grounds near the headwaters of the Yangtze and Yellow Rivers.

Inside the tent, Ba O's wife tosses patties of dried yak dung onto the fire while her four-year-old son plays with a spool of sheep's wool. The family matriarch, Lu Ji, churns yak milk into cheese, rocking back and forth in a hypnotic rhythm. Behind her are two weathered Tibetan chests topped with a small Buddhist shrine: a red prayer wheel, a couple of smudged Tibetan texts, and several yak butter candles whose flames are never allowed to go out. "This is the way we've always done things," Ba O says. "And we don't want that to change."

But it may be too late. The grasslands are dying out, as decades of warming temperatures—exacerbated by overgrazing—turn prairie into desert. Watering holes are drying up, and now, instead of traveling a short distance to find summer grazing for their herds, Ba O and his family must trek more than 30 miles across the high plateau. Even there the grass is meager. "It used to grow so high you could lose a sheep in it," Ba O says. "Now it doesn't reach above their hooves." The family's herd has dwindled from 500 animals to 120. The next step seems inevitable: selling their remaining livestock and moving into a government resettlement camp.

Across Asia the response to climate-induced threats has mostly been slow and piecemeal, as if governments would prefer to leave it up to the industrialized countries that pumped the greenhouse gases into the atmosphere in the first place. There are exceptions. In Ladakh, a bone-dry region in northern India and Pakistan that relies entirely on melting ice and snow, a retired civil engineer named Chewang Norphel has built "artificial glaciers"—simple stone embankments that trap and freeze glacial melt in the fall for use in the early spring growing season. Nepal is developing a remote monitoring system to gauge when glacial lakes are in danger of bursting, as well as the technology to drain them. Even in places facing destructive monsoonal flooding, such as Bangladesh, "floating schools" in the delta enable kids to continue their education—on boats.

But nothing compares to the campaign in China, which has less water than Canada but 40 times more people. In the vast desert in the Xinjiang region, just north of the Tibetan Plateau, China aims to build 59 reservoirs to capture and save glacial runoff. Across Tibet, artillery batteries have been installed to launch rain-inducing silver iodide into the clouds. In Qinghai the government is blocking off degraded grasslands in hopes they can be nurtured back to health. In areas where grasslands have already turned to scrub desert, bales of wire fencing are rolled out over the last remnants of plant life to prevent them from blowing away.

Along the road near the town of Madoi are two rows of newly built houses. This is a resettlement village for Tibetan nomads, part of a massive and controversial program to relieve pressure on the grasslands near the sources of Chinas three major rivers—the Yangtze, Yellow, and Mekong—where nearly half of Qinghai Province's 530,000 nomads have traditionally lived. Tens of thousands of nomads here have had to give up their way of life, and many more—including, perhaps, Ba O—may follow.

The subsidized housing is solid, and residents receive a small annual stipend. Even so, Jixi Lamu, a 33-year-old woman in a traditional embroidered dress, says her family is stuck in limbo, dependent on government handouts. "We've spent the $400 we had left from selling off our animals," she says. "There was no future with our herds, but there's no future here either." Her husband is away looking for menial work. Inside the one-room house, her mother sits on the bed, fingering her prayer beads. A Buddhist shrine stands on the other side of the room, but the candles have burned out.

It is not yet noon in Delhi, just 180 miles south of the Himalayan glaciers. But in the narrow corridors of Nehru Camp, a slum in this city of 16 million, the blast furnace of the north Indian summer has already sent temperatures soaring past 105 degrees Fahrenheit. Chaya, the 25-year-old wife of a fortune-teller, has spent seven hours joining the mad scramble for water that, even today, defines life in this heaving metropolis—and offers a taste of what the depletion of Tibet's water and ice portends.

Chaya's day began long before sunrise, when she and her five children fanned out in the darkness, armed with plastic jugs of every size. After daybreak, the rumor of a tap with running water sent her stumbling in a panic through the slum's narrow corridors. Now, with her containers still empty and the sun blazing overhead, she has returned home for a moment's rest. Asked if she's eaten anything today, she laughs: "We haven't even had any tea yet."

Suddenly cries erupt—a water truck has been spotted. Chaya leaps up and joins the human torrent in the street. A dozen boys swarm onto a blue tanker, jamming hoses in and siphoning the water out. Below, shouting women jostle for position with their containers. In six minutes the tanker is empty. Chaya arrived too late and must move on to chase the next rumor of water.

Delhi's water demand already exceeds supply by more than 300 million gallons a day, a shortfall worsened by inequitable distribution and a leaky infrastructure that loses an estimated 40 percent of the water. More than two-thirds of the city's water is pulled from the Yamuna and the Ganges, rivers fed by Himalayan ice. If that ice disappears, the future will almost certainly be worse. "We are facing an unsustainable situation," says Diwan Singh, a Delhi environmental activist. "Soon—not in thirty years but in five to ten—there will be an exodus because of the lack of water."

The tension already seethes. In the clogged alleyway around one of Nehru Camp's last functioning taps, which run for one hour a day, a man punches a woman who cut in line, leaving a purple welt on her face. "We wake up every morning fighting over water," says Kamal Bhate, a local astrologer watching the melee. This one dissolves into shouting and finger-pointing, but the brawls can be deadly. In a nearby slum a teenage boy was recently beaten to death for cutting in line.

As the rivers dwindle, the conflicts could spread. India, China, and Pakistan all face pressure to boost food production to keep up with their huge and growing populations. But climate change and diminishing water supplies could reduce cereal yields in South Asia by 5 percent within three decades. "We're going to see rising tensions over shared water resources, including political disputes between farmers, between farmers and cities, and between human and ecological demands for water," says Peter Gleick, a water expert and president of the Pacific Institute in Oakland, California. "And I believe more of these tensions will lead to violence."

The real challenge will be to prevent water conflicts from spilling across borders. There is already a growing sense of alarm in Central Asia over the prospect that poor but glacier-heavy nations (Tajikistan, Kyrgyzstan) may one day restrict the flow of water to their parched but oil-rich neighbors (Uzbekistan, Kazakhstan, Turkmenistan). In the future, peace between Pakistan and India may hinge as much on water as on nuclear weapons, for the two countries must share the glacier-dependent Indus.

The biggest question mark hangs over China, which controls the sources of the region's major rivers. Its damming of the Mekong has sparked anger downstream in Indochina. If Beijing follows through on tentative plans to divert the Brahmaputra, it could provoke its rival, India, in the very region where the two countries fought a war in 1962.

For the people in Nehru Camp, geopolitical concerns are lost in the frenzied pursuit of water. In the afternoon, a tap outside the slum is suddenly turned on, and Chaya, smiling triumphantly, hauls back a full, ten-gallon jug on top of her head. The water is dirty and bitter, and there are no means to boil it. But now, at last, she can give her children their first meal of the day: a piece of bread and a few spoonfuls of lentil stew. "They should be studying, but we keep shooing them away to find water," Chaya says. "We have no choice, because who knows if we'll find enough water tomorrow."

Fatalism may be a natural response to forces that seem beyond our control. But Jia Son, the Tibetan farmer watching Mingyong Glacier shrink, believes that every action counts—good or bad, large or small. Pausing on the mountain trail, he makes a guilty confession. The melting ice, he says, may be his fault.

When Jia Son first noticed the rising temperatures—an unfamiliar trickle of sweat down his back about a decade ago—he figured it was a gift from the gods. Winter soon lost some of its brutal sting. The glacier began releasing its water earlier in the summer, and for the first time in memory villagers had the luxury of two harvests a year.

Then came the Chinese tourists, a flood of city dwellers willing to pay locals to take them up to see the glacier. The Han tourists don't always respect Buddhist traditions; in their gleeful hollers to provoke an icefall, they seem unaware of the calamity that has befallen the glacier. Still, they have turned a poor village into one of the region's wealthiest. "Life is much easier now," says Jia Son, whose simple farmhouse, like all in the village, has a television and government-subsidized satellite dish. "But maybe our greed has made Kawagebo angry."

He is referring to the temperamental deity above his village. One of the holiest mountains in Tibetan Buddhism, Kawagebo has never been conquered, and locals believe its summit—and its glacier—should remain untouched. When a Sino-Japanese expedition tried to scale the peak in 1991, an avalanche near the top of the glacier killed all 17 climbers. Jia Son remains convinced the deaths were not an accident but an act of divine retribution. Could Mingyong's retreat be another sign of Kawagebo's displeasure?

Jia Son is taking no chances. Every year he embarks on a 15-day pilgrimage around Kawagebo to show his deepening Buddhist devotion. He no longer hunts animals or cuts down trees. As part of a government program, he has also given up a parcel of land to be reforested. His family still participates in the village's tourism cooperative, but Jia Son makes a point of telling visitors about the glacier's spiritual significance. "Nothing will get better," he says, "until we get rid of our materialistic thinking."

It's a simple pledge, perhaps, one that hardly seems enough to save the glaciers of the Tibetan Plateau—and stave off the water crisis that seems sure to follow. But here, in the shadow of one of the world's fastest retreating glaciers, this lone farmer has begun, in his own small way, to restore the balance.

Critical Thinking

1. Sketch a concept map of the linkages between Asia's population centers, location of glaciers, and their agricultural regions. Locate potential areas for conflict over water resources.

2. If water resources (via glaciers) became too scarce, how will Asia feed its people?

3. What other environmental impacts in the region may also result from glacial loss?

4. How is Asia's potential water scarcity issue different from those described in Article 13 (Middle East and North Africa region)?

5. Identify in this article three key terms, concepts, or principles that are used in your textbook (environmental science, economics, sociology, history, geography, etc.) or employed in the discipline you are currently studying. (Note: The terms, concepts, or principles may be implicit, explicit, implied, or inferred.)

The World's Water Challenge

If oil is the key geopolitical resource of today, water will be as important—if not more so—in the not-so-distant future.

ERIK R. PETERSON AND RACHEL A. POSNER

Historically, water has meant the difference between life and death, health and sickness, prosperity and poverty, environmental sustainability and degradation, progress and decay, stability and insecurity. Societies with the wherewithal and knowledge to control or "smooth" hydrological cycles have experienced more rapid economic progress, while populations without the capacity to manage water flows—especially in regions subject to pronounced flood-drought cycles—have found themselves confronting tremendous social and economic challenges in development.

Tragically, a substantial part of humanity continues to face acute water challenges. We now stand at a point at which an obscenely large portion of the world's population lacks regular access to fresh drinking water or adequate sanitation. Water-related diseases are a major burden in countries across the world. Water consumption patterns in many regions are no longer sustainable. The damaging environmental consequences of water practices are growing rapidly. And the complex and dynamic linkages between water and other key resources—especially food and energy—are inadequately understood. These factors suggest that even at current levels of global population, resource consumption, and economic activity, we may have already passed the threshold of water sustainability.

An obscenely large portion of the world's population lacks regular access to fresh drinking water or adequate sanitation.

A major report recently issued by the 2030 Water Resources Group (whose members include McKinsey & Company, the World Bank, and a consortium of business partners) estimated that, assuming average economic growth and no efficiency gains, the gap between global water demand and reliable supply could reach 40 percent over the next 20 years. As serious as this world supply-demand gap is, the study notes, the dislocations will be even more concentrated in developing regions that account for one-third of the global population, where the water deficit could rise to 50 percent.

It is thus inconceivable that, at this moment in history, no generally recognized "worth" has been established for water to help in its more efficient allocation. To the contrary, many current uses of water are skewed by historical and other legacy practices that perpetuate massive inefficiencies and unsustainable patterns.

The Missing Links

In addition, in the face of persistent population pressures and the higher consumption implicit in rapid economic development among large populations in the developing world, it is noteworthy that our understanding of resource linkages is so limited. Our failure to predict in the spring of 2008 a spike in food prices, a rise in energy prices, and serious droughts afflicting key regions of the world—all of which occurred simultaneously—reveals how little we know about these complex interrelationships.

Without significant, worldwide changes—including more innovation in and diffusion of water-related technologies; fundamental adjustments in consumption patterns; improvements in efficiencies; higher levels of public investment in water infrastructures; and an integrated approach to governance based on the complex relationships between water and food, water and economic development, and water and the environment—the global challenge of water resources could become even more severe.

Also, although global warming's potential effects on watersheds across the planet are still not precisely understood, there can be little doubt that climate change will in a number of regions generate serious dislocations in water supply. In a June 2008 technical paper, the Intergovernmental Panel on Climate Change (IPCC) concluded that "globally, the negative impacts of climate change on freshwater systems are expected to outweigh the benefits." It noted that "higher water temperatures and changes in extremes, including droughts and floods, are projected to affect water quality and exacerbate many forms of water pollution."

Climate change will in a number of regions generate serious dislocations in water supply.

As a result, we may soon be entering unknown territory when it comes to addressing the challenges of water in all their dimensions, including public health, economic development, gender equity, humanitarian crises, environmental degradation, and global security. The geopolitical consequences alone could be profound.

Daunting Trends

Although water covers almost three-quarters of the earth's surface, only a fraction of it is suitable for human consumption. According to the United Nations, of the water that humans consume, approximately 70 percent is used in agricultural production, 22 percent in industry, and 8 percent in domestic use. This consumption—critical as it is for human health, economic development, political and social stability, and security—is unequal, inefficient, and unsustainable.

Indeed, an estimated 884 million people worldwide do not have access to clean drinking water, and 2.5 billion lack adequate sanitation. A staggering 1.8 million people, 90 percent of them children, lose their lives each year as a result of diarrheal diseases resulting from unsafe drinking water and poor hygiene. More generally, the World Health Organization (WHO) estimates that inadequate water, sanitation, and hygiene are responsible for roughly half the malnutrition in the world.

In addition, we are witnessing irreparable damage to ecosystems across the globe. Aquifers are being drawn down faster than they can naturally be recharged. Some great lakes are mere fractions of what they once were.

And water pollution is affecting millions of people's lives. China typifies this problem. More than 75 percent of its urban river water is unsuitable for drinking or fishing, and 90 percent of its urban groundwater is contaminated. On the global scale, according to a recent UN report on world water development, every day we dump some 2 million tons of industrial waste and chemicals, human waste, and agricultural waste (fertilizers, pesticides, and pesticide residues) into our water supply.

Over the past century, as the world's population rose from 1.7 billion people in 1900 to 6.1 billion in 2000, global fresh water consumption increased six-fold—more than double the rate of population growth over the same period. The latest "medium" projections from the UN's population experts suggest that we are on the way to 8 billion people by the year 2025 and 9.15 billion by the middle of the century.

The contours of our predicament are clear-cut: A finite amount of water is available to a rapidly increasing number of people whose activities require more water than ever before. The UN Commission on Sustainable Development has indicated that we may need to double the amount of freshwater available today to meet demand at the middle of the century—after which time demand for water will increase by 50 percent with each additional generation.

Why is demand for water rising so rapidly? It goes beyond population pressures. According to a recent report from the UN Food and Agriculture Organization, the world will require 70 percent more food production over the next 40 years to meet growing per capita demand. This rising agricultural consumption necessarily translates into higher demand for water. By 2025, according to the water expert Sandra Postel, meeting projected global agricultural demand will require additional irrigation totaling some 2,000 cubic kilometers—roughly the equivalent of the annual flow of 24 Nile Rivers or 110 Colorado Rivers.

Consumption patterns aside, climate change will accelerate and intensify stress on water systems. According to the IPCC, in coming decades the frequency of extreme droughts will double while the average length of droughts will increase six times. This low water flow, combined with higher temperatures, not only will create devastating shortages. It will also increase pollution of fresh water by sediments, nutrients, pesticides, pathogens, and salts. On the other hand, in some regions, wet seasons will be more intense (but shorter).

In underdeveloped communities that lack capture and storage capacity, water will run off and will be unavailable when it is needed in dry seasons, thus perpetuating the cycle of poverty.

Climatic and demographic trends indicate that the regions of the world with the highest population growth rates are precisely those that are already the "driest" and that are expected to experience water stress in the future. The Organization for Economic Cooperation and Development has suggested that the number of people in water-stressed countries—where governments encounter serious constraints on their ability to meet household, industrial, and agricultural water demands—could rise to nearly 4 billion by the year 2030.

The Geopolitical Dimension

If oil is the key geopolitical resource of today, water will be as important—if not more so—in the not-so-distant future. A profound mismatch exists between the distribution of the human population and the availability of fresh water. At the water-rich extreme of the spectrum is the Amazon region, which has an estimated 15 percent of global runoff and less than 1 percent of the world's people. South America as a whole has only 6 percent of the world's population but more than a quarter of the world's runoff.

At the other end of the spectrum is Asia. Home to 60 percent of the global population, it has a freshwater endowment estimated at less than 36 percent of the world total. It is hardly surprising that some water-stressed countries in the region have pursued agricultural trade mechanisms to gain access to more water—in the form of food. Recently, this has taken the form of so-called "land grabs," in which governments and state companies have invested in farmland overseas to meet their countries' food security needs. *The Economist* has estimated that, to date, some 50 million acres have been remotely purchased or leased under these arrangements in Africa and Asia.

Although freshwater management has historically represented a means of preventing and mitigating conflict between countries with shared water resources, the growing scarcity of water will likely generate new levels of tension at the local, national, and even international levels. Many countries with limited water availability also depend on shared water, which increases the risk of friction, social tensions, and conflict.

The Euphrates, Jordan, and Nile Rivers are obvious examples of places where frictions already have occurred. But approximately 40 percent of the world's population lives in more than 260 international river basins of major social and economic importance, and 13 of these basins are shared by five or more countries. Interstate tensions have already escalated and could easily intensify as increasing water scarcity raises the stakes.

Within countries as well, governments in water-stressed regions must effectively and transparently mediate the concerns and demands of various constituencies. The interests of urban and rural populations, agriculture and industry, and commercial and domestic sectors often conflict. If allocation issues are handled inappropriately, subnational disputes and unrest linked to water scarcity and poor water quality could arise, as they already have in numerous cases.

Addressing the Challenge

Considering the scope and gravity of these water challenges, responses by governments and nongovernmental organizations have fallen short of what is needed. Despite obvious signs that we overuse water, we continue to perpetuate gross inefficiencies. We continue to skew consumption on the basis of politically charged subsidies or other supports. And we continue to pursue patently unsustainable practices whose costs will grow more onerous over time.

The Colorado River system, for example, is being overdrawn. It supplies water to Las Vegas, Los Angeles, San Diego, and other growing communities in the American Southwest. If demand on this river system is not curtailed, there is a 50 percent chance that Lake Mead will be dry by 2021, according to experts from the Scripps Institution of Oceanography.

Despite constant reminders of future challenges, we continue to be paralyzed by short-term thinking and practices. What is especially striking about water is the extent to which the world's nations are unprepared to manage such a vital resource sustainably. Six key opportunities for solutions stand out.

First, the global community needs to do substantially more to address the lack of safe drinking water and sanitation. Donor countries, by targeting water resources, can simultaneously address issues associated with health, poverty reduction, and environmental stewardship, as well as stability and security concerns. It should be stressed in this regard that rates of return on investment in water development—financial, political, and geopolitical—are all positive. The WHO estimates that the global return on every dollar invested in water and sanitation programs is $4 and $9, respectively.

Consider, for example, how water problems affect the earning power of women. Typically in poor countries, women and girls are kept at home to care for sick family members inflicted with water-related diseases. They also spend hours each day walking to collect water for daily drinking, cooking, and washing. According to the United Nations Children's Fund, water and sanitation issues explain why more than half the girls in sub-Saharan Africa drop out of primary school.

Second, more rigorous analyses of sustainability could help relevant governments and authorities begin to address the conspicuous mismanagement of water resources in regions across the world. This would include reviewing public subsidies—for water-intensive farming, for example—and other supports that tend to increase rather than remove existing inefficiencies.

Priced to Sell

Third, specialists, scholars, practitioners, and policy makers need to make substantial progress in assigning to water a market value against which more sustainable consumption decisions and policies can be made. According to the American Water Works Association, for example, the average price of water in the United States is $1.50 per 1,000 gallons—or less than a single penny per gallon. Yet, when it comes to the personal consumption market, many Americans do not hesitate to pay prices for bottled water that are higher than what they pay at the pump for a gallon of gasoline. What is clear, both inside and outside the United States, is that mechanisms for pricing water on the basis of sustainability have yet to be identified.

Fourth, rapid advances in technology can and should have a discernible effect on both the supply and demand sides of the global water equation. The technology landscape is breathtaking—from desalination, membrane, and water-reuse technologies to a range of cheaper and more efficient point-of-use applications (such as drip

irrigation and rainwater harvesting). It remains to be seen, however, whether the acquisition and use of such technologies can be accelerated and dispersed so that they can have an appreciable effect in offsetting aggregate downside trends.

From a public policy perspective, taxation and regulatory policies can create incentives for the development and dissemination of such technologies, and foreign assistance projects can promote their use in developing countries. Also, stronger links with the private sector would help policy makers improve their understanding of technical possibilities, and public-private partnerships can be effective mechanisms for distributing technologies in the field.

Fifth, although our understanding of the relationship between climate change and water will continue to be shaped by new evidence, it is important that we incorporate into our approach to climate change our existing understanding of water management and climate adaptation issues.

Sixth, the complex links among water, agriculture, and energy must be identified with greater precision. An enormous amount of work remains to be done if we are to appreciate these linkages in the global, basin, and local contexts.

In the final analysis, our capacity to address the constellation of challenges that relate to water access, sanitation, ecosystems, infrastructure, adoption of technologies, and the mobilization of resources will mean the difference between rapid economic development and continued poverty, between healthier populations and continued high exposure to water-related diseases, between a more stable world and intensifying geopolitical tensions.

Critical Thinking

1. In what way will the geopolitical patterns of water resources and consumption be different than the geopolitical patterns of oil resources and consumption?

2. Why does "an obscenely large portion of the world's population lack regular access to fresh water and sanitization"?

3. Refer to your map, and map out the "profound mismatch between the distribution of the human population and the availability of freshwater." Make some critical observations regarding that pattern.

4. How might climate change make the world's future water challenge even more challenging?

5. How might the water challenges of the industrialized world be different than the challenges facing "developing" nations?

6. Identify in this article three key terms, concepts, or principles that are used in your textbook (environmental science, economics, sociology, history, geography, etc.) or employed in the discipline you are currently studying. (Note: The terms, concepts, or principles may be implicit, explicit, implied, or inferred.)

ERIK R. PETERSON is senior vice president of the Center for Strategic and International Studies and director of its Global Strategy Institute. **RACHEL A. POSNER** is assistant director of the CSIS Global Water Futures project.

From *Current History*, January 2010, pp. 31–34. Copyright © 2010 by Current History, Inc. Reprinted by permission.

UNIT 5
Changing Climate

Unit Selections

Learning Outcomes

After reading this Unit, you should be able to:

- Offer insight on the complex linkages that exist between environmental issue #3: climate change, and social, economic, political, and global security issues.
- Outline what are the most common misconceptions regarding climate change.
- Explain what evidence is given to argue that reversing climate change may be mostly an issue of "political will".
- Describe how climate change will affect different world regions differently.
- Discuss the connections between "failed states," geopolitical stability, and climate change.
- Explain how the health of glaciers can affect international conflicts.
- Describe three straightforward ways to reduce carbon emissions.
- Define what it means to "focus on the big wins" with regard to resolving the challenges of climate change.
- Discuss the societal effects of climate change.
- Identify the four key effects climate change may have on the "global resource equation".
- Assess what role variable global patterns of natural resource consumption, resource access, and levels of industrialization may play with regard to global patterns of "security".

Student Website

www.mhhe.com/cls

Internet References

The Climate Project
www.theclimateprojectus.org
Global Climate Change
www.puc.state.oh.us/consumer/gcc/index.html
Global Climate Change Facts: The Truth, The Consensus, and the Skeptics
www.climatechangefacts.info
NASA Global Change site
www.climate.nasa.gov
National Oceanic and Atmospheric Administration (NOAA)
www.noaa.gov
National Operational Hydrologic Remote Sensing Center (NOHRSC)
www.nohrsc.nws.gov
National Security and the Threat of Climate Change
www.npr.org/documents/2007/apr/security_climate.pdf
RealClimate
www.realclimate.org

The system of nature of which man is a part, tends to be self-balancing, self-adjusting, self-cleansing. Not so with technology.

—*E. F. Schumacher,* Small is Beautiful, 1973

For a long time *climate change* was referred to as "global warming," the phrase propagated to a large extent by the media. Fortunately, brighter minds have prevailed, and this particular atmospheric phenomenon of "greenhouse warming" is now addressed as climate change. A changing climate that is evidencing itself not just in higher average temperatures, but in drought, flooding, storms, wind patterns, glacial melting, sea levels rising, crop failures, ecosystem stresses, habit degradation, biodiversity losses, and more. What also has changed is the narrow, sometimes almost myopic, focus on what is causing greenhouse warming and who is to blame. The consensus among nearly all scientists is that climate change (planetary warming) is real, excess carbon emissions (mostly anthropogenic) are the cause, and we can reverse the trend. Some scientists believe we have already crossed a critical threshold with atmospheric degradation, and although we can reverse this event in the long run, they argue there may already be set in motion inevitable environmental consequences. Nevertheless, what is not clear is a precise picture of what will be the *patterns* and *places* of climate change impacts should we not react immediately. Also what is not clear is an accurate map of the complex linkages that exist between climate change and global economics, geopolitics, resource competition, and global security. But in **Unit 5,** we see that what is clear is that the environmental, social, political, economic, agricultural, and biotic impacts will vary across the globe affecting (perhaps catastrophically) different peoples to different degrees. And herein lies the real challenge: predicting, managing, mitigating, and resolving the *where* and *how much* of climate change consequences. Like all environmental issues, the complex linkages involved with climate change will require critical thinking investments from all disciplines and from multiple angles of analysis.

What also makes the climate issue so difficult to contend with is that it is essentially an issue of culture. Why are we in this fix of "climate change"? Because humans consume the Earth. But different cultures, different societies consume the Earth differently. And some cultures benefit more from certain resource consumption patterns, whereas others benefit less, sometimes far less. In fact, some cultures actually suffer socially, economically, and environmentally, as the result of other culture's resource consumption patterns. Articles 17 and 19 suggest that this may very much be the case.

In Article 17, "The Last Straw," the author points out that one in four countries, which include some of the world's most politically unstable and volatile, are at high risk of climate-induced conflict. Article 19 should encourage us to consider further questions. What is or are the origins of this "induced" risk of conflict? Should more pressure be put on the "who" is melting the glaciers on which places like Pakistan and India depend, and less on the "why"? Are "industrialized" cultures responsible? Or will rural, agrarian, emerging economy cultures seeking the

© Design Pics/Carson Ganci

"raising standards of living" development approach discussed in Unit 2 become equally responsible?

Michael Klare asks in Article 19, "What are the societal effects of climate change?" and argues that they will not be limited to humanitarian disasters, but will include ethnic conflicts, insurgencies, and civil violence resulting from increasing resource scarcities and patterns of access. In "Global Warming Battlefields: How Climate Change Threatens Security," the author suggests that that climate change will affect the global resource equation in many ways and can be grouped into four key effects: diminishing rainfall, diminished river flow, rising sea levels, and more frequent and severe storm events. However, more critical than the *effects* will be (1) the *patterns of the event impacts*—the places and peoples who will be affected directly, and (2) which nations will be able to *maintain critical access* to necessary resources or leverage for those they need.

The other two articles in this unit, #16 and #18, focus less on the potential global consequences of climate change and more on actions we can take right now to possibly avoid such dire scenarios. In Article 16, "Climate Change," Bill McKibben believes the only real obstacle to averting a climate catastrophe is simply political will; we know what the job at hand is, and it just needs to get done. What is the alternative to not doing it? But, fixing the problem will require both *cooperation* and *sacrifice* by everyone. And therein may lie the greatest global human challenge to have ever confronted the human race.

Despite such a challenge, like McKibben, author Mark Lynas feels guardedly optimistic in Article 18, "How to Stop Climate Change: The Easy Way." The author argues that first and foremost, we need to stop debating about climate change—it is real. Second, the author believes we need to approach the job at hand with a "win–win" mentality. Who doesn't want to win? According to the author, putting money in the right place can go a long way to creating "win–win" situations. Want to save our rain forests? Then pay the countries who are cutting them down not to cut them down. Make it profitable for them to keep their forests standing. Win–win. Common sense and simple logic can also make for win–win scenarios. Encourage walking, it's good for your health. Enforce speed limits, it saves lives. Encourage quality-of-life

development choices, they require less fossil fuel than amassing material possessions. Localize our lives. It reduces carbon emissions and reconnects us to place and community. Win–win. And third, Lynas says use technology. We do not need new inventions, just a ramping up of the technologies we already know work. But we need to do what we need to do, now.

The reader is particularly encouraged to consider the issues brought forward in this unit and consider them in light of Units 1 to 4. For instance, how will population growth, demographic changes, and consumer megatrends not only be affected by climate change, but act as active agents in contributing to it? What will be the impacts on climate change of raising standards of living versus improving quality of life development choices? What agricultural production approaches should we institute now that will have the best chance of resiliency against climate change? Considering population growth, development, food and freshwater, are there nations or cultures who must take more responsibility for climate change mitigation? Although there are no simple answers, we must come up with answers, nevertheless.

Climate Change

BILL MCKIBBEN

Scientists Are Divided

No, they're not. In the early years of the global warming debate, there was great controversy over whether the planet was warming, whether humans were the cause, and whether it would be a significant problem. That debate is long since over. Although the details of future forecasts remain unclear, there's no serious question about the general shape of what's to come.

Every national academy of science, long lists of Nobel laureates, and in recent years even the science advisors of President George W. Bush have agreed that we are heating the planet. Indeed, there is a more thorough scientific process here than on almost any other issue: Two decades ago, the United Nations formed the Intergovernmental Panel on Climate Change (IPCC) and charged its scientists with synthesizing the peer-reviewed science and developing broad-based conclusions. The reports have found since 1995 that warming is dangerous and caused by humans. The panel's most recent report, in November 2007, found it is "very likely" (defined as more than 90 percent certain, or about as certain as science gets) that heat-trapping emissions from human activities have caused "most of the observed increase in global average temperatures since the mid-20th century."

If anything, many scientists now think that the IPCC has been too conservative—both because member countries must sign off on the conclusions and because there's a time lag. Its last report synthesized data from the early part of the decade, not the latest scary results, such as what we're now seeing in the Arctic.

In the summer of 2007, ice in the Arctic Ocean melted. It melts a little every summer, of course, but this time was different—by late September, there was 25 percent less ice than ever measured before. And it wasn't a one-time accident. By the end of the summer season in 2008, so much ice had melted that both the Northwest and Northeast passages were open. In other words, you could circumnavigate the Arctic on open water. The computer models, which are just a few years old, said this shouldn't have happened until sometime late in the 21st century. Even skeptics can't dispute such alarming events.

We Have Time

Wrong. Time might be the toughest part of the equation. That melting Arctic ice is unsettling not only because it proves the planet is warming rapidly, but also because it will help speed up the warming. That old white ice reflected 80 percent of incoming solar radiation back to space; the new blue water left behind absorbs 80 percent of that sunshine. The process amps up. And there are many other such feedback loops. Another occurs as northern permafrost thaws. Huge amounts of methane long trapped below the ice begin to escape into the atmosphere; methane is an even more potent greenhouse gas than carbon dioxide.

Such examples are the biggest reason why many experts are now fast-forwarding their estimates of how quickly we must shift away from fossil fuel. Indian economist Rajendra Pachauri, who accepted the 2007 Nobel Peace Prize alongside Al Gore on behalf of the IPCC, said recently that we must begin to make fundamental reforms by 2012 or watch the climate system spin out of control; NASA scientist James Hansen, who was the first to blow the whistle on climate change in the late 1980s, has said that we must stop burning coal by 2030. Period.

All of which makes the Copenhagen climate change talks that are set to take place in December 2009 more urgent than they appeared a few years ago. At issue is a seemingly small number: the level of carbon dioxide in the air. Hansen argues that 350 parts per million is the highest level we can maintain "if humanity wishes to preserve a planet similar to that on which civilization developed and to which life on Earth is adapted." But because we're already past that mark—the air outside is currently about 387 parts per million and growing by about 2 parts annually—global warming suddenly feels less like a huge problem, and more like an Oh-My-God Emergency.

Climate Change Will Help as Many Places as It Hurts

Wishful thinking. For a long time, the winners-and-losers calculus was pretty standard: Though climate change will cause some parts of the planet to flood or shrivel up, other frigid, rainy regions would at least get some warmer days every year. Or so the thinking went. But more recently, models have begun to show that after a certain point almost everyone on the planet will suffer. Crops might be easier to grow in some places for a few decades as the danger of frost recedes, but over time the threat of heat stress and drought will almost certainly be stronger.

A 2003 report commissioned by the Pentagon forecasts the possibility of violent storms across Europe, megadroughts across the Southwest United States and Mexico, and unpredictable monsoons causing food shortages in China. "Envision Pakistan, India, and China—all armed with nuclear weapons—skirmishing at their borders over refugees, access to shared rivers, and arable land," the report warned. Or Spain and Portugal "fighting over fishing rights—leading to conflicts at sea."

Of course, there are a few places we used to think of as possible winners—mostly the far north, where Canada and Russia could theoretically produce more grain with longer growing seasons, or perhaps explore for oil beneath the newly melted Arctic ice cap. But even those places will have to deal with expensive consequences—a real military race across the high Arctic, for instance.

Want more bad news? Here's how that Pentagon report's scenario played out: As the planet's carrying capacity shrinks, an ancient pattern of desperate, all-out wars over food, water, and energy supplies would reemerge. The report refers to the work of Harvard archaeologist Steven LeBlanc, who notes that wars over resources were the norm until about three centuries ago. When such conflicts broke out, 25 percent of a population's adult males usually died. As abrupt climate change hits home, warfare may again come to define human life. Set against that bleak backdrop, the potential upside of a few longer growing seasons in Vladivostok doesn't seem like an even trade.

It's China's Fault

Not so much. China is an easy target to blame for the climate crisis. In the midst of its industrial revolution, China has overtaken the United States as the world's biggest carbon dioxide producer. And everyone has read about the one-a-week pace of power plant construction there. But those numbers are misleading, and not just because a lot of that carbon dioxide was emitted to build products for the West to consume. Rather, it's because China has four times the population of the United States, and per capita is really the only way to think about these emissions. And by that standard, each Chinese person now emits just over a quarter of the carbon dioxide that each American does. Not only that, but carbon dioxide lives in the atmosphere for more than a century. China has been at it in a big way less than 20 years, so it will be many, many years before the Chinese are as responsible for global warming as Americans.

What's more, unlike many of their counterparts in the United States, Chinese officials have begun a concerted effort to reduce emissions in the midst of their country's staggering growth. China now leads the world in the deployment of renewable energy, and there's barely a car made in the United States that can meet China's much tougher fuel-economy standards.

For its part, the United States must develop a plan to cut emissions—something that has eluded Americans for the entire two-decade history of the problem. Although the U.S. Senate voted down the last such attempt, Barack Obama has promised that it will be a priority in his administration. He favors some variation of a "cap and trade" plan that would limit the total amount of carbon dioxide the United States could release, thus putting a price on what has until now been free.

Despite the rapid industrialization of countries such as China and India, and the careless neglect of rich ones such as the United States, climate change is neither any one country's fault, nor any one country's responsibility. It will require sacrifice from everyone. Just as the Chinese might have to use somewhat more expensive power to protect the global environment, Americans will have to pay some of the difference in price, even if just in technology. Call it a Marshall Plan for the environment. Such a plan makes eminent moral and practical sense and could probably be structured so as to bolster emerging green energy industries in the West. But asking Americans to pay to put up windmills in China will be a hard political sell in a country that already thinks China is prospering at its expense. It could be the biggest test of the country's political maturity in many years.

Climate Change Is an Environmental Problem

Not really. Environmentalists were the first to sound the alarm. But carbon dioxide is not like traditional pollution. There's no Clean Air Act that can solve it. We must make a fundamental transformation in the most important part of our economies, shifting away from fossil fuels and on to something else. That means, for the United States, it's at least as much a problem for the Commerce and Treasury departments as it is for the Environmental Protection Agency.

And because every country on Earth will have to coordinate, it's far and away the biggest foreign-policy issue we face. (You were thinking terrorism? It's hard to figure out a scenario in which Osama bin Laden destroys Western civilization. It's easy to figure out how it happens with a rising sea level and a wrecked hydrological cycle.)

Expecting the environmental movement to lead this fight is like asking the USDA to wage the war in Iraq. It's not equipped for this kind of battle. It may be ready to save Alaska's Arctic National Wildlife Refuge, which is a noble undertaking but on a far smaller scale. Unless climate change is quickly deghettoized, the chances of making a real difference are small.

Solving It Will Be Painful

It depends. What's your definition of painful? On the one hand, you're talking about transforming the backbone of the world's industrial and consumer system. That's certainly expensive. On the other hand, say you manage to convert a lot of it to solar or wind power—think of the money you'd save on fuel.

And then there's the growing realization that we don't have many other possible sources for the economic growth we'll need to pull ourselves out of our current economic crisis. Luckily, green energy should be bigger than IT and biotech combined.

Almost from the moment scientists began studying the problem of climate change, people have been trying to estimate the costs of solving it. The real answer, though, is that it's such a huge transformation that no one really knows for sure. The bottom line is, the growth rate in energy use worldwide could be cut in half during the next 15 years and the steps would, net, save more money than they cost. The IPCC included a cost estimate in its latest five-year update on climate change and looked a little further into the future. It found that an attempt to keep carbon levels below about 500 parts per million would shave a little bit off the world's economic growth—but only a little. As in, the world would have to wait until Thanksgiving 2030 to be as rich as it would have been on January 1 of that year. And in return, it would have a much-transformed energy system.

Unfortunately though, those estimates are probably too optimistic. For one thing, in the years since they were published, the science has grown darker. Deeper and quicker cuts now seem mandatory.

But so far we've just been counting the costs of fixing the system. What about the cost of doing nothing? Nicholas Stern, a renowned economist commissioned by the British government to study the question, concluded that the costs of climate change could eventually reach the combined costs of both world wars and the Great Depression. In 2003, Swiss Re, the world's biggest reinsurance company, and Harvard Medical School explained why global warming would be so expensive. It's not just the infrastructure, such as sea walls against rising oceans, for example. It's also that the increased costs of natural disasters begin to compound. The diminishing time between monster storms in places such as the U.S. Gulf Coast could eventually mean that parts of "developed countries would experience developing nation conditions for prolonged periods." Quite simply, we've already done too much damage and waited too long to have any easy options left.

We Can Reverse Climate Change

If only. Solving this crisis is no longer an option. Human beings have already raised the temperature of the planet about a degree Fahrenheit. When people first began to focus on global warming (which is, remember, only 20 years ago), the general consensus was that at this point we'd just be standing on the threshold of realizing its consequences—that the big changes would be a degree or two and hence several decades down the road. But scientists seem to have systematically underestimated just how delicate the balance of the planet's physical systems really is.

The warming is happening faster than we expected, and the results are more widespread and more disturbing. Even that rise of 1 degree has seriously perturbed hydrological cycles: Because warm air holds more water vapor than cold air does, both droughts and floods are increasing dramatically. Just look at the record levels of insurance payouts, for instance. Mosquitoes, able to survive in new places, are spreading more malaria and dengue. Coral reefs are dying, and so are vast stretches of forest.

None of that is going to stop, even if we do everything right from here on out. Given the time lag between when we emit carbon and when the air heats up, we're already guaranteed at least another degree of warming.

The only question now is whether we're going to hold off catastrophe. It won't be easy, because the scientific consensus calls for roughly 5 degrees more warming this century unless we do just about everything right. And if our behavior up until now is any indication, we won't.

Critical Thinking

1. What arguments does the author use to dispel the certainties that "climate change will help as many as hurt"?

2. What other population–environment insights can you add to support his argument?

3. The author suggests that climate change is "not really" an environmental problem, but rather a fossil-fuel economy problem. What role does "consumption" play in this fossil fuel–climate change relationship?

4. Refer to the map provided in this text, and locate the nations currently benefitting most from fossil fuels, and which nations' environments suffer most from fossil fuel use today. In 10 years? In 50 years?

5. Why don't all world's nations just "chip in" and resolve climate change issue?

6. Identify in this article three key terms, concepts, or principles that are used in your textbook (environmental science, economics, sociology, history, geography, etc.) or employed in the discipline you are currently studying. (Note: The terms, concepts, or principles may be implicit, explicit, implied, or inferred.)

The Last Straw

If you think these failed states look bad now, wait until the climate changes.

Stephan Faris

Hopelessly overcrowded, crippled by poverty, teeming with Islamist militancy, careless with its nukes—it sometimes seems as if Pakistan can't get any more terrifying. But forget about the Taliban: The country's troubles today pale compared with what it might face 25 years from now. When it comes to the stability of one of the world's most volatile regions, it's the fate of the Himalayan glaciers that should be keeping us awake at night.

> **When it comes to the stability of one of the world's most volatile regions, it's the fate of the Himalayan glaciers that should be keeping us awake at night.**

In the mountainous area of Kashmir along and around Pakistan's contested border with India lies what might become the epicenter of the problem. Since the separation of the two countries 62 years ago, the argument over whether Kashmir belongs to Muslim Pakistan or secular India has never ceased. Since 1998, when both countries tested nuclear weapons, the conflict has taken on the added risk of escalating into cataclysm. Another increasingly important factor will soon heighten the tension: Ninety percent of Pakistan's agricultural irrigation depends on rivers that originate in Kashmir. "This water issue between India and Pakistan is the key," Mohammad Yusuf Tarigami, a parliamentarian from Kashmir, told me. "Much more than any other political or religious concern."

Until now, the two sides had been able to relegate the water issue to the back burner. In 1960, India and Pakistan agreed to divide the six tributaries that form the Indus River. India claimed the three eastern branches, which flow through Punjab. The water in the other three, which pass through Jammu and Kashmir, became Pakistan's. The countries set a cap on how much land Kashmir could irrigate and agreed to strict regulations on how and where water could be stored. The resulting Indus Waters Treaty has survived three wars and nearly 50 years. It's often cited as an example of how resource scarcity can lead to cooperation rather than conflict.

But the treaty's success depends on the maintenance of a status quo that will be disrupted as the world warms. Traditionally, Kashmir's waters have been naturally regulated by the glaciers in the Himalayas. Precipitation freezes during the coldest months and then melts during the agricultural season. But if global warming continues at its current rate, the Intergovernmental Panel on Climate Change estimates, the glaciers could be mostly gone from the mountains by 2035. Water that once flowed for the planting will flush away in winter floods.

Research by the global NGO ActionAid has found that the effects are already starting to be felt within Kashmir. In the valley, snow rarely falls and almost never sticks. The summertime levels of streams, rivers, springs, and ponds have dropped. In February 2007, melting snow combined with unseasonably heavy rainfall to undermine the mountain slopes; landslides buried the national highway—the region's only land connection with the rest of India—for 12 days.

Normally, countries control such cyclical water flows with dams, as the United States does with runoff from the Rocky Mountains. For Pakistan, however, that solution is not an option. The best damming sites are in Kashmir, where the Islamabad government has vigorously opposed Indian efforts to tinker with the rivers. The worry is that in times of conflict, India's leaders could cut back on water supplies or unleash a torrent into the country's fields. "In a warlike situation, India could use the project like a bomb," one Kashmiri journalist told me.

Water is already undermining Pakistan's stability. In recent years, recurring shortages have led to grain shortfalls. In 2008, flour became so scarce it turned into an election issue; the government deployed thousands of troops to guard its wheat stores. As the glaciers melt and the rivers dry, this issue will only become more critical. Pakistan—unstable, facing dramatic drops in water supplies, caged in by India's vastly superior conventional forces—will be forced to make one of three choices. It can let its people starve. It can cooperate with India in building dams and reservoirs, handing over control of its waters to the country it regards as the enemy. Or it can ramp up support for the insurgency, gambling that violence can bleed India's resolve without degenerating into full-fledged war. "The idea of ceding territory to India is anathema," says Sumit Ganguly, a professor of political science at Indiana University. "Suffering, particularly for the elite, is unacceptable. So what's the other option? Escalate."

"It's very bad news," he adds, referring to the melting glaciers. "It's extremely grim."

The Kashmiri water conflict is just one of many climate-driven geopolitical crises on the horizon. These range from possible economic and treaty conflicts that will likely be resolved peacefully—the waters of the Rio Grande and Colorado River have long been a point of contention between the United States and Mexico, for instance—to possible outright wars. In 2007, the London-based NGO International Alert compiled a list of countries with a high risk of armed conflict due to climate change. They cited no fewer than 46 countries, or one in every four, including some of the world's most gravely unstable countries, such as

Life in a Failed State

Pakistan

"I remember being in high school when the Taliban took over in Afghanistan. I remember thinking to myself, 'Oh God, it's so close. Could that ever happen here?' and thinking that it wasn't possible. And now, 10 years later, I see the same thing happening in my own country."

— Fatima Bhutto Pakistani commentator

Somalia, Nigeria, Iran, Colombia, Bolivia, Israel, Indonesia, Bosnia, Algeria, and Peru. Already, climate change might be behind the deep drought that contributed to the conflict in the Darfur region of Sudan and hundreds of thousands of deaths.

Rising global temperatures are putting the whole world under stress, and the first countries to succumb will be those, such as Sudan, that are least able to adapt. Compare the Netherlands and Bangladesh: Both are vulnerable to rises in sea levels, with large parts of their territory near or under the level of the waves. But the wealthy Dutch are building state-of-the-art flood-control systems and experimenting with floating houses. All the impoverished Bangladeshis can do is prepare to head for higher ground. "It's best not to get too bogged down in the physics of climate," says Nils Gilman, an analyst at Monitor Group and the author of a 2006 report on climate change and national security. "Rather, you should look at the social, physical, and political geography of regions that are impacted."

Indeed, with a population half that of the United States crammed into an area a little smaller than Louisiana, Bangladesh might be among the most imperiled countries on Earth. In a normal decade, the country experiences one major flood. In the last 11 years, its rivers have leapt their banks three times, most recently in 2007. That winter, Cyclone Sidr, a Category 5 storm, tore into the country's coast, flattening tin shacks, ripping through paddies, and plunging the capital into darkness. As many as 10,000 people may have died.

Bangladesh's troubles are likely to ripple across the region, where immigration flows have been historically accompanied by rising tensions. In India's northeastern state of Assam, for instance, rapidly changing demographics have led to riots, massacres, and the rise of an insurgency. As global warming tightens its squeeze on Bangladesh, these pressures will mount. And in a worst-case scenario, in which the country is struck by sudden, cataclysmic flooding, the international community will have to cope with a humanitarian emergency in which tens of millions of waterlogged refugees suddenly flee toward India, Burma, China, and Pakistan.

Indeed, the U.S. military has come to recognize that weakened states—the Bangladeshes and Pakistans of the world—are often breeding grounds for extremism, terrorism, and potentially destabilizing conflict. And as it has done so, it has increasingly deployed in response to natural disasters. Such missions often require a warlike scale of forces, if not warlike duration. During the 2004 Indian Ocean tsunami, for instance, the United States sent 15,000 military personnel, 25 ships, and 94 aircraft. "The military brings a tremendous capacity of command-and-control and communications," says retired Gen. Anthony Zinni, the former head of U.S. Central Command. "You

have tremendous logistics capability, transportation, engineering, the ability to purify water."

As the world warms, more years could start to look like 2007, when the U.N. Office for the Coordination of Humanitarian Affairs announced it had responded to a record number of droughts, floods, and storms. Of the 13 natural disasters it responded to, only one—an earthquake in Peru—was not related to the climate.

Worryingly, some analysts have suggested the United States might not fully grasp what it needs to respond to this challenge. The U.S. military has been required by law since 2008 to incorporate climate change into its planning, but though Pentagon strategic documents describe a climate-stressed future, there's little sign the Department of Defense is pivoting to meet it. "Most of the things that the military is requesting are still for a conventional war with a peer competitor," says Sharon Burke, an energy and climate change specialist at the Washington-based Center for a New American Security. "They say they're going to have more humanitarian missions, but there's no discussion at all of 'What do you need?'" The rate at which the war in Iraq has chewed through vehicles and equipment, for instance, has astonished military planners. "Is this a forewarning of what it's like to operate in harsher conditions?" Burke asks.

To be sure, some of the more severe consequences of climate change are expected to unfold over a relatively extended time frame. But so does military development, procurement, and planning. As global warming churns the world's weather, it's becoming increasingly clear that it's time to start thinking about the long term. In doing so, the West may need to adopt an even broader definition of what it takes to protect itself from danger. Dealing with the repercussions of its emissions might mean buttressing governments, deploying into disaster zones, or tamping down insurgencies. But the bulk of the West's effort might be better spent at home. If the rivers of Kashmir have the potential to plunge South Asia into chaos, the most effective response might be to do our best to ensure the glaciers never melt at all.

Critical Thinking

1. Describe what a "failed state" is and locate examples of them on the map provided in this text. List some of your observations regarding the geographic pattern of these states.

2. Make three lists: environmental characteristics, population characteristics, and economic characteristics of the "failed state" places referred to in the article. Now look at the lists side by side, and write down your critical observations.

3. Based on your textbooks and any number of articles in this reader, make some predictions of other countries you think may be "at risk for climate change induced conflict" in the next 20 years.

4. What do you think needs to be done to prevent such potential conflicts?

5. Identify in this article three key terms, concepts, or principles that are used in your textbook (environmental science, economics, sociology, history, geography, etc.) or employed in the discipline you are currently studying. (Note: The terms, concepts, or principles may be implicit, explicit, implied, or inferred.)

STEPHAN FARIS is the author of *Forecast: The Consequences of Climate Change, from the Amazon to the Arctic, from Darfur to Napa Valley,* from which reporting for this article is drawn.

How to Stop Climate Change: The Easy Way

Changing your light bulbs may not be enough to save a single polar bear, but there *are* things we can do collectively—and easily—that will really make a measurable difference in the battle against global warming.

MARK LYNAS

We have about 100 months left. If global greenhouse gas emissions have not begun to decline by the end of 2015, then our chances of restraining climate change to within the two degrees "safety line"—the level of warming below which the impacts are severe but tolerable—diminish day by day thereafter. This is what the latest science now demands: the peaking of emissions within eight years, worldwide cuts of 60 percent by 2030, and 80 percent or more by 2050. Above two degrees, our chances of crossing "tipping points" in the earth's system—such as the collapse of the Amazon rainforest, or the release of methane from thawing Siberian permafrost—is much higher.

Despite this urgent timetable, our roads continue to heave with traffic. Power companies draft blueprints for new coal-fired plants. The skies over England are criss-crossed with vapour trails from aircraft travelling some of the busiest routes in the world. Global emissions, far from decreasing, remain on a steep upward curve of almost exponential growth.

Sure, there are some encouraging signs. Media coverage of climate change remains high, and a worldwide popular movement—now perhaps upwards of a million people—is mobilising. But with so little time left, we must recognise that most people won't do anything to save the planet unless we make it much, much easier for them. This essay outlines my three-part strategy for stopping climate change—the easy way.

Step One: Stop Debating, Start Doing

Although there is now a very broad consensus on climate in the media and politics, opinion polls show that many people still harbour doubts about climate change. One of the peculiarities of the climate debate is that although more than 99 percent of international climate change scientists agree on the causes of global warming, the denial lobby still only has to produce one contrarian to undermine the consensus in the public mind. Similarly, changes in our understanding can be magnified and distorted to suggest that, because we don't know everything, therefore we must know nothing. Thus, data from one glacier that apparently bucks the global trend can be wielded as a trump card against all the accumulated knowledge of climate science.

This partly reflects a perhaps healthy scepticism in the public mind about believing "experts". But there is also a darker force at work: doubt undermines responsibility for action. If you don't know for sure that global warming isn't caused by sunspots or cosmic rays, then it's OK to go on driving and flying without feeling as if you're doing something bad. When it comes to global warming, many people—subconsciously at least—actually *want* to be lied to.

This is where the psychology gets interesting. Most green campaigners assume that information leads to action, and that deeper knowledge will undermine denial. Actually, the reverse may well be true: the more disempowered that people feel about a huge, scary issue like climate change, the more unwilling they may be to believe it is a problem. This sounds illogical, but it makes sense. If people don't feel they can do very much about climate change, they will prefer to cling to any tempting doubts that are dangled their way. Presenting people with more gloom-and-doom scenarios, however true they might be, may thus serve to reinforce denial.

Most campaigners try to mitigate this by also offering people easy things they can do: the "just change your light bulbs" approach. However, most people intuitively understand that an enormous problem cannot be solved by a tiny solution; that changing your light bulbs will not save a single polar bear. They are right, of course. So how can we mobilise collective action on a sufficiently grand scale to make a measurable contribution to solving the problem?

The American political strategists Ted Nordhaus and Michael Shellenberger make a specific proposal in a recent paper, and this forms the first plank of my three-part strategy to tackle global warming. Stop debating, they say, and start doing. Instead of confronting deeply established patterns of behaviour head on, let's start focusing on preparing for the impacts of global warming that are already inevitable. That means working on flood defences for vulnerable towns, helping to drought-proof agriculture and population centres, and adapting to sea-level rise in low-lying areas.

By sidestepping the tedious causality argument (is it us or natural cycles?), focusing on global warming preparedness can also help reopen the mitigation agenda. Shifting sandbags is empowering because you feel as if you're doing something tangible and useful. But accepting the need for adaptation and preparation implicitly involves accepting the reality of global warming, and therefore the eventual need to cut emissions. Many more people may be prepared to accept the change—the introduction of personal carbon allowances, for example—that this will inevitably mean.

In any case, adaptation is now essential because of the one degree or so of additional global warming that is already locked into the system thanks to past emissions. With proper planning, we can not only save thousands of human lives, but also try to protect natural ecosystems by establishing new "refuge" coral reefs in cooler waters or helping species to migrate as temperature zones shift.

Step Two: Focus on the Big Wins

But this is a long-term agenda, and we don't have much time. Hence my second proposal, which is for a much clearer focus on win–win strategies for immediate emissions reductions. These are things we would want to be doing anyway, even if global warming had never been thought of. Reducing deforestation in the tropics is a big win–win. Inherently desirable, this by itself would reduce global carbon emissions by 10 percent or more. All it takes is money: we have to pay countries such as Brazil and Indonesia to leave their forests alone rather than chop them down to sell to us as plywood and furniture.

There are obvious win–win strategies in the domestic sector. Better insulation makes living conditions more comfortable and reduces fuel bills. Even without climate change we'd still want to be getting cars out of town centres to reduce air pollution and improve the urban experience. Getting more children to walk and cycle to school improves their physical health and helps to tackle obesity. Enforcing speed limits (and reducing them further) would save hundreds of lives a year, and give some respite from the incessant noise pollution of speeding traffic.

Quality-of-life issues are by their nature subjective, so we need to focus on things that most people will agree on. Partly, this depends on how an issue is framed: most people don't want motorists to be unjustifiably hounded, but nor are they likely to oppose a measure that is about saving children's lives. The ban on smoking in public, for instance, was accepted precisely because the issue was correctly framed, and quickly became imbued with a sense of inevitability.

There is also a high degree of consensus about the desirability of localisation: protecting and encouraging small shops and local businesses, privileging farmers' markets over supermarkets, helping build stronger and more cohesive communities by reducing the need for travel, and so on. The fact that all of these measures will also reduce carbon emissions simply underlines the need for a more determined approach to their implementation. A much longer-term agenda here might be the reconnecting of people with their place and surroundings, helping them feel more rooted in their communities and proud of what is distinctive about their own areas. We are bringing up children who often have no direct experience of nature any more. Tree houses are replaced with Nintendos, the unsupervised exercise of playing outdoors replaced with structured exercise of sporting events. The author Richard Louv terms this "nature deficit disorder" and asks whether this disconnection might have something to do with the alienation and boredom that many youngsters feel today.

Step Three: Use Technology

But there are some areas of high-carbon behaviour that people will always be reluctant to give up, and this brings me to the third and final part of my strategy to deal with global warming—technology.

Today we face a situation where a global population of potentially nine billion or so by 2050 continues to demand a steadily increasing consumer lifestyle. There is nothing we can do to stop this, and nor should we try. But it does put humanity on a very real collision course with the planet, so we are going to have to throw every technological tool we have at the problem to try to meet people's aspirations without worsening our climatic predicament. Some of this will involve technology leapfrogging: helping developing countries skip over our dirty phase of industrialisation, by installing solar power in remote, off-grid areas of Africa and Asia, for example. We also need to help developing countries make choices that put fossil fuels at the bottom of the energy shopping list, by helping them use carbon capture and storage technology as well as nuclear power. Both have obvious drawbacks, but I would rather see China building two nuclear reactors a week than two coal-fired plants.

The localisation agenda can only go so far: in an age of carbon-fuelled globalisation, we need to figure out ways to transport people and goods long distances without increasing emissions. Aviation in particular is crying out for a techno-fix. Humanity went from the first manned flight in 1903 to putting a man on the moon in 1969. I think we should give the aviation industry 15 years to find a low-carbon way to shuttle people between continents—or get taxed out of existence. I believe with this kind of incentive, designers would come up with ideas none of us today could even conceive of.

The technological challenge is not just to come up with new inventions, but—in the words of Robert Socolow and Stephen Pacala from Princeton University—"to scale up what we already know how to do". In their concept of "stabilisation wedges," each wedge represents a billion tonnes of carbon shaved off the upward trend of emissions over the next 50 years. Building two million one-megawatt wind turbines, for example, is a wedge, as are two million hectares of solar panels, a 700-fold increase from today's deployment. There are many more wedges in the fields of transport, power generation and energy efficiency. As the two researchers say, this reduces a "heroic challenge" merely to a set of "monumental tasks". No one said it would be easy.

Perhaps the most controversial technological option of all is one that we need to keep strictly in reserve for real emergencies—geo-engineering. Here, some proposals have more merit than others, whether they be seeding the oceans with iron filings or putting up solar mirrors in space. None of them is an alternative to reducing emissions, but one just might be a valuable piece of insurance against the worst-case climate change scenarios. Believe me, pretty much anything is better than five or six degrees of global warming.

This may seem like a depressing conclusion, but it's really an optimistic one. If we fail to reduce emissions quickly enough and find ourselves frying, we must throw everything we possibly can at the problem to counteract the warming process, however temporarily. At no point—I repeat, at no point—do we give up and admit that all is lost. If we go over two degrees, then we have to try and stop ourselves going over three. If we fail to stabilise emissions by 2015, then we have to try and stabilise them by 2016 or 2020. If people continue to demand economic growth, then we have to try to deliver than growth in a low-carbon way. It will never be too late. As long as people and nature remain alive on this planet, we will still have everything to fight for.

Critical Thinking

1. Why are three steps outlined by the author to stop climate change "the easy way"?

2. When the author says, "We have about 100 months left," who is the "we" he is referring to? Locate this "we" on the world map in your text. Make some observations of what the map is saying.

3. The article was published in 2007. One hundred months equals 8.3 years, which equals the year 2015, which equals 3 years from now. Assess the progress of the "we."

4. The author states: "Today we face a situation where a global population of potentially nine billion or so by 2050 continues to demand a steadily increasing consumer lifestyle. There is nothing we can do to stop this, nor should we." Appraise this statement in light of the concept "environmental sustainability."

5. Identify in this article three key terms, concepts, or principles that are used in your textbook (environmental science, economics, sociology, history, geography, etc.) or employed in the discipline you are currently studying. (Note: The terms, concepts, or principles may be implicit, explicit, implied, or inferred.)

MARK LYNAS is the New Statesman's environment correspondent, and author of "*Six Degrees: Our Future on a Hotter Planet,*" published by Fourth Estate.

Global Warming Battlefields: How Climate Change Threatens Security

Michael T. Klare

By any reckoning, global climate change poses a threat to world security writ large. Because it will imperil food production around the world and could render many heavily populated areas uninhabitable, it has the potential to endanger the lives and livelihoods of hundreds of millions of people. So far, most experts' warnings have naturally tended to focus on the large-scale, non-traditional security implications of global warming: mass starvation resulting from persistent drought, humanitarian disasters caused by severe hurricane and typhoon activity, the inundation of coastal cities, and so on. Just as likely, however, is an increase in more familiar security threats: war, insurgency, ethnic conflict, state collapse, and civil violence. The Nobel committee affirmed as much in October when it awarded the Peace Prize to former Vice President Al Gore and the Intergovernmental Panel on Climate Change for their efforts to raise awareness about global warming. The prize committee cited "increased danger of violent conflicts and wars, within and between states."

Climate change will increase the risk of conflict because it is almost certain to diminish the supply of vital resources—notably food, water, and arable land—in areas of the planet that already are suffering from resource scarcity, thus increasing the risk that desperate groups will fight among themselves for whatever remains of the means of survival. In wealthier societies, such conflicts can be mitigated by food and housing subsidies provided by the central governments and by robust schemes for relocation and reconstruction. In poorer countries, where little or no such capacity exists, the conflicts are more likely to be decided by ethnic or religious militias and the power of the gun.

Violent conflict over vital resources has, of course, been a characteristic of the human condition since very ancient times. Archaeological remains and the oldest written records attest to the fact that early human communities fought for control over prime growing areas, hunting zones, timber stands, and so on. A growing body of evidence also suggests that severe climate changes—for example, the "little Ice Age" of circa AD 1300–1700—have tended to increase the risk of resource-related conflict. Steven A. LeBlanc of the Peabody Museum of Archaeology and Ethnology at Harvard has noted, for example, that conflict among the Anasazi people of the American Southwest appears to have increased substantially with the cooling trend (and reduced food output) of the early 1300s, as indicated by the abandonment of exposed valley-floor settlements in favor of more defensible cliff dwellings.

Resource conflict has continued into more recent times, growing even more pronounced as European adventurers and settlers invaded Africa, Asia, and the Americas in search of gold, furs, spices, timber, land, human chattel, iron, copper, and oil—often encountering fierce resistance in the process. Today, indigenous peoples are still battling to preserve their lands and traditional means of livelihood in the few remaining unexploited tropical forests, mountain highlands, and other wilderness areas left on the planet.

As climate change kicks in, the risk of resource wars will grow many times over.

Elsewhere, many of those on the bottom rungs of the socioeconomic ladder—especially those who depend on agriculture or herding for their livelihoods—are also caught up in perennial conflict over access to land, water, energy, and other resources. Even as inter-state wars have diminished in number worldwide, conflicts between various groups in recent times have been exacerbated by rapid population growth; increased competition from agribusiness and cheap imported foodstuffs; the growing popularity of militant ethnic, religious, and political ideologies; and other exogenous factors. Even without global warming, these factors will continue to increase the likelihood of intergroup conflict. As climate change kicks in, the risk of resource wars will grow many times over.

The Hardest Hit

Accelerated by the greenhouse effect (the warming produced as greenhouse gases such as carbon dioxide trap heat in the atmosphere), climate change will affect the global resource equation in many ways. Essentially, these can be grouped into four key effects: (1) *diminished rainfall* in many tropical and temperate areas, leading to more frequent and prolonged droughts; (2) *diminished river flow* in many of these same areas as a result of reduced rainfall or the shrinking of mountain glaciers, producing greater water scarcity in food-producing regions; (3) *a rising sea level,* leading to the inundation of coastal cities and farmlands; and (4) *more frequent and severe storm events,* producing widespread damage to farms, factories, and villages. These effects will vary in their application to different parts of the globe, with some areas experiencing greater trauma than others, but the net result will be a substantial reduction in life-sustaining resources for a good part of the earth's population.

The particular impacts of these global warming effects on various communities have been studied in a piecemeal fashion for some time, but were given their most systematic examination in the Fourth Assessment Report of the UN-sponsored Intergovernmental Panel on Climate Change (IPCC), released in April 2007. As part of this study—the most comprehensive of global warming yet conducted—the IPCC convened a task force on "Impacts, Adaptation, and Vulnerability," called Working Group II. Ecosystem by ecosystem, region by region, this group's report provides an extraordinary overview of what can be expected from global warming's long-term impact on natural habitats and human communities around the planet. Although dry and dispassionate in tone, the report of Working Group II is devastating in its conclusions. Among its principal findings:

On water scarcity. "By mid-century, annual average river runoff and water availability are projected to . . . decrease by 10 to 30 percent over some dry regions at mid-latitudes and in the dry tropics, some of which are presently water-stressed areas. . . . In the course of the century, water supplies stored in glaciers and snow cover are projected to decline, reducing water availability in regions supplied by meltwater from major mountain ranges, where more than one-sixth of the world population currently lives."

On food availability. "At lower latitudes, especially seasonally dry and tropical regions, crop productivity is projected to decrease for even small local temperature increases (1 to 2 degrees centigrade), which would increase the risk of hunger. . . . Increases in the frequency of droughts and floods are projected to affect local crop production negatively, especially in subsistence sectors at low latitudes.

On coastal inundation. "Many millions more people are projected to be flooded every year due to sea-level rise by the 2080s. Those densely populated and low-lying areas where adaptive capacity is relatively low, and which already face other challenges such as tropical storms or local coastal subsidence, are especially at risk."

Most of these effects will be felt across the entire planet, but the degree to which they produce death, injury, and suffering will vary with the relative wealth and resiliency of the societies involved. In general, affluent and well-governed societies will be better able to cope with trauma and provide for the minimum needs of their affected citizens; poor and inadequately governed nations will be much less able to cope. And it is in the latter countries where conflict is most likely to arise over the allocation of relief supplies and relocation options.

The pivotal relationship between climate change and the coping capacity of affected states will be especially pronounced in Africa. That continent is expected to suffer disproportionately from the direst effects of global warming—especially from prolonged drought and water scarcity—and it possesses the least capacity to mitigate these impacts. According to Working Group II, as early as 2020, between 75 million and 250 million Africans are expected to face increased water scarcity as a result of climate change; by the 2050s, this number is projected to range between 350 and 600 million people. Because food production in Africa is already stretched to the limit, the decline in water availability will reduce crop yields and greatly increase the risk of hunger and malnutrition. According to the Working Group II report, yields from rain-fed agriculture in some African countries could be reduced by as much as half by 2020. Increased rural unrest and conflicts over land are a likely result.

Parts of South and Central Asia could also suffer from violence related to global warming. A rise in average temperatures and a decline in water supplies are expected to produce a sharp reduction in cereal production throughout this vast region, which contains some of the most heavily populated countries in the world. In Bangladesh, for example, wheat production could decline by 32 percent by 2050 and rice production by 8 percent. For all of South Asia, according to the IPCC report, "net cereal production . . . is projected to decline at least between 4 to 10 percent by the end of this century under the most conservative climate change scenario." With many millions of subsistence farmers in these countries already struggling to survive, production declines on this scale will prove catastrophic. Large numbers of rural residents forced into destitution will no doubt migrate to cities in search of jobs. But some may be attracted to radical sects or ethnic bands that promise salvation through the seizure of less-affected lands held by wealthy landowners or other ethnic groups.

In general, adverse effects from global warming likely will produce suffering on an unprecedented scale. Many will starve; many more will perish from disease, flooding, or fire. Others, however, will attempt to survive in the same manner as their predecessors: by fighting among themselves for whatever food and water remains; by invading more favorable locales; or by migrating to distant lands, even in the face of violent resistance.

Resource Wars

The onset of severe climate change will increase the frequency and intensity of certain familiar types of conflict and also introduce some new or largely unfamiliar forms. Two kinds of conflicts—resource wars and ethnic warfare attendant on state collapse—are among the more familiar of these. A third, less familiar type of violence likely to increase as a result of global warming might best be described as migratory conflict.

Resource wars arise when competing states or ethnic enclaves fight over the possession of key resources—particularly water supplies, oil reserves, diamond fields, timber stands, and mineral deposits. Such conflicts, as noted, have been a feature of human behavior since time immemorial. Conflicts over resources were less prevalent during the cold war era, when ideological antagonisms were the driving force in world affairs, but they have become more conspicuous since the demise of the Soviet Union and the outbreak of fresh disputes in the developing world. Though often characterized as ethnic and religious wars, many of these newer conflicts have been, at root, disputes over the allocation of land, water, timber, or other valuable commodities. Bitter fighting in Angola and Sierra Leone, for example, was principally driven by competition over the illicit trade in diamonds. Struggles over diamonds, timber, and coltan (a critical ingredient in the manufacture of cell phones) have fueled the ongoing violence in Congo. Wars in Somalia, Ethiopia, and the Darfur region of Sudan have largely been sparked by disputes over land and water rights.

Even without global warming, the incidence of intergroup wars like these is likely to increase because the demand for key resources is growing while supplies, in many cases, are shrinking. On the demand side of the ledger, many developing countries are expected to experience a sharp increase in population over the next several decades along with a steady increase in per capita consumption levels. On the supply side, many once-lucrative sources of oil, natural gas, uranium, copper, timber, fish, and underground water (aquifers) are expected to be depleted, producing significant scarcities of these materials. Virtually all states and societies are likely to experience some traumas and hardships as a result, but some groups will suffer far more than others. And because these disparities are likely to coincide with national, ethnic, and religious distinctions, they will provide ample fodder for those who seek justifications for waging war on "others" who can be portrayed as the cause of one's own hardships and misfortunes.

Add climate change to the equation, and the picture becomes much, much worse. While most of the world's regions are likely to experience a reduction in the supply of at least some critical resources, it is true that a few could see limited gains from global warming. Some countries in the far north, for example, could benefit from more rainfall and longer growing seasons, allowing for increased food output. Russia also hopes to benefit from the melting of the Arctic ice cap, which theoretically would allow oil and natural gas drilling in areas now covered year-round by thick ice. But even if these hypothetical advantages are not outweighed by other, less desirable consequences of global warming, any perception of a widening chasm between the "winners" and "losers" of climate change—when the overwhelming majority of the world's population is likely to fall in the latter category—could direct angry and potentially lethal attention toward the former.

Climate change will increase pressure on nearly every key resource used by humans, but land and fresh water will probably experience the greatest effects. Conflict over arable land has been one of the most persistent causes of warfare throughout history, and it is hard to imagine that global warming will not increase the likelihood of this type of conflict. If the projections by the IPCC's Working Group II prove accurate, vast inland areas of North and South America, Africa, and South and Central Asia are likely to suffer from diminished rainfall and recurring drought, turning once productive croplands into lifeless dustbowls. At the same time, many once reliable river systems will offer sharply reduced water flows, as the glaciers and snowpacks that feed them melt and recede or disappear. Again, not *every* area will suffer in this fashion: Some coastal highlands (for example, in the Horn of Africa) could experience increased precipitation and longer growing seasons, allowing greatly increased food production. Under these circumstances, those who feel cheated by the vagaries of climate change may feel impelled to invade and occupy the lands of those who, in their view, are unjustly blessed by the same fickle forces.

An area of particular concern among many climatologists is the Sahel region of Central Africa. The Sahel—the southern fringe of the Sahara Desert—stretches clear across Africa at its widest point from Senegal and Mauritania in the west through Mali, Niger, Chad, and Sudan (notably Darfur), to Eritrea and Ethiopia in the east. This is an area historically inhabited by Muslim pastoralists (mainly cattle herders) who are now being driven south by the steady advance of the Sahara into lands occupied by non-Muslim farmers—often provoking conflict in the process. The southward advance of the Sahara is believed to be one of the earliest observable effects of climate change and, as global temperatures rise, its rate of expansion is expected to increase. This, in turn, is likely to trigger intensified conflict from one end of the Sahel to the other.

Watching the River Flow

Global climate change is also likely to increase the risk of conflict over vital supplies of fresh water. Although states have rarely gone to war over disputed water supplies in recent times, they have often threatened to do so, and the risk factors appear to be growing. Water scarcity and stress are already a significant problem in many parts of the world, and are expected to become more so as a result of population growth, urbanization, and industrialization. Furthermore, many of the countries with the greatest exposure to water scarcity are highly dependent on river systems that arise outside their territory and pass through nations with which they have poor or unfriendly relations. Egypt, for example, is almost entirely dependent for its fresh water on the Nile, which arises in Central Africa (in the case of the White Nile) and Ethiopia (in the case of the Blue Nile). Iraq and Syria both depend for much of their water on the Tigris and Euphrates, which originate in Turkey. Israel relies on the Jordan River, which originates, in part, in Lebanon and Syria.

When an upstream country in any of these trans-boundary systems decides to dam a river and use it for domestic irrigation, the downstream country will inevitably experience a reduced flow, and this can be seen by that country as a significant threat to its well-being. It is not surprising, then, that downstream states like Egypt and Israel have threatened to go to war against any upstream state that endangers its water supply in this manner. "Water for Israel is not a luxury," former Prime Minister Moshe Sharett once proclaimed. "It is not just a desirable and helpful addition to our natural resources. Water is life itself."

How, exactly, global warming will affect any particular river system cannot be predicted with absolute assurance. However, it is clear from the IPCC's Fourth Assessment Report that many of the world's most important trans-boundary river systems in tropical and mid-temperate areas are likely to experience reduced flows as a result of climate change. This will significantly increase the competition among states for the ever-diminishing supply, thus increasing the risk of conflict.

Egypt is a source of particular concern in this regard, both because of its extreme reliance on the Nile—which originates far from its own territory—and because of its repeated threats to attack any upstream country that attempts to interfere with the river's flow. The Nile is the world's longest river and travels for more than a thousand miles through nearly cloudless desert before reaching Cairo, making it especially vulnerable to higher evaporation rates as global temperatures rise. Any reduction in upstream rainfall would also reduce its flow, increasing Egypt's vulnerability. Meanwhile, as water scarcity grows, so will Egypt's population, which is projected to increase from 80 million today to between 115 million and 179 million by 2050. Under these circumstances, any efforts by Ethiopia, Sudan, Uganda, or any of the other upstream states to divert the Nile's flow to meet the needs of their own soaring populations would almost assuredly trigger a panicky and quite possibly violent response from Egypt.

The Mogadishu Effect

A second type of violence that likely will increase as a result of global climate change is the sort of militia rule and gang warfare that prevail today in Mogadishu, the Somali capital. This is a condition in which the established government no longer exists or exercises effective authority; as a consequence, armed bands of one kind or another control access to critical resources, and these gangs or militias constantly fight among themselves over what little remains. Such violence has become familiar in the post–cold war era, as once vigorous states have collapsed and ethnic militias—often allied with or built around criminal associations—have arisen in their place. Multinational peacekeeping forces have confronted such bands in Somalia, Sierra Leone, Congo, Bosnia, Haiti, Sudan, Rwanda, and elsewhere, often with discouraging results.

Much research has been devoted to the causes of state collapse and the rise of ethnic militias in the post–cold war era, but this research has not managed to identify a clear, consistent set of precipitating factors. Corruption, authoritarianism, endemic poverty, ethnic favoritism, and poor social services are often common factors, but each case has its own distinctive features. What is true in all of these instances, however, is that the central government proves incapable of coping with an onslaught of powerful socioeconomic forces and either disintegrates entirely (as in Somalia) or loses control of large regions of the country—sometimes everything but the capital itself. Looking around the world, one can identify any number of countries that could, under existing circumstances, fall prey to these types of forces. Add global warming to the mix, and the pressures on these already vulnerable states grow much stronger.

Global warming will contribute to the propensity for weak states to collapse and give rise to militia rule and ethnic conflict.

Climate change will contribute to the propensity for weak states to collapse and give rise to militia rule and ethnic conflict for a variety of reasons. Consider any nation in the tropical or sub-temperate regions that depends for a significant share of its gross domestic product on farming, herding, forestry, and fishing, and that encompasses within its population more than one major ethnic, religious, or linguistic community. As indicated in the report of Working Group II, global warming is likely to harm some if not all of these livelihoods, though not to the same extent and not all at once. Also, some outlying parts of the country may become virtually uninhabitable, forcing people to migrate to the major city (or cities), often the capital or major port; or to areas more fortunate, which may be occupied by people of a different ethnicity (or religion, language group, and so forth). The decline in farming, fishing, and other livelihoods will contribute to a reduction in GDP, diminishing the revenues of the central government and thus its ability to shoulder additional burdens. Meanwhile, the movement of desperate refugees to the cities or other areas will produce an enormous need for relief services and exacerbate inter-group tensions.

All this would be a Herculean challenge for even the most affluent and capable governments, as the aftermath of Hurricane Katrina showed Americans. For poor, weak, and divided governments, the challenge could well prove insurmountable. As states collapse under the strain, what might be called the "Mogadishu effect" will kick in. Armed groups will coalesce around clans, tribes, village ties, and so on, as each group strives to ensure its own survival, at whatever price in bullets and blood. It is in precisely these circumstances, moreover, that extremist movements take root. With food and housing in short supply and city streets clogged with refugees, it is easy for a demagogue to blame another group or tribe for his own group's misfortunes and to call for violent action to redress grievances.

Because existing models cannot pinpoint future climate trends at the local level, it is not possible to predict where this combination of effects is most likely to result in state collapse, militia rule, and warfare among armed bands in the years ahead. Nonetheless, in its carefully worded manner, the IPCC in its Fourth Assessment Report provides some revealing hints. Speaking of the "negative effects" of climate change on human populations, it concludes: "The most vulnerable industries, settlements, and societies are generally those in coastal and river flood plains, those whose economies are closely linked with climate-sensitive resources, and those in areas prone to extreme weather events, especially where rapid urbanization is occurring." In most nations, moreover, vulnerable areas are not isolated but are linked in extensive and complex ways to other parts of their countries and to the surrounding regions.

As the impacts from climate change spread and state systems collapse in their wake, giving rise to gang and militia rule, violence might take many forms. Street-to-street combat for control over particular neighborhoods (and the distribution of relief supplies or vital commodities) is one common form. Fighting for control over desert oases, as in Darfur, is another. Impoverished farmers driven from their own territories by drought may invade neighboring lands. All of this will cause immense suffering and generate more refugees, creating ripple effects and sparking calls for international humanitarian intervention, as in the case of Darfur. This could lead as well to the deployment of international peacekeeping forces, and so result in clashes of yet another sort. The breakdown of state rule in affected areas will also facilitate the activities of criminals, mercenaries, drug traffickers, and others who flourish in an atmosphere of chaos—including international terrorists.

Migratory Conflicts

Yet a third type of conflict arising from global climate change might best be described as migratory warfare. This is armed violence provoked by efforts by large groups of people to migrate from environmentally devastated areas to less affected regions in the face of armed resistance by those inhabiting the more privileged locales. Of course, undocumented migrants from drought-prone areas of Mexico and Africa are already meeting significant resistance in their efforts to enter the United States and Europe, respectively. Many have perished in desperate attempts to circumvent the fences, border patrols, and other means employed to impede them.

But these struggles take place largely on an individual basis, or in groups of ten or a hundred at a time. Once global warming occurs on a massive scale, it is possible to conceive of migratory movements encompassing entire communities or regions, involving tens of thousands of people—some equipped with arms or formed into militias. In such scenarios, the struggle to cross national borders and settle in new lands could take on the form of ragged military campaigns—not unlike the travails experienced by the Israelites on entering the "Promised Land" (the Jordan River valley) as recounted in the Jewish scriptures.

The growing prevalence of migratory pressures like these could produce a number of worrisome phenomena. On one hand, destination countries like the United States, Spain, France, and Italy could choose to place even greater emphasis on the physical sealing-off of their borders and the use of military or paramilitary forces to stem the flood of immigrants. Even as experts debate the utility of such measures—especially in the face of vastly increased migratory pressures—more forcible means are certain to produce higher levels of death (whether through drowning, heat prostration, or shootouts with border guards), raising difficult moral questions. On the other hand, the growing intrusion of people with alien backgrounds could further inflame anti-immigrant sentiment in these countries, possibly leading to anti-immigrant violence or the ascendancy of ultranationalistic or even neo-Nazi political parties (with attendant social disorder and the possibility of increased militarism). However this plays out, migratory pressures on this scale are certain to prove highly destabilizing in many parts of the world.

Migratory pressures are likely to be among the most destabilizing impacts of global warming.

The global security implications of such phenomena were first accorded serious consideration in a 2007 study conducted by a group of retired US military officers convened by the CNA Corporation, a not-for-profit military contractor. In *National Security and the Threat of Climate Change,* the officers, led by former Army Chief of Staff General Gordon R. Sullivan, concluded that migratory pressures are likely to be among the most destabilizing impacts of global warming. For the United States, the greatest national security problem arising out of climate change "may be an increased flow of migrants northward into the U.S." Despite vigorous efforts to stem the flood of illegal immigrants, the rate of immigration is likely to rise "because the water situation in Mexico is already marginal and could worsen with less rainfall and more droughts." This, in turn, is likely to produce increased political outrage among long-term residents of border states, along with intensified efforts to exclude illegal migrants—with questionable results.

A similar dynamic is developing in the wealthier parts of Europe. "The greater threat to Europe lies in migration of people from across the Mediterranean, from the Maghreb, the Middle East, and sub-Saharan Africa," the CNA study observed. "As more people migrate from the Middle East because of water shortages and loss of their already marginal agricultural lands (as, for instance, if the Nile Delta disappears under the rising sea level), the social and economic stress on European nations will grow." Such stress can take many forms, but anti-immigrant violence is sure to be among the more likely outcomes.

A sharp increase in international migrations brought about by climate change is likely to have other baneful effects for world stability. As countries act to seal their borders and prevent the in-migration of environmental refugees, they are far less likely to participate in inclusionary political projects like the further expansion of the European Union—which, by its nature, would facilitate migration. Indeed, one might expect the EU to fall apart, with the original core of Western European countries breaking away from Greece and newer members in south-eastern Europe, which are expected to suffer from drought and water scarcity as a result of global warming. This, in turn, could give rise to the emergence of new military blocs, with Russia keen to restore its influence in east-central Europe. Other efforts to build and expand regional partnerships—for example, the Association of Southeast Asian Nations—could suffer a similar fate, as politically insecure states seek to drive off migrants from less-fortunate neighboring nations. Isolationism and xenophobia historically have been harbingers of conflict.

Looking Ahead

As the effects of global warming become more pronounced, the world community will have to cope with a wide range of extreme environmental perils: prolonged droughts, intense storms, extensive coastal flooding, and so on. In practical terms, however, it may well be that the most costly and challenging consequence of climate change will be an increase in violent conflict and all the humanitarian trauma this brings with it. The war in Darfur provides a sobering indication of what this is likely to entail: Although this tragic conflict, affecting hundreds of thousands of still-vulnerable civilians, cannot be attributed to climate change alone, it comes as close as any contemporary contest to the global warming battlefields of the future.

It may well be that the most costly and challenging consequence of climate change will be an increase in violent conflict and all the humanitariantrauma this brings with it.

All this should lead us to think differently about both "national security" and global warming itself. In traditional national security discourse, the overarching challenge is how best to protect the nation against identifiable armed adversaries. But in a world of severe climate change, the greatest challenge is not likely to come from well-equipped hostile powers. More likely it will emerge from chaos and violence and civil conflicts arising from the breakdown of states and the ensuing struggles over scarce and diminishing resources.

Under these circumstances, national security will take on an entirely new meaning—requiring, for one thing, a corresponding transformation of the role and organization of a nation's armed forces. One would expect, for instance, a much greater emphasis on civil-defense functions: flood control, emergency response in times of disaster, anti-looting operations, refugee protection and resettlement, and so forth. The international community will also need to be much better prepared for humanitarian interventions and peacekeeping in the event of environmental-related conflicts.

Finally, it should be apparent that a world of unceasing resource wars, state disintegration, militia rule, and migratory conflicts will not be a safe or desirable world for anyone. These may not be the images one most associates with global warming, but they may, in fact, prove to be the most immediate and concrete consequences of climate change. As if there were not already enough good reasons to take swift action to curb emissions of greenhouse gases, the prospect of increased armed conflict should, one would hope, convince those still unaware of the magnitude of the danger.

Critical Thinking

1. The author argues that global climate change will increase the risk conflict, but will some world regions be more at risk than other regions? Why or why not?

2. What "unfamiliar forms" of conflict may we be confronting in the future? Describe some connections these conflicts might have with global environmental resources.

3. Pick five countries in the world (other than the United States) that you think will suffer least from climate change and explain.

4. Locate the high-risk conflict areas on your textbook world map. Do some "critical" viewing of the global patterns and record your observations.

5. How do you think the United States will fare in these global warming battlefields?

6. Identify in this article three key terms, concepts, or principles that are used in your textbook (environmental science, economics, sociology, history, geography, etc.) or employed in the discipline you are currently studying. (Note: The terms, concepts, or principles may be implicit, explicit, implied, or inferred.)

MICHAEL T. KLARE, a *Current History* contributing editor, is a professor at Hampshire College and author of the forthcoming *Rising Powers, Shrinking Planet: The New Geopolitics of Energy* (Metropolitan Books, 2008).

From *Current History*, vol. 106, 2007, pp. 355–361. Copyright © 2007 by Current History, Inc. Reprinted by permission.

UNIT 6

Endangered Diversity

Unit Selections

Learning Outcomes

After reading this Unit, you should be able to:

- Provide examples of the biological, ecological, cultural, and resource management complexities/linkages involved with environmental issue #6: endangered diversity.

- Identify the primary indicators of biodiversity decline.

- Describe the connections between human well-being and protecting biodiversity.

- Discuss where biodiversity is most threatened and describe what actions need to be implemented to reduce both direct and indirect drivers of biodiversity loss.

- Offer insights into connections between illegal hunting and endangered species.

- Elaborate on the linkages between ecological diversity and human cultural diversity.

- Explain the complexities of ecosystem diversity management when people, animals, culture, and government agencies are interlinked.

- Assess the current relation between human and biodiversity, and offer some predictions for the future.

Student Website

www.mhhe.com/cls

Internet References

Endangered Species
www.endangeredspecie.com
Friends of the Earth
www.foe.co.uk/index.html
Greenpeace
www.greenpeace.org
Natural Resources Defense Council
http://nrdc.org
Smithsonian Institution Website
www.si.edu
World Wildlife Federation (WWF)
www.wwf.org

If all mankind were to disappear, the world would regenerate back to the rich state of equilibrium that existed ten thousand years ago. If insects were to vanish, the environment would collapse into chaos.

— Edward O. Wilson

According to scientists, our fossil record suggests more than 99 percent of all species ever in existence are now extinct. To look at it in a different way, that means that the living world you see out there each and every day represents a less than 1 percent success rate in the Great Living Planet Experiment. With an apparent 99+percent planetary organic failure rate, how could anyone deny that the Earth and its myriad permutations of life forms are beyond priceless. Nothing short of a mind-boggling miracle. But what may prove to be hardest on the future success rate of the nonhuman species world is not meteors or volcanic activity or climate change or some other cataclysmic natural event, but the human species. And that fact may be our undoing, as well. **Unit 6** considers the current state of *Top Ten Environmental Issue 6:* Endangered Diversity

The extinction of plant and animal species is nothing new. In fact, it is a natural phenomenon that has been occurring since the beginning of life on Earth. But just because it is "natural" does not mean it does not come without its costs. With each disappearance of a species there is information loss—genetic information loss. And it's the kind of information that helps living organisms make the necessary adjustments and adaptations to a changing world to ensure its survival. Furthermore, life forms, species, do not live in a mason jar—at least not for long. They are the cogs in the wheels of the ecosystems of a planet that taken together have had a synergistic effect creating a planet that we fondly refer to as Earth. These cogs are known more formally in the scientific world as species. Scientists estimate there are 1.6 million species presently known. Taxonomists estimate there may be somewhere between 3 million and 50 million species alive today. And that is a lot of biological diversity—the spice of life that keeps ecosystems working and humans alive.

So, when too many cogs begin to disappear and cannot be replaced, ecosystems can begin to break down, and falter. The great wheel of life—the living earth—screeches to a halt. Humans, lest we forget, have been hitching a ride on that great wheel of life. Unfortunately, though, we humans have not only been contributing significantly to the loss of cogs, but we are doing so at an accelerating rate as the result of our increasingly cavalier attitude toward maintaining of our planet's wheels of life. To address this attitude, the World Summit on Sustainable Development convened in 2002 in Johannesburg, South Africa, and set a goal "to achieve by 2010 a significant reduction of the current rate of biodiversity loss at the global, regional and national level as a contribution to poverty alleviation and to the benefit of all life on Earth." It's now 2012, and we didn't make it. Why?

Perhaps a primary reason is that the linkages between biodiversity, ecosystem stability, and human-well-being are not readily recognizable. Not by biologists and environmental scientists, but by people (but not all people). Article 20, "Executive Summary from the Secretariat of the Convention on Biological Diversity," clearly substantiates this observation. The linkages are not recognized clearly by those human groups that are the least "intimate" and most disconnected with the ecosystems they depend on and the organisms responsible for the health of those ecosystems. The industrialized nations are home to many of these people, and the nations on which the industrialized world depends oftentimes bear an unfair share of the biodiversity loss burden.

It should be no surprise that the more intimately a person lives with their environment, the more aware they are of the linkages mentioned earlier. They must have intimate appreciation or they would not survive. Ask any African peasant farmer, Amazon River fisherman, or Mongolian herder. This may be said as well with subgroups even within industrialized nations. Take, for example, a Midwest mixed-grain-and-livestock farmer or a Louisiana commercial fisherman. Ask these "ecosystem" people about the condition of their "environment," and you will get intimate details. Ask the average person walking down the street in Manhattan, Chicago, or Los Angeles about the health of the ecosystems that have contributed to their walking on that street, and unless they are a student of environmental studies, their appreciation of the biodiversity–ecosystem–human well-being connection will be woefully limited.

But what of the behavior of local peoples demonstrated in Article 23, "Cry of the Wild"? Here, the impact of biodiversity loss is from the direct hand of the people who live there—the killing of apes and other endangered species. We would think the local people would be more "aware," but their actions are heavily influenced by the profiteering motives of individuals thousands of miles away. Or consider the situation in Indonesia, where dragons and humans are coming into a conflict that never existed until recently. In Article 22, "When Good Lizards Go Bad: Komodo Dragons Take Violent Turn," diversity is being threatened because a fundamental ecological principle—resource partitioning—is being destabilized by resource mismanagement (nonlocal) policies and incursion into the Komodo dragons' habitat by village population growth rates, which are eroding their formerly small village "ecosystem" intimacy.

Finally, Article 21, "When Diversity Vanishes," is included in this unit on diversity to remind the reader that it is in fact "diversity" (not just biological) that is being threatened and that must be protected. Human and cultural diversity is just as important to maintaining our social, political, economic, and food production ecosystems as biological diversity is to the preservation of Earth's life support systems. Scientists note that cultural and biological diversity often occur in similar geographic "spaces." It's reported that seven of the countries with the highest cultural diversity in the world are also on the list of "megadiversity" countries with the highest number of unique biological organisms (Indonesia, New Guinea, Mexico, China, Brazil, United States, Philippines). Human systems, like biological systems, characterized by diversity are better able to respond and adapt to changing environmental conditions. Author Don Monroe may well be quite correct that to avoid the Earth's "tipping point" both human–cultural and biological diversity must each be protected.

Executive Summary from Secretariat of the Convention on Biological Diversity

GLOBAL BIODIVERSITY OUTLOOK 3

The target agreed by the world's Governments in 2002, "to achieve by 2010 a significant reduction of the current rate of biodiversity loss at the global, regional and national level as a contribution to poverty alleviation and to the benefit of all life on Earth," has not been met.

There are multiple indications of continuing decline in biodiversity in all three of its main components—genes, species and ecosystems—including:

- Species which have been assessed for extinction risk are on average moving closer to extinction. Amphibians face the greatest risk and coral species are deteriorating most rapidly in status. Nearly a quarter of plant species are estimated to be threatened with extinction.
- The abundance of vertebrate species, based on assessed populations, fell by nearly a third on average between 1970 and 2006, and continues to fall globally, with especially severe declines in the tropics and among freshwater species.
- Natural habitats in most parts of the world continue to decline in extent and integrity, although there has been significant progress in slowing the rate of loss for tropical forests and mangroves, in some regions. Freshwater wetlands, sea ice habitats, salt marshes, coral reefs, seagrass beds and shellfish reefs are all showing serious declines.
- Extensive fragmentation and degradation of forests, rivers and other ecosystems have also led to loss of biodiversity and ecosystem services.
- Crop and livestock genetic diversity continues to decline in agricultural systems.
- The five principal pressures directly driving biodiversity loss (habitat change, overexploitation, pollution, invasive alien species and climate change) are either constant or increasing in intensity.
- The ecological footprint of humanity exceeds the biological capacity of the Earth by a wider margin than at the time the 2010 target was agreed.

The loss of biodiversity is an issue of profound concern for its own sake. Biodiversity also underpins the functioning of ecosystems which provide a wide range of services to human societies. Its continued loss, therefore, has major implications for current and future human well-being. The provision of food, fibre, medicines and fresh water, pollination of crops, filtration of pollutants, and protection from natural disasters are among those ecosystem services potentially threatened by declines and changes in biodiversity. Cultural services such as spiritual and religious values, opportunities for knowledge and education, as well as recreational and aesthetic values, are also declining.

The existence of the 2010 biodiversity target has helped to stimulate important action to safeguard biodiversity, such as creating more protected areas (both on land and in coastal waters), the conservation of particular species, and initiatives to tackle some of the direct causes of ecosystem damage, such as pollution and alien species invasions. Some 170 countries now have national biodiversity strategies and action plans. At the international level, financial resources have been mobilized and progress has been made in developing mechanisms for research, monitoring and scientific assessment of biodiversity.

Many actions in support of biodiversity have had significant and measurable results in particular areas and amongst targeted species and ecosystems. This suggests that with adequate resources and political will, the tools exist for loss of biodiversity to be reduced at wider scales. For example, recent government policies to curb deforestation have been followed by declining rates of forest loss in some tropical countries. Measures to control alien invasive species have helped a number of species to move to a lower extinction risk category. It has been estimated that at least 31 bird species (out of 9,800) would have become extinct in the past century, in the absence of conservation measures.

However, action to implement the Convention on Biological Diversity has not been taken on a sufficient scale to address the pressures on biodiversity in most places. There has been insufficient integration of biodiversity issues into broader policies, strategies and programmes, and the under-lying drivers of biodiversity loss have not been addressed significantly. Actions to promote the conservation and sustainable use of biodiversity receive a tiny fraction of funding compared to activities aimed at promoting infrastructure and industrial developments. Moreover, biodiversity considerations are often ignored when such developments are designed, and opportunities to plan in ways that minimize unnecessary negative impacts on biodiversity are missed. Actions to address the underlying drivers of biodiversity loss, including demographic, economic, technological, socio-political and cultural pressures, in meaningful ways, have also been limited.

Most future scenarios project continuing high levels of extinctions and loss of habitats throughout this century, with associated decline of some ecosystem services important to human well-being.
For example:

- Tropical forests would continue to be cleared in favour of crops and pastures, and potentially for biofuel production.
- Climate change, the introduction of invasive alien species, pollution and dam construction would put further pressure on freshwater biodiversity and the services it underpins.
- Overfishing would continue to damage marine ecosystems and cause the collapse of fish populations, leading to the failure of fisheries.

Changes in the abundance and distribution of species may have serious consequences for human societies. The geographical distribution of species and vegetation types is projected to shift radically due to climate change, with ranges moving from hundreds to thousands of kilometres towards the poles by the end of the 21st century. Migration of marine species to cooler waters could make tropical oceans less diverse, while both boreal and temperate forests face widespread dieback at the southern end of their existing ranges, with impacts on fisheries, wood harvests, recreation opportunities and other services.

There is a high risk of dramatic biodiversity loss and accompanying degradation of a broad range of ecosystem services if ecosystems are pushed beyond certain thresholds or tipping points. The poor would face the earliest and most severe impacts of such changes, but ultimately all societies and communities would suffer.
Examples include:

- The Amazon forest, due to the interaction of deforestation, fire and climate change, could undergo a widespread dieback, with parts of the forest moving into a self-perpetuating cycle of more frequent fires and intense droughts leading to a shift to savanna-like vegetation. While there are large uncertainties associated with these scenarios, it is known that such dieback becomes much more likely to occur if deforestation exceeds 20–30% (it is currently above 17% in the Brazilian Amazon). It would lead to regional rainfall reductions, compromising agricultural production. There would also be global impacts through increased carbon emissions, and massive loss of biodiversity.

- The build-up of phosphates and nitrates from agricultural fertilizers and sewage effluent can shift freshwater lakes and other inland water ecosystems into a long-term, algae dominated (eutrophic) state. This could lead to declining fish availability with implications for food security in many developing countries. There will also be loss of recreation opportunities and tourism income, and in some cases health risks for people and livestock from toxic algal blooms. Similar, nitrogen-induced eutrophication phenomena in coastal environments lead to more oxygen-starved dead zones, with major economic losses resulting from reduced productivity of fisheries and decreased tourism revenues.

- The combined impacts of ocean acidification, warmer sea temperatures and other human induced stresses make tropical coral reef ecosystems vulnerable to collapse. More acidic water—brought about by higher carbon dioxide concentrations in the atmosphere—decreases the availability of the carbonate ions required to build coral skeletons. Together with the bleaching impact of warmer water, elevated nutrient levels from pollution, overfishing, sediment deposition arising from inland deforestation, and other pressures, reefs worldwide increasingly become algae-dominated with catastrophic loss of biodiversity and ecosystem functioning, threatening the livelihoods and food security of hundreds of millions of people.

There are greater opportunities than previously recognized to address the biodiversity crisis while contributing to other social objectives. For example, analyses conducted for this Outlook identified scenarios in which climate change is mitigated while maintaining and even expanding the current extent of forests and other natural ecosystems (avoiding additional habitat loss from the widespread deployment of biofuels). Other opportunities include "rewilding" abandoned farmland in some regions, and the restoration of river basins and other wetland ecosystems to enhance water supply, flood control and the removal of pollutants.

Well-targeted policies focusing on critical areas, species and ecosystem services are essential to avoid the most dangerous impacts on people and societies. Preventing further human-induced biodiversity loss for the near term future will be extremely challenging, but biodiversity loss may be halted and in some aspects reversed in the longer term, if urgent, concerted and effective action is initiated now in support of an agreed long-term vision. Such action to conserve biodiversity and use its components sustainably will reap rich rewards—through better health, greater food security, less poverty and a greater capacity to cope with, and adapt to, environmental change.

Placing greater priority on biodiversity is central to the success of development and poverty-alleviation measures. It is clear that continuing with "business as usual" will jeopardize the future of all human societies, and none more so than the poorest who depend directly on biodiversity for a particularly high proportion of their basic needs. The loss of biodiversity is frequently linked to the loss of cultural diversity, and has an especially high negative impact on indigenous communities.

The linked challenges of biodiversity loss and climate change must be addressed by policymakers with equal priority and in close co-ordination, if the most severe impacts of each are to be avoided. Reducing the further loss of carbon storing ecosystems such as tropical forests, salt marshes and peatlands will be a crucial step in limiting the build-up of greenhouse gases in the atmosphere. At the same time, reducing other pressures on ecosystems can increase their resilience, make them less vulnerable to those impacts of climate change which are already unavoidable, and allow them to continue to provide services to support people's livelihoods and help them adapt to climate change.

Better protection of biodiversity should be seen as a prudent and cost-effective investment in risk-avoidance for the global community. The consequences of abrupt ecosystem changes on a large scale affect human security to such an extent, that it is rational to minimize the risk of triggering them—even if we are not clear about the precise probability that they will occur. Ecosystem degradation, and the consequent loss of ecosystem services, has been identified as one of the main sources of disaster risk. Investment in resilient and diverse ecosystems, able to withstand the multiple pressures they are subjected to, may be the best-value insurance policy yet devised.

Scientific uncertainty surrounding the precise connections between biodiversity and human well-being, and the functioning of ecosystems, should not be used as an excuse for inaction. No one can predict with accuracy how close we are to ecosystem tipping points, and how much additional pressure might bring them about. What is known from past examples, however, is that once an ecosystem shifts to another state, it can be difficult or impossible to return it to the former conditions on which economies and patterns of settlement have been built for generations.

Effective action to address biodiversity loss depends on addressing the underlying causes or indirect drivers of that decline.

This will mean:

- Much greater efficiency in the use of land, energy, fresh water and materials to meet growing demand.
- Use of market incentives, and avoidance of perverse subsidies to minimize unsustainable resource use and wasteful consumption.
- Strategic planning in the use of land, inland waters and marine resources to reconcile development with conservation of biodiversity and the maintenance of multiple ecosystem services. While some actions may entail moderate costs or tradeoffs, the gains for biodiversity can be large in comparison.

- Ensuring that the benefits arising from use of and access to genetic resources and associated traditional knowledge, for example through the development of drugs and cosmetics, are equitably shared with the countries and cultures from which they are obtained.
- Communication, education and awareness raising to ensure that as far as possible, everyone understands the value of biodiversity and what steps they can take to protect it, including through changes in personal consumption and behaviour.

The real benefits of biodiversity, and the costs of its loss, need to be reflected within economic systems and markets. Perverse subsidies and the lack of economic value attached to the huge benefits provided by ecosystems have contributed to the loss of biodiversity. Through regulation and other measures, markets can and must be harnessed to create incentives to safeguard and strengthen, rather than to deplete, our natural infrastructure. The re-structuring of economies and financial systems following the global recession provides an opportunity for such changes to be made. Early action will be both more effective and less costly than inaction or delayed action.

Urgent action is needed to reduce the direct drivers of biodiversity loss. The application of best practices in agriculture, sustainable forest management and sustainable fisheries should become standard practice, and approaches aimed at optimizing multiple ecosystem services instead of maximizing a single one should be promoted. In many cases, multiple drivers are combining to cause biodiversity loss and degradation of ecosystems. Sometimes, it may be more effective to concentrate urgent action on reducing those drivers most responsive to policy changes. This will reduce the pressures on biodiversity and protect its value for human societies in the short- to medium-term, while the more intractable drivers are addressed over a longer time-scale. For example the resilience of coral reefs—and their ability to withstand and adapt to coral bleaching and ocean acidification—can be enhanced by reducing overfishing, land-based pollution and physical damage.

Direct action to conserve biodiversity must be continued, targeting vulnerable as well as culturally-valued species and ecosystems, combined with steps to safeguard key ecosystem services, particularly those of importance to the poor. Activities could focus on the conservation of species threatened with extinction, those harvested for commercial purposes, or species of cultural significance. They should also ensure the protection of functional ecological groups—that is, groups of species that collectively perform particular, essential roles within ecosystems, such as pollination, control of herbivore numbers by top predators, cycling of nutrients and soil formation.

Increasingly, restoration of terrestrial, inland water and marine ecosystems will be needed to re-establish ecosystem functioning and the provision of valuable services. Economic analysis shows that ecosystem restoration can give good economic rates of return. However the biodiversity and associated services of restored ecosystems usually remain below the levels of natural ecosystems. This reinforces the argument that, where

111

possible, avoiding degradation through conservation is preferable (and even more cost-effective) than restoration after the event.

Better decisions for biodiversity must be made at all levels and in all sectors, in particular the major economic sectors, and government has a key enabling role to play. National programmes or legislation can be crucial in creating a favourable environment to support effective "bottom-up" initiatives led by communities, local authorities, or businesses. This also includes empowering indigenous peoples and local communities to take responsibility for biodiversity management and decision-making; and developing systems to ensure that the benefits arising from access to genetic resources are equitably shared.

We can no longer see the continued loss of and changes to biodiversity as an issue separate from the core concerns of society: to tackle poverty, to improve the health, prosperity and security of our populations, and to deal with climate change. Each of those objectives is undermined by current trends in the state of our ecosystems, and each will be greatly strengthened if we correctly value the role of biodiversity in supporting the shared priorities of the international community. Achieving this will involve placing biodiversity in the mainstream of decision-making in government, the private sector, and other institutions from the local to international scales.

The action taken over the next decade or two, and the direction charted under the Convention on Biological Diversity, will determine whether the relatively stable environmental conditions on which human civilization has depended for the past 10,000 years will continue beyond this century. If we fail to use this opportunity, many ecosystems on the planet will move into new, unprecedented states in which the capacity to provide for the needs of present and future generations is highly uncertain.

Critical Thinking

1. In a brief paragraph, summarize the current state of global biodiversity.

2. If there seems to be plenty of evidence regarding the importance of biodiversity and human well-being, then why are we not succeeding in protecting biodiversity?

3. Do ecosystem resources benefit all people in all parts of the world similarly? Explain.

4. Why might wealthy people and wealthy nations not feel the urgency to protect biodiversity?

5. Do plants, animals, and other organisms that threaten human well-being deserve the same protection as the ones that benefit humans? Explain.

6. Identify in this article three key terms, concepts, or principles that are used in your textbook (environmental science, economics, sociology, history, geography, etc.) or employed in the discipline you are currently studying. (Note: The terms, concepts, or principles may be implicit, explicit, implied, or inferred.)

When Diversity Vanishes

Don Monroe

Complex systems, from ecologies to economies, do interesting and unexpected things. Much of this rich behavior can be traced to the networks through which the underlying "agents" affect each other. Often, however, diversity of the agents themselves is essential. If they act too similarly, the entire system can cease to function. At the annual symposium of the Santa Fe Institute Business Network, November 1–3, 2007, an array of experts explored this "diversity collapse" in contexts ranging from ecology and the food we eat, to finance and organizational structure.

Ecological Collapse

The most dramatic diversity collapses are mass extinctions, which have wiped out much of life five times in Earth's history. Doug Erwin, of the National Museum of Natural History and SFI, said that one of these, the end-Permian extinction, wiped out "90 to 95 percent of everything in the oceans, about 70 percent of everything on land, and by all accounts was about the best thing that ever happened to life on Earth." The extinctions made room for later innovation—but not right away. "Eventually the diversity got bigger than before, but it took four million years to even get started."

In contrast to some observers, Erwin does not believe that we are entering a "sixth wave" of extinction. "At least if we're lucky, we're not," he said. Nonetheless, "the crisis is real." Erwin emphasized that there are many types of diversity, which do not have the same impact. For example, individual species on different branches of the tree of life forms can become extinct without substantial effect, but losing the same number of species on a single branch could eliminate that entire branch.

Global extinction reflects the combined changes in smaller, individual ecosystems around the world. Andrew Dobson of Princeton University described what he called "probably the best-studied" ecosystem: the Serengeti National Park in Tanzania. Established in 1951, this park and the surrounding areas provide "a natural example of what happens when we perturb an ecosystem," he observed. Outside the park, the ecology changes dramatically because of farming and grazing. The difference is most notable at the highest trophic levels in the food chain, Dobson observed.

Dobson contributed to the Millennium Ecosystem Assessment, which framed the contributions of healthy ecosystems, at least in part, in terms of the economic "services" they provide to people. Almost half of the value, Dobson said, comes from the most basic level, including bacteria, and another one third from plants. These lower levels also tend to be more resilient. Higher trophic levels, including grazers and predators, are more visible but provide less value, he said. They are also more sensitive to changes, so "monitoring these brittle species gives an early warning" of damage.

"We have a scarily short time scale to understand how ecosystems collapse," Dobson commented. "Most large natural ecosystems will be destroyed in the next 30 to 50 years. The quality of human life on this planet is dependent on the economic services supplied by those webs."

Historically, said Mercedes Pascual of the University of Michigan and SFI, ecologists viewed complexity in food webs as an essential feature of healthy ecosystems that helps them to resist disruption. In contrast, monocultures, such as the endless fields of U.S. Midwestern corn, can succumb to a single pest.

The important work of Robert May in the 1970s, however, showed that complexity actually reduces stability in some mathematical models. Ever since, Pascual said, ecologists have tried to understand "how more realistic structures lead to higher stability."

Instead of studying small perturbations as May did, Stefano Allesina, of NCEAS, and Pascual looked at major shifts such as the disappearance of prey causing a predator to become extinct. They also used food webs taken directly from ecological studies instead of mathematically generated networks. They then determined which pathways are functional and which are redundant, from the perspective of secondary extinction, and found that in real webs about 90 percent of connections are functional, independent of the size of the ecosystem.

As a result, Pascual observed, "Even when secondary extinctions are not observed, the loss of species makes ecosystems more fragile to further extinctions." There may be little warning of an approaching "tipping point," in which the entire ecosystem collapses.

Pascual suggested that work by SFI External Professor Ricard Solé may clarify the dynamics of interacting populations contributing to extinctions in these ecological networks. In his model, as species continually become extinct and new ones immigrate into a region, the food web forms a self-organized state, with many of the features observed in real webs.

Although the system as a whole seems static, individual species are not. "Individual populations . . . are all going up and down like crazy," Pascual said. These fluctuations at one level may even enhance the stability at a higher level. The extinction of individual species may therefore be a misleading measure of the loss of diversity.

The question of the best level for gauging diversity also arose in work by Katia Koelle (Penn State), Sarah Cobey (University of Michigan), Bryan Grenfell (Penn State), and Pascual on the evolution of flu. Genes evolve continuously, but often with no effect on the "phenotype": the surface proteins that determine immune response. The researchers modeled genetic evolution coupled with the prevalence in the human population of immunity to particular variants. In this model, viruses multiply rapidly whenever they take on a new phenotype, quickly crowding out other variants. "This pattern of boom and bust is explained by an interaction of genetic drift and selection, and not exclusively one or the other," Pascual said.

Managed Ecosystems

If the ecologists are right, natural ecosystems, which have evolved complex webs of interactions, are ideally "managed" by leaving them alone—when possible. In stark contrast, agricultural crop management resembles control theory—an engineering tool developed for much simpler systems—and features simplified ecosystems that depend heavily on external inputs. Much of the corn in the U.S. is grown with petroleum-derived pesticides and fertilizer, and fed to cattle whose excrement then becomes toxic waste rather than nutrition for plants.

An alternative was described by Joel Salatin, who recovered marginal land in Virginia by using cow manure to fertilize the grasses that are the natural food for cattle. But Salatin complained that the many regulations aimed at industrial-scale production present formidable barriers to small farms like his.

In spite of such efforts, local production is likely to be an anomaly in an industrial food system that prizes cheap and abundant food. But Cary Fowler of the Global Crop Diversity Trust asserted that this system depends on underappreciated diversity of plant varieties.

Even as agriculture has focused on fewer species over the past 12,000 to 15,000 years, Fowler said, "diversity in some sense was increasing," as farmers in different regions selected variants with different traits. "There are about 120,000 different varieties of rice, each as distinct one from the other as a Great Dane from a Chihuahua," he commented.

Industrial agriculture, Fowler said, threatens this variation within species, although scientists still cannot agree on how to measure it. In response to new challenges, he wondered, "can we continue to develop our agriculture without diversity?" His answer: "Obviously we can't."

The expected global warming in coming decades makes these issues especially urgent. "My guess is we are ill-prepared for this kind of change," Fowler said. "But if we are prepared, it will be because of the gene banks and the diversity they contain."

> **"My guess is we are ill-prepared for this kind of change," Fowler said. "But if we are prepared, it will be because of the gene banks and the diversity they contain."**

Numerous gene banks have been storing seeds for crops and other plants around the world. Unfortunately, Fowler said, many of them are poorly funded and maintained. He stressed that for a modest cost—an endowment of some $250 million—"we can conserve the gene pool of our major crops in perpetuity." As a start in this direction, Fowler's organization is funding a facility above the Arctic Circle to provide a global seed repository, sometimes called the "doomsday vault."

Valuing Diversity

No such vault exists to preserve human culture. "The forces of homogenization are rampant," said Suzanne Romaine of the University of Oxford. She described the rapid extinction of languages, as large ones like Mandarin, Spanish, and English spread at the expense of smaller ones.

As a linguist, Romaine values languages as data for her own work. But she sees linguistic extinction as part of a larger problem. "It's not just languages that are at stake, but forms of knowledge," she said. "They can't be separated from people, their identities, their cultural heritage, their well-being and their rights." She also stressed that language diversity and biodiversity often disappear together.

In a similar way, shared communication drives a homogenization in computer systems, said Gabriela Barrantes, of the University of Costa Rica and SFI. The dominance of Microsoft in personal computer software is only the most visible example of this diversity collapse, she said.

This uniform computing environment is sensitive to threats, just as monocultures are vulnerable to agricultural pests. Barrantes and Stephanie Forrest, of the University of New Mexico and SFI, have been exploring how artificial variability in computer systems can slow the spread of malware. In any successful scheme, she stressed, computers must still interoperate with similar performance and cost.

Diversity can be introduced at many levels, Barrantes said. For example, the well-known "buffer overflow" attacks rely on long data spilling into areas of memory intended for programs. Varying the locations of these segments can often thwart the spread of infection between different machines. This kind of artificial diversification is "currently being used in two major operating systems," Barrantes said.

Diversity collapse in computer systems is probably well ahead of that in ecosystems, Forrest suggested. But she noted that "adding diversity back in is much easier than it would be in the natural world."

The dominance of Google, eBay, and others shows that "online niches are often winner-take all," notes Virgil Griffith of the California Institute of Technology and SFI. But Griffith claimed that "promiscuous interoperability" can promote diversity by allowing people to use data for new purposes. "When all-powerful monocultures make data available, diversity flourishes," Griffith claimed. "The users diversify the monoculture, not the other way around."

Diverse Perspectives

In finance, diversification reduces risk by spreading money among assets that respond differently during market moves. But during the 1998 international financial crisis, Long-Term Capital Management suffered enormous losses when its ostensibly diversified investments began to react similarly as other investors desperately sold the same assets. "Diversity collapses are really the source of inefficiencies in markets," asserted Michael Mauboussin, Chief Investment Strategist at Legg Mason Capital Management and an SFI trustee.

Mauboussin reviewed three theories for how markets become efficient, meaning that prices reflect value. The first, in which all investors behave rationally, is unrealistic. A second explanation, which requires only that some investors exploit—and thereby remove—arbitrage opportunities, "has failed us in critical junctures," he asserted.

Mauboussin contrasted these models with a view of "markets as a complex adaptive system, where prices essentially emerge from the interaction of many agents." In this view, also called "the wisdom of crowds," three conditions assure efficiency: diversity among investors, an aggregation mechanism, and financial incentives. Of these assumptions, he said, "the most likely to be violated is diversity," which may decline imperceptibly until it suddenly collapses.

Scott Page of the University of Michigan and SFI, has compiled many ways that diverse groups outperform individuals. In expert judgment, for example, as in diversified portfolios, the average judgment of a group is always better. "This is not a feel-good statement, this is a mathematical theorem," Page said. For problem solving, having multiple strategies can help a group evade roadblocks that hamper any one approach.

"In human systems," Page said, "the thing that really works against cognitive diversity is selection." The increasingly global marketplace of ideas selects the current "best practices" at the expense of other approaches. "If the world is flat, we cannot count on the right amount of diversity existing," Page said.

Page warned that merely recognizing the advantages of cognitive diversity might not be enough to preserve it. "Diversity's benefits may be public goods that are not in any one's interest to maintain."

One way to maintain diversity is through time-varying selective pressures that prevent one idea from dominating. However, Page showed a simple model in which such churn did not prevent uniformity. He suggested that maintaining diversity also requires diverse selective processes or richer networks, so that the criteria for picking winners varies.

The broad range of speakers at this symposium shows that the Santa Fe Institute is in no danger of diversity collapse, although similar principles apply in very different fields. Still, in the world outside, increasing interconnectedness seriously threatens diversity in both human organizations and ecosystems.

Critical Thinking

1. If industrial agriculture threatens biodiversity, then critically assess the arguments made in Article 9 in support of commercial agriculture and genetic modification.

2. Referring to the text world map, identify locations where you think biological diversity is most threatened. Locate places where you think cultural/human diversity is most threatened. Offer some insights regarding your observations.

3. Do you think there is more diversity protection in "industrialized" regions, or "rural" regions of the world? Explain.

4. Describe the "ecosystem" within which you live each day, and describe its diversity. What do you see or think?

5. Identify in this article three key terms, concepts, or principles that are used in your textbook (environmental science, economics, sociology, history, geography, etc.) or employed in the discipline you are currently studying. (Note: The terms, concepts, or principles may be implicit, explicit, implied, or inferred.)

DON MONROE (www.donmonroe.info) writes on biology, physics, and technology from Berkeley Heights, New Jersey. Prior to 2003, he spent 18 years in basic and applied research at Bell Laboratories.

When Good Lizards Go Bad: Komodo Dragons Take Violent Turn

Villagers Blame Environmentalists for Reptiles' Mood; Ban on Goat Sacrifice

YAROSLAV TROFIMOV

A t least once a week, an unwelcome intruder crawls under a clapboard wall and, forked tongue darting, lumbers its way into Syarif Maulana's classroom.

"Then, everyone screams, there is no more school, and we all run away very fast," says the 10-year-old boy. "We are very afraid."

The intruder, a Komodo dragon, is the world's largest lizard, an ancient, fierce carnivore found only on a handful of remote islands in eastern Indonesia. Reaching 10 feet in length, the dragons feed on buffaloes, deer and an occasional human. Just a year ago, a boy about Syarif's age died in a dragon's jaws, his bones smashed against rocks to facilitate reptilian digestion.

That killing, and a spate of other close encounters, has fanned a panic in the dragons' main habitat, the Komodo National Park. Touted by Indonesia as its "Jurassic Park," this rocky, barren archipelago is home to some 2,500 dragons and nearly 4,000 people, clustered in four fishing villages of wooden stilt houses.

These locals have long viewed the dragons as a reincarnation of fellow kinsfolk, to be treated with reverence. But now, villagers say, the once-friendly dragons have turned into vicious man-eaters. And they blame policies drafted by American-funded environmentalists for this frightening turn of events.

"When I was growing up, I felt the dragons were my family," says 55-year-old Hajji Faisal. "But today the dragons are angry with us, and see us as enemies." The reason, he and many other villagers believe, is that environmentalists, in the name of preserving nature, have destroyed Komodo's age-old symbiosis between dragon and man.

For centuries, local tradition required feeding the dragons —which live more than 50 years, can recognize individual humans and usually stick to fairly small areas. Locals say they always left deer parts for the dragons after a hunt, and often tied goats to a post as sacrifice. Island taboos strictly prohibited hurting the giant reptiles, a possible reason why the dragons have survived in the Komodo area despite becoming extinct everywhere else.

'Sacred Duty'

"For us, giving food to the dragons is an obligation, our sacred duty," says Hajji Adam, headman of the park's biggest village, Kampung Komodo.

Indonesia invited the Nature Conservancy, a Virginia-based environment protection group, to help manage the park in 1995. An Indonesian subsidiary of the group, called Putri Naga Komodo, gained a tourism concession for the park in 2005 and is investing in the conservation effort some $10 million of its own money and matching financing from international donors.

With this funding and advice, park authorities put an end to villagers' traditional deer hunting, enforcing a prohibition that had been widely disregarded. They declared canines an alien species, and outlawed the villagers' dogs, which used to keep dragons away from homes. Park authorities banned the goat sacrifices, previously staged on Komodo for the benefit of picture-snapping tourists.

"We don't want the Komodo dragon to be domesticated. It's against natural balance," says Widodo Ramono, policy director of the Nature Conservancy's Indonesian branch and a former director of the country's national park service. "We have to keep this conservation area for the purpose of wildlife. It is not for human beings."

When people hunt deer, it poses a mortal threat to the dragons, which disappeared from a small island near Komodo after poachers decimated deer stocks there, officials say. "If we let the locals hunt again, the dragons will be gone," says Vinsensius Latief, the national park's chief for Komodo island. "If we are not strict in enforcing the ban, everything here will be destroyed."

But, while the deer population remains stable in the park, many dragons these days prefer to seek easier prey in the vicinity of humans. They frequently descend from the hills to the villages, hiding under stilt houses and waiting for a chance to snap at passing chicken or goats. Much to the fury of villagers, park authorities, while endorsing the idea in principle, so far haven't acted on repeated requests to build dragon-proof fences around the park's inhabited areas. The measure is estimated to cost about $5,000 per village.

"People are scared because, every day, the dragons come down and eat our goats," complains Ibrahim Hamso, secretary of the Kampung Rinca village. "Today it's a goat, and tomorrow it can be our child."

'Crazed with Blood'

A year ago, a 9-year-old named Mansur was one such victim. The boy went to answer the call of nature behind a bush near his home in Kampung Komodo. In broad daylight, as terrified relatives looked on, a dragon lunged from his hideout, took a bite of the boy's stomach and chest, and started crushing his skull.

"We threw branches and stones to drive him away, but the dragon was crazed with blood, and just wouldn't let go," says the boy's father, Jamain, who, like many Indonesians, goes by only one name.

Unlike in the U.S. and many other Western countries, park rangers here don't routinely put down animals that develop a taste for human flesh.

A few months later, Jamain's neighbor Mustaming Kiswanto, a 38-year-old who makes a living selling dragon woodcarvings to tourists, and whose son had been bitten by a dragon, was attacked by another giant lizard after falling asleep. In June, five European divers, stranded in an isolated part of the park, said they successfully fended off an aggressive dragon by throwing their weight belts at it.

One of the most famous lizard attacks occurred half a world away in 2001, when a Komodo dragon kept by the Los Angeles Zoo tried to ingest the foot of Phil Bronstein, then editor of the San Francisco Chronicle and husband of actress Sharon Stone.

To the villagers in Komodo, the recent incidents provide clear evidence of an ominous change in reptile behavior. "I don't blame the dragons for my boy's death. I blame those who forbade us from following custom and feeding them," says Jamain. "If it weren't for them, my boy would still be alive."

Officials at the Nature Conservancy's Indonesian headquarters in Bali dismiss such widespread belief about a connection between the attacks and the ban on feeding the dragons as "superstition." The group and its Komodo subsidiary reject any responsibility for Mansur's death.

The boy "shouldn't have crouched like a prey species in a place where dragons live," says Marcus Matthews-Sawyer, tourism, marketing and communications director at Putri Naga Komodo. "You've got to be very careful about extrapolating and drawing any conclusions."

Tackling Fears

Despite such disbelief in the Komodo villagers' theories, executives at the Nature Conservancy's headquarters in the U.S. pledge to reach out and tackle local fears. "Any concern expressed by the villagers will be taken seriously and we will address it if we can," says Chief Communications Officer James R. Petterson. "The Komodo effort is a work in progress."

Dragon and man could coexist here in harmony in the past, Komodo park officials add, because at the time the area's human population was a fraction of today's size. Now, with local villages pushing deeper inland and attracting new settlers from elsewhere in Indonesia, conflict may be inevitable—and even a fence won't be able to prevent dragon infiltrations.

"The smell of the village—goats, chicken, drying fish—all this invites the dragons," says Mr. Latief. "And if the dragons can't grab the animals, they will bite the villagers."

Critical Thinking

1. Draw a concept map of the biological and social components and then indicate their linkages within the "ecosystem" of Komodo National Park. What is the "diversity issue" in the park?

2. Why do villages believe environmentalists have destroyed the age-old symbiosis between dragon and man?

3. What do park officials say has disrupted the balance of this ecosystem?

4. Who is right? Is it okay to endanger humans in order to protect biological diversity and species protection? Explain.

5. Identify in this article three key terms, concepts, or principles that are used in your textbook (environmental science, economics, sociology, history, geography, etc.) or employed in the discipline you are currently studying. (Note: The terms, concepts, or principles may be implicit, explicit, implied, or inferred.)

From *Wall Street Journal.com*, August 25, 2008, pp. A1. Copyright © 2008 by Wall Street Journal. Reprinted by permission via Rightslink.

Cry of the Wild

Last week four gorillas were slaughtered in Congo. With hunting on the rise, our most majestic animals are facing a new extinction crisis.

SHARON BEGLEY

On the lush plains of Congo's Virunga National Park last week, the convoy of porters rounded the final hill and trooped into camp. They gently set down the wooden frame they had carried for miles, and with it the very symbol of the African jungle: a 600-pound silverback mountain gorilla. A leader of a troop often visited by tourists, his arms and legs were lashed to the wood, his head hanging low and spots of blood speckling his fur. The barefoot porters, shirts torn and pants caked with dust from their trek, lay him beside three smaller gorillas, all females, who had also been killed, then silently formed a semicircle around the bodies. As the stench of death wafted across the camp in the waning afternoon light, a park warden stepped forward. "What man would do this?" he thundered. He answered himself: "Not even a beast would do this."

Park rangers don't know who killed the four mountain gorillas found shot to death in Virunga, but it was the seventh killing of the critically endangered primates in two months. Authorities doubt the killers are poachers, since the gorillas' bodies were left behind and an infant—who could bring thousands of dollars from a collector—was found clinging to its dead mother in one of the earlier murders. The brutality and senselessness of the crime had conservation experts concerned that the most dangerous animal in the world had found yet another excuse to slaughter the creatures with whom we share the planet. "This area must be immediately secured," said Deo Kujirakwinja of the Wildlife Conservation Society's Congo Program, "or we stand to lose an entire population of these endangered animals."

Back when the Amazon was aflame and the forests of Southeast Asia were being systematically clear-cut, biologists were clear about what posed the greatest threat to the world's wildlife, and it wasn't men with guns. For decades, the chief threat was habitat destruction. Whether it was from impoverished locals burning a forest to raise cattle or a multinational denuding a tree-covered Malaysian hillside, wildlife was dying because species were being driven from their homes. Yes, poachers killed tigers and other trophy animals—as they had since before Theodore Roosevelt—and subsistence hunters took monkeys for bushmeat to put on their tables, but they were not a primary danger.

That has changed. "Hunting, especially in Central and West Africa, is much more serious than we imagined," says Russell Mittermeier, president of Conservation International. "It's huge," with the result that hunting now constitutes the pre-eminent threat to some species. That threat has been escalating over the past decade largely because the opening of forests to logging and mining means that roads connect once impenetrable places to towns. "It's easier to get to where the wildlife is and then to have access to markets," says conservation biologist Elizabeth Bennett of the Wildlife Conservation Society. Economic forces are also at play. Thanks to globalization, meat, fur, skins and other animal parts "are sold on an increasingly massive scale across the world," she says. Smoked monkey carcasses travel from Ghana to New York and London, while gourmets in Hanoi and Guangzhou feast on turtles and pangolins (scaly anteaters) from Indonesia. There is a thriving market for bushmeat among immigrants in Paris, New York, Montreal, Chicago and other points in the African diaspora, with an estimated 13,000 pounds of bushmeat—much of it primates—arriving every month in seven European and North American cities alone. "Hunting and trade have already resulted in widespread local extinctions in Asia and West Africa," says Bennett. "The world's wild places are falling silent."

When a company wins a logging or mining concession, it immediately builds roads wide enough for massive trucks where the principal access routes had been dirt paths no wider than a jaguar. "Almost no tropical forests remain across Africa and Asia which are not penetrated by logging or other roads," says Bennett. Hunters and weapons follow, she notes, "and wildlife flows cheaply and rapidly down to distant towns where it is either sold directly or links in to global markets." How quickly can opening a forest ravage the resident wildlife? Three weeks after a logging company opened up one Congo forest, the density of animals fell more than 25 percent; a year after a logging road went into forest areas in Sarawak, Malaysia, in 2001, not a single large mammal remained.

A big reason why hunting used to pale next to habitat destruction is that as recently as the 1990s animals were killed mostly for subsistence, with locals taking only what they needed to live. Governments and conservation groups helped reduce even that through innovative programs giving locals an economic stake in the preservation of forests and the survival of wildlife. In the mountains of Rwanda, for instance, tourists pay $500 to spend an hour with the majestic mountain gorillas, bolstering the economy of the surrounding region. But recent years have brought a more dangerous kind of hunter, and not only because they use AK-47s and even land mines to hunt.

The problem now is that hunting, even of supposedly protected animals, is a global, multimillion-dollar business. Eating bushmeat "is now a status symbol," says Thomas Brooks of Conservation International. "It's not a subsistence issue. It's not a poverty issue. It's considered supersexy to eat bushmeat." Exact figures are hard to come by, but what conservation groups know about is sobering. Every year a single province in Laos exports $3.6 million worth of wildlife, including pangolins, cats, bears and primates. In Sumatra, about 51 tigers were killed each year between 1998 and 2002; there are currently an estimated 350 tigers left on the island (down from 1,000 or so in the 1980s) and fewer than 5,000 in the world.

If a wild population is large enough, it can withstand hunting. But for many species that "if" has not existed for decades. As a result, hunting in Kilum-Ijim, Cameroon, has pushed local elephants, buffalo, bushbuck, chimpanzees, leopards and lions to the brink of extinction. The common hippopotamus, which in 1996 was classified as of "least concern" because its numbers seemed to be healthy, is now "vulnerable": over the past 10 years its numbers have fallen as much as 20 percent, largely because the hippos are illegally hunted for meat and ivory. Pygmy hippos, classified as "vulnerable" in 2000, by last year had become endangered, at risk of going extinct. Logging has allowed bushmeat hunters to reach the West African forests where the hippos live; fewer than 3,000 remain.

Setting aside parks and other conservation areas is only as good as local enforcement. "Half of the major protected areas in Southeast Asia have lost at least one species of large mammal due to hunting, and most have lost many more," says Bennett. In Thailand's Doi Inthanon and Doi Suthep National Parks, for instance, elephants, tigers and wild cattle have been hunted into oblivion, as has been every primate and hornbill in Sarawak's Kubah National Park. The world-famous Project Tiger site in India's Sariska National Park has no tigers, biologists

announced in 2005. Governments cannot afford to pay as many rangers as are needed to patrol huge regions, and corruption is rife. The result is "empty-forest syndrome": majestic landscapes where flora and small fauna thrive, but where larger wildlife has been hunted out.

Which is not to say the situation is hopeless. With governments and conservationists recognizing the extinction threat posed by logging and mining, they are taking steps to ensure that animals do not come out along with the wood and minerals. In one collaboration, the government of Congo and the WCS work with a Swiss company, Congolaise Industrielle des Bois—which has a logging concession near Nouabalé-Ndoki National Park—to ensure that employees and their families hunt only for their own food needs; the company also makes sure that bushmeat does not get stowed away on logging trucks as illegal hunters try to take their haul to market. Despite the logging, gorillas, chimps, forest elephants and bongos are thriving in the park.

Anyone who thrills at the sight of man's distant cousins staring silently through the bush can only hope that the executions of Virunga's gorillas is an aberration. At the end of the week, UNESCO announced that it was sending a team to investigate the slaughter.

Critical Thinking

1. What exactly is the environmental issue in the Congo according to this article?

2. If gorillas weren't an endangered species, would it be okay to hunt them?

3. Although not addressed directly in the article, reflect on why eating bushmeat "is now a status symbol" and "it's super sexy to eat bushmeat."

4. What are the benefits to human well-being of protecting gorillas?

5. Read Article 33 and explain why a human would kill gorillas.

6. Identify in this article three key terms, concepts, or principles that are used in your textbook (environmental science, economics, sociology, history, geography, etc.) or employed in the discipline you are currently studying. (Note: The terms, concepts, or principles may be implicit, explicit, implied, or inferred.)

UNIT 7

Degrading Ecosystems

Unit Selections

Learning Outcomes

After reading this Unit, you should be able to:

- Describe the key features of ecosystem functions and explain why/how human well-being is ultimately linked to these ecosystems.

- Outline the arguments used to emphasize the connections between human well-being and the earth's ecosystems.

- Explain why it is difficult to manage ecosystems sustainably.

- Provide examples of effective responses to better managing of human–ecosystem relationships.

- Explain why "geography matters" in the valuation of ecosystem goods and services.

- Provide examples of how people benefit from nature.

- Discuss how people from different geographic regions of the world might have varying levels of dependency on ecosystem services.

- Offer insights on how industrialized, wealthy nations use ecosystem services compared to developing and/or poor countries.

Student Website

www.mhhe.com/cls

Internet References

National Center for Electronics Recycling
www.electronicsrecycling.org/Public/default.aspx
Terrestrial Sciences
www.cgd.ucar.edu/tss
World Health Organization
www.who.int
The World Watch Institute
www.worldwatch.org

And Man created the plastic bag and the tin and aluminum can and the cellophane wrapper and the paper plate, and this was good because Man could then take his automobile and buy all his food in one place and He could save that which was good to eat in the refrigerator and throw away that which had no further use. And soon the earth was covered with plastic bags and aluminum cans and paper plates and disposable bottles and there was nowhere to sit down or walk, and Man shook his head and cried: "Look at this God-awful mess."

—*Art Buchwald*, 1970

Which came first, the chicken or the egg? Likewise, which environmental problem arrived on the scene first: *Top Ten Environmental Issue 6,* Endangered Diversity, or *Issue 7,* Degrading Ecosystems? Probably what came first was modern man, followed by the invention of agriculture, which supported accelerating human population growth in those agricultural regions. Then the Industrial Revolution tamed more of the Earth and cleared the way for even more explosive growth of the human population in those regions. However, to meet the growing demands these revolutions and populations required for more food, freshwater, timber, fiber, minerals, ores, and fuel, Earth was consumed at faster and faster rates, and into ever-increasing hinterlands beyond the hearths of human communities. Whatever obstacles—ecosystems, habitats, nonhuman (and human) species—that lay in the path of this tsunami of humanity crashing over the planet were in dire straits. Fortunately, though, the nature of geography provides places of refuge and safety with its variable topographies of peaks and highlands and distances from epicenters. At least for now, sufficient ecosystems, habitats, and nonhuman species remain, if not endangered. Otherwise, our global ecosystems would be near collapse, or collapsing, as would the human race.

Unit 7 explores the question: Why the concern? Because the human population depends on Earth's ecosystems and the services they provide. And despite our transformation (frequently abuse) of many of these ecosystems over the past 50 years, these ecosystems have contributed to our making substantial gains in human well-being and socioeconomic development. However, such gains have not come without their costs. First, the negative externalities and nonlinear changes we have created through our transformation of ecosystems to meet our needs are not clearly understood. And second, perhaps more important, is that the harmful effects of ecosystem degradation are too often instituted in one place by people who are not intimately connected to the ecosystems but who benefit from its services, while at the same time the harmful effects are borne disproportionately in another place by people who, more often than not, are not just "benefiting" from an ecosystem, but have their lives depend on it. The irony is that those people who are most affected by ecosystem degradation are the ones who can least afford it. For example, wealthy, industrialized societies have the capital and technology to build extensive and complex linkages to capture the benefits of ecosystem services located great distances from their own degraded habitats and/or faltering

© Alan Morgan

ecosystems. Poor, agrarian, nonindustrialized economies (ecosystem peoples) have no such capabilities to reach that far for ecosystem services should their systems destabilize, and they have even far less agency to protect those systems. When was the last time (in recent history) that a drought in modern industrialized Europe or the United States resulted in death by starvation of millions of people? When the glaciers of the American Rocky Mountains dry up, what will be the death toll in Montana, Colorado, or Idaho? What catastrophic political conflicts will emerge in the United States? India? Pakistan? The Western Alpine System of South America?

According to Article 24, "Ecosystems and Human Well-being Synthesis," approximately 60 percent of the ecosystem services examined during the assessment are being degraded or used unsustainably. Most notable, however, are three emerging observations: First, many ecosystem services are being degraded as the result of our need to increase the supply of other services— like producing food or impounding freshwater. Such trade-offs often shift the costs of degradation from one group to another (or one nation to another), or to future generations. Second, the costs are shifted disproportionally to the poor (groups or nations). And third, the degradation of ecosystem services is proving to be a significant barrier to achieving the Millennium Development Goals. Finally, the article report includes a set of four scenarios that explore plausible futures (as of the 2005 writing) for ecosystems and human well-being based on assumptions about driving forces of change. The reader is encouraged to review those scenarios and assess them in the light of our present 2012 planetary environment and make future predictions.

Throughout this text, there has been a consistent theme that "geography matters" in addressing all environmental issues, particularly with regard to place. For example, the political economy issues of freshwater shortages in the United States will be different than in India. In Article 25, "The Geography of Ecosystem Services," James Boyd believes geography also matters because nature moves. And nature's ecosystem services are also dynamic because they result from the interactions of myriad biophysical and economic components and processes. They are, in a very real sense, *spatial production functions*. Boyd's observations regarding the connections between ecological features, their scarcity, and the size of population benefiting from it, and value, are particularly noteworthy in light of the ecological valuation discussion of Articles 24 and 26. If ecological services benefits were available to everyone everywhere in the same amounts and for the same costs, we would not have an issue.

But scarcity and demand are both key elements of ecosystem benefits and key realities of ecosystem geographies: patterns of people, resources, access, and consumption.

Finally, Rebecca L. Goldman's Article 26, "Ecosystem Services: How People Benefit from Nature," sheds further light on demonstrating how ecosystem services are not only the link between people and the natural world, but also critical link to ensuring our future well-being. However, the author argues that it will become increasingly important to assess real value to these systems and evaluate the costs/benefits of the trade-offs we choose to make in our accounting of these values. But most important, her article emphasizes how this appeal for an appreciation of this link can be employed to encourage new and increased interest in our conservation of natural resources across a wide range of resource management issues.

There appears solid consensus regarding the benefits to human well-being that ecosystem services provide. Where there will not be consensus, but instead contention and perhaps even conflict, is determining: Who will bear the costs for maintaining our ecosystems? Who will be compensated and how much for having their local and regional ecosystems degraded for the benefits of nonlocal/regional peoples? Who will decide what ecosystem services should be used to provide for human *needs* and which ones are used to provide for human *wants*? How will our industrialized consumption patterns have to change to protect our ecosystems?

Ecosystems and Human Well-Being

MILLENNIUM ECOSYSTEM ASSESSMENT, 2005

Summary for Decision-Makers

Everyone in the world depends completely on Earth's ecosystems and the services they provide, such as food, water, disease management, climate regulation, spiritual fulfillment, and aesthetic enjoyment. Over the past 50 years, humans have changed these ecosystems more rapidly and extensively than in any comparable period of time in human history, largely to meet rapidly growing demands for food, freshwater, timber, fiber, and fuel. This transformation of the planet has contributed to substantial net gains in human well-being and economic development. But not all regions and groups of people have benefited from this process—in fact, many have been harmed. Moreover, the full costs associated with these gains are only now becoming apparent.

Three major problems associated with our management of the world's ecosystems are already causing significant harm to some people, particularly the poor, and unless addressed will substantially diminish the long-term benefits we obtain from ecosystems:

- First, approximately 60% (15 out of 24) of the ecosystem services examined during the Millennium Ecosystem Assessment are being degraded or used unsustainably, including freshwater, capture fisheries, air and water purification, and the regulation of regional and local climate, natural hazards, and pests. The full costs of the loss and degradation of these ecosystem services are difficult to measure, but the available evidence demonstrates that they are substantial and growing. Many ecosystem services have been degraded as a consequence of actions taken to increase the supply of other services, such as food. These trade-offs often shift the costs of degradation from one group of people to another or defer costs to future generations.
- Second, there is *established but incomplete* evidence that changes being made in ecosystems are increasing the likelihood of nonlinear changes in ecosystems (including accelerating, abrupt, and potentially irreversible changes) that have important consequences for human well-being. Examples of such changes include disease emergence, abrupt alterations in water quality, the creation of "dead zones" in coastal waters, the collapse of fisheries, and shifts in regional climate.

- Third, the harmful effects of the degradation of ecosystem services (the persistent decrease in the capacity of an ecosystem to deliver services) are being borne disproportionately by the poor, are contributing to growing inequities and disparities across groups of people, and are sometimes the principal factor causing poverty and social conflict. This is not to say that ecosystem changes such as increased food production have not also helped to lift many people out of poverty or hunger, but these changes have harmed other individuals and communities, and their plight has been largely overlooked. In all regions, and particularly in sub-Saharan Africa, the condition and management of ecosystem services is a dominant factor influencing prospects for reducing poverty.

The degradation of ecosystem services is already a significant barrier to achieving the Millennium Development Goals agreed to by the international community in September 2000 and the harmful consequences of this degradation could grow significantly worse in the next 50 years. The consumption of ecosystem services, which is unsustainable in many cases, will continue to grow as a consequence of a likely three- to sixfold increase in global GDP by 2050 even while global population growth is expected to slow and level off in mid-century. Most of the important direct drivers of ecosystem change are unlikely to diminish in the first half of the century and two drivers—climate change and excessive nutrient loading—will become more severe.

Already, many of the regions facing the greatest challenges in achieving the MDGs coincide with those facing significant problems of ecosystem degradation. Rural poor people, a primary target of the MDGs, tend to be most directly reliant on ecosystem services and most vulnerable to changes in those services. More generally, any progress achieved in addressing the MDGs of poverty and hunger eradication, improved health, and environmental sustainability is unlikely to be sustained if most of the ecosystem services on which humanity relies continue to be degraded. In contrast, the sound management of ecosystem services provides cost-effective opportunities for addressing multiple development goals in a synergistic manner.

There is no simple fix to these problems since they arise from the interaction of many recognized challenges, including climate change, biodiversity loss, and land degradation, each of which is complex to address in its own right. Past actions to slow or reverse the degradation of ecosystems have yielded significant benefits, but these improvements have generally not kept pace with growing pressures and demands. Nevertheless, there is tremendous scope for action to reduce the severity of these problems in the coming decades. Indeed, three of four detailed scenarios examined by the MA suggest that significant changes in policies, institutions, and practices can mitigate some but not all of the negative consequences of growing pressures on ecosystems. But the changes required are substantial and are not currently under way.

An effective set of responses to ensure the sustainable management of ecosystems requires substantial changes in institutions and governance, economic policies and incentives, social and behavior factors, technology, and knowledge. Actions such as the integration of ecosystem management goals in various sectors (such as agriculture, forestry, finance, trade, and health), increased transparency and accountability of government and private-sector performance in ecosystem management, elimination of perverse subsidies, greater use of economic instruments and market-based approaches, empowerment of groups dependent on ecosystem services or affected by their degradation, promotion of technologies enabling increased crop yields without harmful environmental impacts, ecosystem restoration, and the incorporation of nonmarket values of ecosystems and their services in management decisions all could substantially lessen the severity of these problems in the next several decades.

The remainder of this Summary for Decision-makers presents the four major findings of the Millennium Ecosystem Assessment on the problems to be addressed and the actions needed to enhance the conservation and sustainable use of ecosystems.

> Finding #1: Over the past 50 years, humans have changed ecosystems more rapidly and extensively than in any comparable period of time in human history, largely to meet rapidly growing demands for food, freshwater, timber, fiber, and fuel. This has resulted in a substantial and largely irreversible loss in the diversity of life on Earth.

The structure and functioning of the world's ecosystems changed more rapidly in the second half of the twentieth century than at any time in human history.

More land was converted to cropland in the 30 years after 1950 than in the 150 years between 1700 and 1850. Cultivated systems (areas where at least 30% of the landscape is in croplands, shifting cultivation, confined livestock production, or freshwater aquaculture) now cover one quarter of Earth's terrestrial surface.

Approximately 20% of the world's coral reefs were lost and an additional 20% degraded in the last several decades of the twentieth century, and approximately 35% of mangrove area was lost during this time (in countries for which sufficient data exist, which encompass about half of the area of mangroves).

The amount of water impounded behind dams quadrupled since 1960, and three to six times as much water is held in reservoirs as in natural rivers. Water withdrawals from rivers and lakes doubled since 1960; most water use (70% worldwide) is for agriculture.

Since 1960, flows of reactive (biologically available) nitrogen in terrestrial ecosystems have doubled, and flows of phosphorus have tripled. More than half of all the synthetic nitrogen fertilizer, which was first manufactured in 1913, ever used on the planet has been used since 1985.

Since 1750, the atmospheric concentration of carbon dioxide has increased by about 32% (from about 280 to 376 parts per million in 2003), primarily due to the combustion of fossil fuels and land use changes. Approximately 60% of that increase (60 parts per million) has taken place since 1959.

Humans are fundamentally, and to a significant extent irreversibly, changing the diversity of life on Earth, and most of these changes represent a loss of biodiversity.

More than two thirds of the area of 2 of the world's 14 major terrestrial biomes and more than half of the area of 4 other biomes had been converted by 1990, primarily to agriculture.

Across a range of taxonomic groups, either the population size or range or both of the majority of species is currently declining.

The distribution of species on Earth is becoming more homogenous; in other words, the set of species in any one region of the world is becoming more similar to the set in other regions primarily as a result of introductions of species, both intentionally and inadvertently in association with increased travel and shipping.

The number of species on the planet is declining. Over the past few hundred years, humans have increased the species extinction rate by as much as 1,000 times over background rates typical over the planet's history (*medium certainty*). Some 10–30% of mammal, bird, and amphibian species are currently threatened with extinction (*medium to high certainty*). Freshwater ecosystems tend to have the highest proportion of species threatened with extinction.

Genetic diversity has declined globally, particularly among cultivated species.

Most changes to ecosystems have been made to meet a dramatic growth in the demand for food, water, timber, fiber, and fuel. Some ecosystem changes have been the inadvertent result of activities unrelated to the use of ecosystem services, such as the construction of roads, ports, and cities and the discharge of pollutants. But most ecosystem changes were the direct or indirect result of changes made to meet growing demands for ecosystem services, and in particular growing demands for food, water, timber, fiber, and fuel (fuelwood and hydropower). Between 1960 and 2000, the demand for ecosystem services grew significantly as world population doubled to 6 billion people and the global economy increased more than sixfold. To meet this demand, food production increased by roughly two-and-a-half times, water use doubled, wood harvests for pulp and paper production tripled, installed hydropower capacity doubled, and timber production increased by more than half.

The growing demand for these ecosystem services was met both by consuming an increasing fraction of the available supply (for example, diverting more water for irrigation or capturing more fish from the sea) and by raising the production of some services, such as crops and livestock. The latter has been accomplished through the use of new technologies (such as new crop varieties, fertilization, and irrigation) as well as through increasing the area managed for the services in the case of crop and livestock production and aquaculture.

> Finding #2: The changes that have been made to ecosystems have contributed to substantial net gains in human well-being and economic development, but these gains have been achieved at growing costs in the form of the degradation of many ecosystem services, increased risks of nonlinear changes, and the exacerbation of poverty for some groups of people. These problems, unless addressed, will substantially diminish the benefits that future generations obtain from ecosystems.

In the aggregate, and for most countries, changes made to the world's ecosystems in recent decades have provided substantial benefits for human well-being and national development. Many of the most significant changes to ecosystems have been essential to meet growing needs for food and water; these changes have helped reduce the proportion of malnourished people and improved human health. Agriculture, including fisheries and forestry, has been the mainstay of strategies for the development of countries for centuries, providing revenues that have enabled investments in industrialization and poverty alleviation. Although the value of food production in 2000 was only about 3% of gross world product, the agricultural labor force accounts for approximately 22% of the world's population, half the world's total labor force, and 24% of GDP in countries with per capita incomes of less than $765 (the low-income developing countries, as defined by the World Bank).

These gains have been achieved, however, at growing costs in the form of the degradation of many ecosystem services, increased risks of nonlinear changes in ecosystems, the exacerbation of poverty for some people, and growing inequities and disparities across groups of people. . . .

> Finding #3: The degradation of ecosystem services could grow significantly worse during the first half of this century and is a barrier to achieving the Millennium Development Goals.

The MA developed four scenarios to explore plausible futures for ecosystems and human well-being. (See Box 1.) The scenarios explored two global development paths, one in which the world becomes increasingly globalized and the other in which it becomes increasingly regionalized, as well as two different approaches to ecosystem management, one in which actions are reactive and most problems are addressed only after they become obvious and the other in which ecosystem management is proactive and policies deliberately seek to maintain ecosystem services for the long term.

Most of the direct drivers of change in ecosystems currently remain constant or are growing in intensity in most ecosystems. In all four MA scenarios, the pressures on ecosystems are projected to continue to grow during the first half of this century.

The most important direct drivers of change in ecosystems are habitat change (land use change and physical modification of rivers or water withdrawal from rivers), overexploitation, invasive alien species, pollution, and climate change. These direct drivers are often synergistic. For example, in some locations land use change can result in greater nutrient loading (if the land is converted to high-intensity agriculture), increased emissions of greenhouse gases (if forest is cleared), and increased number of invasive species (due to the disturbed habitat). . . .

The degradation of ecosystem services poses a significant barrier to the achievement of the Millennium Development Goals and the MDG targets for 2015. The eight Millennium Development Goals adopted by the United Nations in 2000 aim to improve human well-being by reducing poverty, hunger, child and maternal mortality, by ensuring education for all, by controlling and managing diseases, by tackling gender disparity, by ensuring environmental sustainability, and by pursuing global partnerships. Under each of the MDGs, countries have agreed to targets to be achieved by 2015. Many of the regions facing the greatest challenges in achieving these targets coincide with regions facing the greatest problems of ecosystem degradation.

Although socioeconomic policy changes will play a primary role in achieving most of the MDGs, many of the targets (and goals) are unlikely to be achieved without significant improvement in management of ecosystems. The role of ecosystem changes in exacerbating poverty (Goal 1, Target 1) for some groups of people has been described already, and the goal of environmental sustainability, including access to safe drinking water (Goal 7, Targets 9, 10, and 11), cannot be achieved as long as most ecosystem services are being degraded. Progress toward three other MDGs is particularly dependent on sound ecosystem management:

- Hunger (Goal 1, Target 2): All four MA scenarios project progress in the elimination of hunger but at rates far slower than needed to attain the internationally agreed target of halving, between 1990 and 2015, the share of people suffering from hunger. Moreover, the improvements are slowest in the regions in which the problems are greatest: South Asia and sub-Saharan Africa. Ecosystem condition, in particular climate, soil degradation, and water availability, influences process toward this goal through its effect on crop yields as well as through impacts on the availability of wild sources of food.
- Child mortality (Goal 4): Undernutrition is the underlying cause of a substantial proportion of all child deaths. Three of the MA scenarios project reductions in child undernourishment by 2050 of between 10% and 60% but undernourishment increases by 10% in *Order from Strength* (*low certainty*). Child mortality is also strongly influenced by diseases associated with water quality. Diarrhea is one of the predominant causes of infant deaths worldwide. In sub-Saharan Africa, malaria additionally plays an important part in child mortality in many countries of the region.

Box 1.
MA Scenarios

The MA developed four scenarios to explore plausible futures for ecosystems and human well-being based on different assumptions about driving forces of change and their possible interactions:

Global Orchestration—This scenario depicts a globally connected society that focuses on global trade and economic liberalization and takes a reactive approach to ecosystem problems but that also takes strong steps to reduce poverty and inequality and to invest in public goods such as infrastructure and education. Economic growth in this scenario is the highest of the four scenarios, while it is assumed to have the lowest population in 2050.

Order from Strength—This scenario represents a regionalized and fragmented world, concerned with security and protection, emphasizing primarily regional markets, paying little attention to public goods, and taking a reactive approach to ecosystem problems. Economic growth rates are the lowest of the scenarios (particularly low in developing countries) and decrease with time, while population growth is the highest.

Adapting Mosaic—In this scenario, regional watershed-scale ecosystems are the focus of political and economic activity. Local institutions are strengthened and local ecosystem management strategies are common; societies develop a strongly proactive approach to the management of ecosystems. Economic growth rates are somewhat low initially but increase with time, and population in 2050 is nearly as high as in *Order from Strength.*

TechnoGarden—This scenario depicts a globally connected world relying strongly on environmentally sound technology, using highly managed, often engineered, ecosystems to deliver ecosystem services, and taking a proactive approach to the management of ecosystems in an effort to avoid problems. Economic growth is relatively high and accelerates, while population in 2050 is in the midrange of the scenarios.

The scenarios are not predictions; instead they were developed to explore the unpredictable features of change in drivers and ecosystem services. No scenario represents business as usual, although all begin from current conditions and trends.

Both quantitative models and qualitative analyses were used to develop the scenarios. For some drivers (such as land use change and carbon emissions) and ecosystem services (water withdrawals, food production), quantitative projections were calculated using established, peer-reviewed global models. Other drivers (such as rates of technological change and economic growth), ecosystem services (particularly supporting and cultural services, such as soil formation and recreational opportunities), and human well-being indicators (such as human health and social relations) were estimated qualitatively. In general, the qualitative models used for these scenarios addressed incremental changes but failed to address thresholds, risk of extreme events, or impacts of large, extremely costly, or irreversible changes in ecosystem services. These phenomena were addressed qualitatively by considering the risks and impacts of large but unpredictable ecosystem changes in each scenario.

Three of the scenarios—*Global Orchestration, Adapting Mosaic,* and *TechnoGarden* incorporate significant changes in policies aimed at addressing sustainable development challenges. In *Global Orchestration* trade barriers are eliminated, distorting subsidies are removed, and a major emphasis is placed on eliminating poverty and hunger. In *Adapting Mosaic,* by 2010, most countries are spending close to 13% of their GDP on education (as compared to an average of 3.5% in 2000), and institutional arrangements to promote transfer of skills and knowledge among regional groups proliferate. In *TechnoGarden* policies are put in place to provide payment to individuals and companies that provide or maintain the provision of ecosystem services. For example, in this scenario, by 2015, roughly 50% of European agriculture, and 10% of North American agriculture is aimed at balancing the production of food with the production of other ecosystem services. Under this scenario, significant advances occur in the development of environmental technologies to increase production of services, create substitutes, and reduce harmful trade-offs.

• Disease (Goal 6): In the more promising MA scenarios, progress toward Goal 6 is achieved, but under *Order from Strength* it is plausible that health and social conditions for the North and South could further diverge, exacerbating health problems in many low-income regions. Changes in ecosystems influence the abundance of human pathogens such as malaria and cholera as well as the risk of emergence of new diseases. Malaria is responsible for 11% of the disease burden in Africa, and it is estimated that Africa's GDP could have been $100 billion larger in 2000 (roughly a 25% increase) if malaria had been eliminated 35 years ago. The prevalence of the following infectious diseases is particularly strongly influenced by ecosystem change: malaria, schistosomiasis, lymphatic filariasis, Japanese encephalitis, dengue fever, leishmaniasis, Chagas disease, meningitis, cholera, West Nile virus, and Lyme disease.

Three of the four MA scenarios show that significant changes in policies, institutions, and practices can mitigate many of the

Finding #4: The challenge of reversing the degradation of ecosystems while meeting increasing demands for their services can be partially met under some scenarios that the MA considered, but these involve significant changes in policies, institutions, and practices that are not currently under way. Many options exist to conserve or enhance specific ecosystem services in ways that reduce negative trade-offs or that provide positive synergies with other ecosystem services.

negative consequences of growing pressures on ecosystems, although the changes required are large and not currently under way. All provisioning, regulating, and cultural ecosystem services are projected to be in worse condition in 2050 than they are today in only one of the four MA scenarios (*Order from Strength*). At least one of the three categories of services is in better condition in 2050 than in 2000 in the other three scenarios. The scale of interventions that result in these positive outcomes are substantial and include significant investments in environmentally sound technology, active adaptive management, proactive action to address environmental problems before their full consequences are experienced, major investments in public goods (such as education and health), strong action to reduce socioeconomic disparities and eliminate poverty, and expanded capacity of people to manage ecosystems adaptively. However, even in scenarios where one or more categories of ecosystem services improve, biodiversity continues to be lost and thus the long-term sustainability of actions to mitigate degradation of ecosystem services is uncertain.

Past actions to slow or reverse the degradation of ecosystems have yielded significant benefits, but these improvements have generally not kept pace with growing pressures and demands.

Although most ecosystem services assessed in the MA are being degraded, the extent of that degradation would have been much greater without responses implemented in past decades. For example, more than 100,000 protected areas (including strictly protected areas such as national parks as well as areas managed for the sustainable use of natural ecosystems, including timber or wildlife harvest) covering about 11.7% of the terrestrial surface have now been established, and these play an important role in the conservation of biodiversity and ecosystem services (although important gaps in the distribution of protected areas remain, particularly in marine and freshwater systems). Technological advances have also helped lessen the increase in pressure on ecosystems caused per unit increase in demand for ecosystem services.

Substitutes can be developed for some but not all ecosystem services, but the cost of substitutes is generally high, and substitutes may also have other negative environmental consequences. For example, the substitution of vinyl, plastics, and metal for wood has contributed to relatively slow growth in global timber consumption in recent years. But while the availability of substitutes can reduce pressure on specific ecosystem services, they may not always have positive net benefits

Box 2.
Examples of Promising and Effective Responses for Specific Sectors

Illustrative examples of response options specific to particular sectors judged to be promising or effective are listed below. A response is considered effective when it enhances the target ecosystem services and contributes to human well-being without significant harm to other services or harmful impacts on other groups of people. A response is considered promising if it does not have a long track record to assess but appears likely to succeed or if there are known ways of modifying the response so that it can become effective.

Agriculture

Removal of production subsidies that have adverse economic, social, and environmental effects.

Investment in, and diffusion of, agricultural science and technology that can sustain the necessary increase of food supply without harmful tradeoffs involving excessive use of water, nutrients, or pesticides.

Use of response polices that recognize the role of women in the production and use of food and that are designed to empower women and ensure access to and control of resources necessary for food security.

Application of a mix of regulatory and incentive- and market-based mechanisms to reduce overuse of nutrients.

Fisheries and Aquaculture

Reduction of marine fishing capacity.

Strict regulation of marine fisheries both regarding the establishment and implementation of quotas and steps to address unreported and unregulated harvest. Individual transferable quotas may be appropriate in some cases, particularly for cold water, single species fisheries.

Establishment of appropriate regulatory systems to reduce the detrimental environmental impacts of aquaculture.

Establishment of marine protected areas including flexible no-take zones.

Water

Payments for ecosystem services provided by watersheds.

Improved allocation of rights to freshwater resources to align incentives with conservation needs.

Increased transparency of information regarding water management and improved representation of marginalized stakeholders.

Development of water markets.

Increased emphasis on the use of the natural environment and measures other than dams and levees for flood control.

Investment in science and technology to increase the efficiency of water use in agriculture.

Forestry

Integration of agreed sustainable forest management practices in financial institutions, trade rules, global environment programs, and global security decision-making.

Empowerment of local communities in support of initiatives for sustainable use of forest products; these initiatives are collectively more significant than efforts led by governments or international processes but require their support to spread.

Reform of forest governance and development of country-led, strategically focused national forest programs negotiated by stakeholders.

on the environment. Substitution of fuelwood by fossil fuels, for example, reduces pressure on forests and lowers indoor air pollution but it also increases net greenhouse gas emissions. Substitutes are also often costlier to provide than the original ecosystem services.

Ecosystem degradation can rarely be reversed without actions that address the negative effects or enhance the positive effects of one or more of the five indirect drivers of change: population change (including growth and migration), change in economic activity (including economic growth, disparities in wealth, and trade patterns), sociopolitical factors (including factors ranging from the presence of conflict to public participation in decision-making), cultural factors, and technological change. Collectively these factors influence the level of production and consumption of ecosystem services and the sustainability of the production. Both economic growth and population growth lead to increased consumption of ecosystem services, although the harmful environmental impacts of any particular level of consumption depend on the efficiency of the technologies used to produce the service. Too often, actions to slow ecosystem degradation do not address these indirect drivers. For example, forest management is influenced more strongly by actions outside the forest sector, such as trade policies and institutions, macroeconomic policies, and policies in other sectors such as agriculture, infrastructure, energy, and mining, than by those within it.

An effective set of responses to ensure the sustainable management of ecosystems must address the indirect and drivers just described and must overcome barriers related to:

- Inappropriate institutional and governance arrangements, including the presence of corruption and weak systems of regulation and accountability.
- Market failures and the misalignment of economic incentives.
- Social and behavioral factors, including the lack of political and economic power of some groups (such as poor people, women, and indigenous peoples) that are particularly dependent on ecosystem services or harmed by their degradation.
- Underinvestment in the development and diffusion of technologies that could increase the efficiency of use of ecosystem services and could reduce the harmful impacts of various drivers of ecosystem change.

- Insufficient knowledge (as well as the poor use of existing knowledge) concerning ecosystem services and management, policy, technological, behavioral, and institutional responses that could enhance benefits from these services while conserving resources.

All these barriers are further compounded by weak human and institutional capacity related to the assessment and management of ecosystem services, underinvestment in the regulation and management of their use, lack of public awareness, and lack of awareness among decision-makers of both the threats posed by the degradation of ecosystem services and the opportunities that more sustainable management of ecosystems could provide. . . .

Note

1. The eight Millennium Development Goals are: eradicating extreme poverty and hunger, achieving universal primary education, promoting gender equality and empowering women, reducing child mortality, improving maternal health, combatting HIV/AIDS, malaria, and other diseases, ensuring environmental sustainability, and developing a global partnership for development. See www.un.org/millenniumgoals/.

Critical Thinking

1. According to the Millennium Ecosystem Assessment, approximately 60 percent of ecosystem services examined are being degraded or used unsustainably. By who? Where?

2. What are some of the connections between the MDGs and the earth's ecosystems?

3. Which of the four MA scenarios do you see as most realistic by the year 2025? Explain.

4. How might different MA scenarios unfold in different places? Give some examples.

5. What impacts on the health of ecosystem services do you think the population megatrends described in Article 4 might have?

6. Identify in this article three key terms, concepts, or principles that are used in your textbook (environmental science, economics, sociology, history, geography, etc.) or employed in the discipline you are currently studying. (Note: The terms, concepts, or principles may be implicit, explicit, implied, or inferred.)

The Geography of Ecosystem Services

James Boyd

The study of ecosystem services involves two broad missions. The first is a *biophysical* one associated with ecology, hydrology, and the other natural sciences. How can we protect—or, ideally, enhance—the biophysical goods and services necessary to our well-being? If we want clean air and water, healthy and abundant species populations, pollination, irrigation, protection from floods and fires, how can we take action to preserve these things?

The second is an *economic* mission to measure and communicate the value of those goods and services. Quantitative measures help justify interventions to protect natural resources and systems. They also spur government and other decisionmakers to take ecological gains and losses into account.

Geography is essential to both missions. Ecologists and economists of ecosystem services are scrambling to develop skills in mapping, visualization, and the manipulation of data via geospatial information systems. These skills aren't optional. We eventually need to be able to see and manage what can be called the "missing economy of nature," which is absent for several reasons. In general, markets and business activity do not produce and trade ecosystem goods and services. Consequently, the information we use to measure the conventional economy doesn't capture the free public goods provided by natural systems. Besides, nature is inherently complex. How does an action taken in one place affect conditions in another?

In Nature, Some Things Move, Others Stay Put

From an ecological perspective, geography matters because *nature moves*. Air circulates. Water runs downhill. Species migrate. Seeds and pollen disperse. Not only that, the movement of one thing—say water—tends to trigger the movement of other things, like birds and fish. With the goal of managing and protecting ecosystem goods and services, we must understand this web of movement. You could say that in nature, *nothing stays put*. Ecologically, the constant movement and mixing of natural systems is what generates the need for geographic science.

Interestingly, you could also say that in nature, *everything stays put*—an apparent contradiction. A distinctive feature of ecosystem goods and services—once produced—is that they are unmovable. You can't move a lake, river basin, or forest. You

can't ship clean air from one city to another. Birds will migrate where birds migrate. Beautiful mountain trails and scenery can be found in Colorado. Too bad, Kansas. To economists, it is this property of ecosystem goods and services that triggers the need for geography. As any realtor will tell you, three things matter: location, location, location. The same is true for ecosystem goods and services. They're just like houses: if you want to know their value, it's all about the neighborhood.

The Production of Ecosystem Services: Nature in Motion

Think about anything in nature you care about. It could be the beauty of a park, a species you fish for or hunt, or the quality of the air you breathe and water you drink. Now ask the following question: what do those things depend on?

Downstream water quality depends on upstream land uses. The health of Gulf of Mexico fisheries, for example, depends on agricultural practices in the upper Midwest. Air quality in the Adirondacks depends on pollution emissions from the Midwest. Coastal cities and towns depend on nearby wetlands to absorb flood pulses. The point is that the ecosystem goods and services we care about often depend on physical conditions at a great distance from the thing we actually care about. This is a consequence of the continual movement of nature's components.

Accordingly, the biophysical analysis of ecosystem goods and services must be geographic. Treating an ecological problem at the point where it occurs usually doesn't work. It's like putting a band-aid on a lesion caused by an underlying disease. Our ecological diseases—and their cures—are geographic, because ecological systems are geographic.

The challenge for ecosystem scientists and managers is to scientifically relate cause and effect when the cause-and-effect relationship is spatial. We call these relationships *spatial production functions,* because they tell us how an action (good or bad) in one place affects the production of ecosystem goods and services in another. Broadly, we need spatial production functions that describe the dependence of:

- species on the configuration of lands needed for their reproduction, forage, and migration;
- surface and aquifer water volumes and quality on land cover configurations and land uses;

- flood and fire protection services on land cover configurations;
- soil quality on climate variables and land uses; and
- air quality on pollutant emissions, atmospheric processes, and natural sequestration.

The science of these effects is already well underway. For example, we know that stream bank vegetation can improve water quality, help prevent soil erosion, and provide desirable habitat for certain species. But much more remains to be done. We know much less about the exact, empirical relationship between vegetation and water quality.

Why is it such a challenge? First, nature is a highly complex and non-uniform system. Complexity means that causal relationships can only be tested using rigorous, data-intensive empirical and scientific methods that are difficult and costly to perform. Second, non-uniformity means that even if you establish a causal relationship in one location, that relationship may not hold in other locations. Third, empirical analysis of causality requires collaboration between different disciplines (ecology and hydrology, for example). Cross-disciplinary collaboration in any scientific inquiry is always a practical barrier. Finally, the biophysical scientists have many other things to study and have limited financial support for all they are asked to do.

However, deeper understanding of these production functions is necessary if the ecosystem services agenda is to be taken seriously. Ecosystem protection and management will be ineffective at best, and dangerous at worst, if we cannot make credible claims about ecological cause and effect. And the only way to test ecological cause and effect is with spatial— that is, geographic—understanding of biophysical production functions.

The good news is that maps and mapping technology are increasingly capable of capturing and manipulating this data. Water-sheds can now be categorized on the basis of their adjoining land uses. Geographic information system (GIS) tools allow us to "see" migratory pathways and design protections accordingly. As ecology becomes ever more sophisticated in its use of spatial science and data the practical ability to measure cause and effect will become more and more possible.

The Value of Ecosystem Services: Nature's Social Neighborhood

When McDonald's wants to open yet another McDonald's, the first thing the company does is look at a map. Where are the customers? How many competitors are in the vicinity? Do people have easy access from the highway? When economists value ecosystem services, the same kind of things matter. How many people can enjoy the service? Are there other ways to get the service in that neighborhood? Do we have easy access to the service?

Ecosystem goods and services are like houses and fast food outlets because we can't have them shipped to us. They don't move to be near us, we move to be near them. This is most obvi-

ous when we talk about recreation. Usually, outdoor recreation requires us to travel to a park, stream, or forest. But backyard ecosystem services are the same. Chances are you chose your house based in part on its proximity to large trees, open space, clean air, and the likelihood someone interesting might show up at the birdfeeder.

We can make several broad statements about the value of ecosystem goods and services and all of them relate to geography:

- The scarcer an ecological feature, the greater its value.
- The scarcer the substitutes for an ecological feature, the greater its value.
- The more abundant the complements to an ecological feature, the greater its value.
- The larger the population benefiting from an ecological feature, the greater its value.
- The larger the economic value protected or enhanced by the feature, the greater its value.

New York's Central Park makes this point clearly. It is one of the most valuable sources of ecosystem services in the world. Central Park isn't particularly desirable ecologically, but it is nevertheless valuable because so many people live near it and have so few substitutes within walking distance. Geography tells us about all of the factors noted above. We can map population densities, measure distances to similar parks, and easily detect the presence of other types of recreational open space and forms of access like roads.

The general proposition holds for most kinds of ecosystem services. The value of irrigation and drinking water quality depends on how many people depend on the water—which is a function of where they are in relation to the water. Flood damage avoidance services are more valuable the larger the value of lives, homes, and businesses protected from flooding. Species important to recreation (for anglers, hunters, birders, and the like) are more valuable when more people can enjoy them.

Placing a value on ecosystem goods and services also requires us to analyze the presence of substitutes for the good. The value of any good or service is higher the scarcer it is. How do you measure the scarcity of an ecosystem good? If recreation is the source of benefits, substitutes depend on travel times. What are walkable substitutes? Driveable substitutes? The value of irrigation water depends on the availability (and hence location) of alternative water sources. If wetlands are plentiful in an area, then a given wetland may be less valuable as a source of flood pulse attenuation than it might be in a region in which it is the only such resource. In all of these cases, geography is necessary to evaluate the presence of scarcity and substitutes.

Finally, many ecosystem goods and services are valuable only if they are bundled with certain manmade assets. These assets are called "complements" because they complement the value of the ecosystem service. Recreational fishing and kayaking require docks or other forms of access. A beautiful vista yields social value when people have access to it. Access may require infrastructure—roads, trails, parks, housing. Note that these complements may themselves not be transportable. Again, neighborhood matters.

There are exceptions, in which geography is less important to valuation. For example, many of us value the existence of species and wild places *wherever they are*. When it comes to these kinds of ecosystem goods and services, location doesn't matter to our enjoyment, as long as the services exist somewhere. Another important clarification is that everything in nature is valuable if it contributes to the health of the overall system. Here, though, the value arises from the way nature produces services (the realm of the biophysical sciences). When it comes to the consumption of ecosystem goods and services, value tends to be determined by the social neighborhood.

Geographic Information as Technological Revolution

Geographic science will be challenging for both ecologists and economists of ecosystem services. The good news is that our technologies, data, and culture are becoming rapidly more map-focused. Armchair cartographers can already do amazing things with application platforms such as Google Earth. Government agencies and conservancies are making maps available that allow us to see both natural and social landscapes with remarkable detail. This technological revolution is having a cultural effect: maps are everywhere, changing the way we communicate and helping develop our spatial understanding of social and natural phenomena.

The growing deployment of geographic information systems is not without teething problems, however. This is particularly true when it comes to government creation and distribution of geographic information. The U.S. Census Bureau, for example, produces massive quantities of geospatial information on households and businesses. The integration of this information into widely shared, open-source software applications remains awkward, however. Private individuals are stepping in to help solve these problems, but much more could be done by government providers to aid the distribution of geographic information.

Nature is as important to our economy as are farms, factories, and multi-national corporations.

A larger worry is the lack of systematically and consistently tracked environmental information by our government trustees—a worry amply documented by the Government Accountability Office and other watchdog organizations. The greatest need facing us is to understand how we can protect and enhance ecosystem services and predict their loss. Geographic analysis of biophysical production functions is the key. But geographic analysis will rely on detailed ecological information tracked consistently over time. Unfortunately, agencies like the U.S. Environmental Protection Agency, NASA, the U.S. Geological Survey, and the Department of the Interior, among others, are given scant resources and authority to gather such information. Nature is as important to our economy as are farms, factories, and multi-national corporations. Geography is the key to understanding that economy.

Critical Thinking

1. Why is geography essential to the study of ecosystem services?

2. How can geography be used to better understand other environmental issues like population growth, climate change, food/hunger, or diversity?

3. Explain the idea of "spatial production functions" and give examples of how this concept can be applied to the environmental issue of global freshwater shortages.

4. Give three reasons why we cannot really talk about environmental issues and ecosystem services without also including the idea of "place."

5. Identify in this article three key terms, concepts, or principles that are used in your textbook (environmental science, economics, sociology, history, geography, etc.) or employed in the discipline you are currently studying. (Note: The terms, concepts, or principles may be implicit, explicit, implied, or inferred.)

From *Resources*, Fall 2008, pp. 11–15. Copyright © 2008 by Resources for the Future, an independent research institute focused on environmental, energy, and natural resource policies. Reprinted by permission. www.rff.org

Ecosystem Services
How People Benefit from Nature

REBECCA L. GOLDMAN

What do the blue jeans you wear, the hamburger you have for lunch, and the sheet you make your bed with have in common? They all take copious amounts of water to produce. One pair of blue jeans takes 2,900 gallons or about 78 bathtubs of water. Even your morning cup of coffee takes 37 gallons (about one bathtub) of water—not just the one cup you consume. But we don't pay for all the water that goes into our morning cup of coffee. The price of the coffee is based on production and transportation costs (among other costs), but it's much more difficult to value where all the water in one cup of coffee comes from. This difficulty arises from the fact that natural ecosystems are responsible for the retention, release, and regulation of water, but how does a person value a natural ecosystem and the services it provides and put that into the cost of a cup of coffee?

Ecosystem services, or the benefits that nature provides to people, have, in the past decade or two, become a growing focus for the conservation movement, both its science and its policy; see, for example, the Millennium Ecosystem Assessment, the launching of The Ecosystem Marketplace (www.ecosystem marketplace.com), DIVERSITAS (see www.diversitas-inter national.org), among many others. Uses and definitions of ecosystem services vary, but in general, ecosystem services help demonstrate the link between people and nature and the interdependence of our lives on ecosystem-based processes that create the products we need and use every day. Some examples of ecosystem services are water purification, water retention, soil fertility, carbon sequestration, and coastal protection, among many others.

What are some examples of how ecosystem services are already a part of our lives, how might ecosystem service considerations change daily decisions, and why is this behavior change important? In this article, I answer these questions using three examples—pollination services, flood and natural disaster protection services, and water services—to illustrate the interrelationship between nature and people. With each example I provide a policy or personal decision-making context into which this link could be relevant and/or could lead to a different choice or behavior.

I do this for three reasons. First, I want to introduce the concept of ecosystem services in a more tangible way to demonstrate that ecosystem services are not just relevant to academics and conservation practitioners but to everyone. Second, I want to demonstrate the impact that behaviors and choices have on service provision. Finally, I want to underscore the importance of ecosystem services for a sustainable future and how, by fully considering the tradeoffs associated with daily choices, people can, with small changes, make a potentially large difference. I conclude by illustrating some programs, policies, and reports that have truly enveloped this approach and demonstrate the large impact the "ecosystem services movement" can have on our future. First, however, I begin with some general background about ecosystem services and their importance to sustainable development and to conservation.

Why are Ecosystem Services Important for Sustainable Development?

The human population is expected to reach 9 billion people by 2050, and with that increase will come a greater demand for many natural resources. Look at freshwater needs, for example. Research has estimated per person per day dietary needs of 2,000–5,000 liters of water, and this does not include water needed for cleaning and other activities. Hand in hand with this growing demand for resources is the conversion of native ecosystems to meet growing needs; this is where a tradeoff assessment in terms of ecosystem services might be useful.

Agricultural and pasture lands represent about 40 percent of global land surface. If people continue to depend on agricultural products as they have in the past, then by 2050, scholars estimate that 10 hectares of natural ecosystems will be converted to agriculture. This conversion would include a 2.4–2.7-fold increase in nitrogen- and phosphorus-driven eutrophication of numerous waters with similar increases in pesticide use. Agriculture already accounts for 70 percent of water withdrawals from lakes, rivers, and aquifers.

This dependence and use pattern we have with our land is no different in the oceans. One clear example is oysters. Oysters have been consumed for sustenance for millennia. Reports from the 1800s in England indicate that in one year, 700 million

European flat oysters were consumed, a process that employed about 120,000 people. In the Chesapeake Bay on the Eastern Shore of the United States, oyster reefs used to extend for miles. By the 1940s, these reefs had largely disappeared. This is, sadly, true of many ocean creatures, and it underscores our dependence on nature and how a growing demand can affect nature's ability to support itself and us.

We are at a critical juncture. Making choices that can benefit both us and nature may be our best option for securing our livelihoods. Ecosystem services provide a means for people to understand the link between their choices and the natural world. But what exactly are ecosystem services, and how does nature "create" these services? It is to the answers of these questions that I now turn.

What are Ecosystem Services?

Underlying all the resources we use, the species we see, and the foods we eat are ecosystem processes: the biological, chemical, and physical interactions between components of an ecosystem (e.g., soil, water, species). These processes produce benefits to people in the form of clean water, carbon sequestration, and reductions in erosion, among others. These benefits are ecosystem services. The ability of nature to help filter, regulate the release of, and capture and store water allows us to wear blue jeans, drink coffee, and eat a hamburger, but we rarely think about the true origin of the products we use every day.

The distinction between ecosystem process, services, and goods is one that has received a lot of attention in the literature. The key points emerging from these discussions are that ecosystem processes create our natural world. Ecosystem services are the link between this natural world and people, that is, the specific processes that benefit people. (There can be processes that are not services if there is no person to value that particular process.) Ecosystem goods are created from processes and services and are the tangible, material products we are familiar with, but again the distinction between goods and services is complicated and interrelated.

The appeal of ecosystem services for conservation is the connection to people and people's well-being and how that appeal translates into new and increased interest in conservation across a wide range of resource management issues. Ecosystem services can provide a means to value people's well-being in conservation projects and can help advance a set of on-the-ground actions that are equitable, just, and moral. Ecosystem services can be a basis for sustainable development by providing a means to think through how to retain our natural resources for people and for nature with a growing population and therefore an ever-increasing demand for them.

Ecosystem services, since they are the benefits from *nature,* are often discussed in the context of conservation, but in our daily lives we make choices that depend on and affect flows of services from nature, since all goods and products we use today originate from nature and its services. Each choice we make—drive or ride a bus, buy organic or buy regular vegetables, turn on the heat or put on an extra sweatshirt—has tradeoffs. Conserving nature or converting nature does too, but tradeoffs associated with nature's values are often harder to assess. Not understanding nature's role in the products we use means we won't conserve nature sufficiently; this in turn will compromise our ability to access products we need, or we will have to find sometimes costly alternatives for what nature could otherwise provide to us. Incorporating the full suite of costs and benefits into decision-making means evaluating all costs and benefits associated with nature, too. Economists refer to this full valuation as shadow pricing, but even an informal, "back-of-the-envelope" calculation of all values can help to illustrate the importance of ecosystem services in our daily lives.

Ecosystem services help connect people to nature and allow us to make more informed decisions by underscoring all the component pieces of the products we value. How might considering these multiple cost and benefit streams alter people's behaviors? How are people's daily choices linked to ecosystem services? I provide concrete examples to answer these questions.

How Might Ecosystem Services Change Which Products we Purchase at the Supermarket?

To provide a more tangible understanding of the behind-the-scene tradeoffs affecting choices we make, I will use the example of pollination services from native pollinators supported by natural ecosystems. In 2006, around the world there was a great deal of news and concern about the sudden and extreme disappearance of the honey bee, *Apis mellifera,* due to colony collapse disorder (CCD). The disappearance of the honey bee would have catastrophic financial outcomes, since it is the most economically valuable pollinator worldwide. About 90 percent of commercially grown field crops, citrus and other fruit crops, vegetables, and nut crops currently depend on honey bee pollination services. These crops in the United States are valued at $15–20 billion and include Pennsylvania's apple harvest, which is the fourth largest in the nation with an estimated worth of about $45 million per year and California's almond harvest, which accounts for 80 percent of the world's market share of almonds.

The honey bee most common in the United States is native to Europe, and many of the crops pollinated by honey bees today such as watermelon, almonds, blackberries, and raspberries, among others, could be pollinated by bees native to the landscape (often more efficiently), allowing an ecosystem to generate these important services rather than having to import honey bees from elsewhere. Bees generally require food (flower pollen) and habitat (native vegetation) to survive in a landscape. In addition, bees do not fly great distances, so they require patches of native habitat they can easily fly between. The good news is that studies have shown that retaining even small patches of native habitat provides native bees a home and best promotes pollination ecosystem services. If, however, we can ship bees in to provide pollination, why should farmers or the general public care if we eliminate native landscapes and their native pollinators?

One good reason to care is that using nature to provide services can reduce costs for farmers, since wild bee pollinators can be as efficient as managed bees. Farmers wouldn't have to pay the costs of transporting and caring for honey bees to successfully produce their crops. In a coffee plantation in Costa Rica, for example, Ricketts et al. found that services from wild pollinators living in native rainforest patches embedded in the coffee farm were worth $60,000/year, and this is just one service provided by the forest. Other services such as carbon sequestration, soil stabilization, flood mitigation, and water purification could also have added value for the farmer (soil stabilization), for other people in the region (water purification and flood mitigation), and for people around the world (carbon sequestration). Loss of this rainforest, thus, has equivalent costs to society. Yet, right now, despite the plethora of studies that demonstrate which farming practices can greatly impact native bee populations, either encouraging their survival or leading to their demise, many coffee farms depend on the honey bee and not wild pollinators for crop production, making them very vulnerable to extreme consequences of honey bee decline. Why?

The use of native pollinators for crop pollination is not ubiquitous, because this ecosystem service does not come without tradeoffs. Keeping patches of native vegetation on an agricultural landscape means less space for crop production and potentially a reduction in yield. Changing farming practices might mean changing machinery or learning new techniques, which can take time and might have other associated costs. As with any change in practice, there is a risk, and just as consumers have an incentive to buy the least costly vegetables in the market so they can save more money, farmers have an incentive to produce crops as cheaply as possible. Changes in practices sometimes include large, upfront costs that are not offset by benefits or not offset immediately enough, even if in the long run production costs are actually less.

As alluded to, there are tradeoffs associated with destroying native ecosystems and therefore eliminating pollination services. What if more crises occur such as the CCD outbreak leading to further loss of honey bees, and suddenly the cost of importing and keeping these bees becomes unsustainable? Keeping patches of natural ecosystems on the landscape can support a diversity of bee species, making pollinators less vulnerable to being completely annihilated by one disease. What if it would be cheaper to restore some cropland now to ensure some native pollinators are retained, since studies show native bee communities might be an insurance policy in the event of honey bee decline? Who should have to pay for these costs? Should consumers pay $.50 more for a cup of coffee, for example, to help offset potential upfront costs to farmers for using ecosystem pollination services?

According to the soon to be released United Nations TEEB report, conserving nature and ecosystem services might be 10–100 times more valuable than the cost of saving the habitats and species associated with the provision of the services; this demonstrates the potentially major impact of including nature's values. As consumers, it might cost us more at first to buy coffee pollinated by native bees, but in the long run it might be much less costly. This matters for policy, too. For example,

should the U.S. farm bill provide financial incentives for farmers to continue growing crops as we do traditionally, or should it subsidize the same crops but provide additional incentives to farmers to restore patches of native habitat to secure native pollination services in appropriate cases? These types of questions not only get at the complexities associated with ecosystem services and how to extract their exact value (how do you know how much extra to charge on the cup of coffee?), but also underscore the importance of broadening our understanding of all the tradeoffs associated with our everyday choices.

How Might Ecosystem Services Save Your Life or Affect the House You Buy?

As with the food we buy in the market, the houses and property we buy may be more or less valuable depending on the impact we have on nature. Nature can provide services that help to mitigate or at least diminish some potentially catastrophic impacts from weather events. Perhaps the most recent and often discussed example of this is the value that mangrove ecosystems have in protecting against coastal flooding and storms. These protection services can enhance or detract from the value of coastal property, and in the case of severe storms like tsunamis, can help save people's lives.

Mangroves are coastal forest systems and make up about 0.4 percent of the world's forests. They are among the most endangered ecosystems on the planet, yet they are frequently cleared so people can make use of the space they occupy for rice paddies, shrimp farms, or other productive activities. Mangroves provide numerous ecosystem services to people, including nurseries for the young of about 30 percent of commercial fishes, coastal protection to prevent erosion and loss of coastal lands, carbon sequestration that helps to reduce the concentration of carbon dioxide in the atmosphere, waste processing, food production, recreation, and protection against large storm surges. For example, a recent study demonstrated that larger mangrove ecosystems led to significantly greater fish catches without having to increase fishing efforts.

Perhaps one of the most significant, yet undervalued, services from mangroves is their ability to help reduce damage caused by tsunamis and tropical storms. Das and Vincent demonstrated that mangrove forests can save lives in the context of large storms like the tsunami of 1999. Even the minimal coverage of mangroves around coastal villages in India reduced the death toll by one-third. Das and Vincent took the study one step further and estimated that the mangroves could have sold for about $67 million (total of 44,000 acres), but their value in terms of life saving services was at least $80 million. Prior to the tsunami, these life saving services may not have been valued, and the consequences could have been grave.

Coastal protection is another important ecosystem service provided by mangroves, which can have important implications for a more basic choice: whether or not to buy a house. Property values are assessed based on a wide range of factors including taxes, location, views, noise pollution, etc., but without factoring in ecosystem services, or the lack thereof, one might pay an inappropriate amount for a house. For example, coastal properties are often the most costly because they have lovely views. If the coastal areas have coral reefs or mangroves, this is no less true; however, through destructive fishing practices and/or overuse, many of these reefs and forests are being destroyed. What once may have been prime real estate right on the water is now a home that is threatened with destruction at any time since it can be easily flooded by a strong storm surge or the ground below it could erode away. The loss of protective ecosystem services from the natural ecosystems may actually mean the coastal home you just bought was vastly overpriced.

Again, the choices are not so black and white. Denying poor coastal communities' access to mangroves for income purposes has costs, just as preserving the mangroves has benefits for human lives. Tradeoffs must be evaluated, and the range of costs and benefits must be included, but without accounting for ecosystem services values, you can't be sure if you should buy the coastal home or if you should cut down the mangroves to be able to sell more fish. More broadly, governments can't assess what policies might be needed to secure the well-being of their citizens. For example, should the Philippine or Indian governments regulate mangrove cutting more stringently while channeling funds into providing incentives to poor coastal dwellers to compensate them for lost income and further decrease the threat to mangrove systems?

What do Ecosystem Services Mean for the Water you Drink?

People rely on clean, regular supplies of water for survival, whether for drinking or for the production of other goods and services (agriculture, electricity, etc.). Water users have an incentive to find the lowest cost options for accessing clean water. Interestingly, nature may provide the lowest-cost, longest-term means of providing such water services. Conservationists are increasingly recognizing the value in thinking about these low-cost approaches for financing conservation. Demonstrating the links between nature and people is a way to engage new stakeholders in conservation and potentially find new ways to finance activities.

Increasing numbers of case studies and research studies are emerging about the benefits of payment for ecosystem service approaches as a means to finance conservation. While there is still uncertainty about what factors are likely to contribute to successful ecosystem service projects, these approaches continue to proliferate. Payments for watershed service projects make up a significant portion of implemented ecosystem services schemes (many others relate to carbon). These schemes often involve water users paying "suppliers" for the delivery of clean, consistent water supplies. Using a payment for watershed services approach, called "water funds," developed by The Nature Conservancy (TNC) and several partners, I will provide a tangible example of an ecosystem services approach that has changed the way users are securing access to water.

Water funds are a public/private partnership focused around a long-term, sustainable finance source for conservation. The partnership determines how to fund conservation of the watershed in order to protect valuable biodiversity and to generate vital water services (namely a clean, regular supply of water) upon which large groups of downstream people depend. For context, water funds are proliferating throughout the Andean region of South America, particularly in Colombia and Ecuador though they are not only in this region. The headwaters of important rivers originate in higher altitude natural ecosystems (composed of native grasslands, páramo, and mixed forest) that serve as the hydrologic regulators for the entire water system.

The problems in these Andean systems are three-fold: growing populations of downstream users require increased flows of services, the natural ecosystems that provide the services are not sufficiently protected, and the human communities that threaten the natural ecosystems are poor and depend upon these ecosystems for their livelihoods. In addition, their land use practices can have their own consequences for service provision, as farming and ranching can lead to reduced water retention and increased water pollution. But, water services can't be preserved solely by keeping people out of the natural areas and restoring the working landscapes, as this would compromise many livelihoods. Using an ecosystem services framework, TNC and partners used the dependence of downstream people on the services provided by natural ecosystems and restored working landscapes to finance conservation and livelihood projects to secure water services sustainably.

In a water fund, water users voluntarily invest money in a trust fund, and the revenue (interest and sometimes part of the principal) from it is used to finance conservation projects in the watershed, which are decided upon by the public/private partnership of users and key stakeholders that oversee the fund. These projects take steps to address the needs of preserving natural ecosystems and maintaining the well-being of watershed communities. Activities and projects can include hiring community-based park guards to maintain the natural areas (to maintain the natural hydrologic regulation of the system, which helps maintain a regular base flow), protecting riparian areas (putting fences up to keep crops and cows away from river banks), re-vegetating riparian areas (to provide a natural filter for sediments and other pollutants), planting live tree fences to delineate property boundaries, and isolating/fencing off headwaters and steep slopes. These practices can have major impacts on water quality, on the timing and volume of water flows (particularly floods), on fires, and on freshwater biodiversity. One study demonstrated that just maintaining natural vegetation on the landscape can decrease sedimentation tenfold compared with converting the area to cropland.

Such management is not without costs, however, and thus the water fund not only finances conservation management projects but also supports community projects to compensate for impacts on livelihoods. Ideally, conservation management activities will enhance farm/ranch productivity through the production of on-farm ecosystem services such as soil stabilization and enhanced soil fertility, but these benefits will not be immediate and are not guaranteed. In the shorter term, conservation management agreements include livelihood investments such as environmental education programs, additional income sources such as guinea pig farms, alternative food sources such as organic vegetable gardens, and expanded capacity for the production of goods such as providing communities with ovens to make the drying of fruit and herbs they sell on the market more efficient and effective.

The premise of water funds, therefore, is that securing natural ecosystems and improving management of farms and ranches in the watershed will help to ensure users have clean water available to them year round . . .

In this case, taking an ecosystem services approach meant being holistic and recognizing that the downstream users of water have an incentive to find the lowest cost option for continued access to clean water and that nature can provide that service potentially more cheaply and for longer than built infrastructure. The premise of water funds, therefore, is that securing natural ecosystems and improving management of farms and ranches in the watershed will help to ensure users have clean water available to them year round, and in return for these services, users pay for the upstream conservation management. This approach has had tremendous replication success in this region of South America, with the implementation of seven water funds in the last decade serving cities with combined populations of over 11 million people and helping protect over 1.6 million hectares of land.

How are Ecosystem Service Approaches Being Leveraged?

The last five years have seen the proliferation of ecosystem services strategies, not just in on-the-ground actions but also in the emergence of new offices, new projects, and new strategies within conservation NGOs, governments, and multilateral donor agencies. This increased attention started with books such as *The New Economy of Nature* and interdisciplinary scholarly investigations such as the *Millennium Ecosystem Assessment* (MA) demonstrating the ecosystem alternatives to resource problems. The MA was called for by the United Nations secretary general in 2000 as a way to assess the impact of ecosystem change on human well-being including a scientific assessment of how to increase the conservation and sustainable use of these ecosystems to secure well-being. Over 1,360 scholars worldwide collaborated on the study publishing the findings in five technical volumes and six synthesis reports (see www.millennium assessment.org/en/index.aspx for more information).

The concept has now been integrated into funding criteria by donor agencies. The World Bank, for example, has an entire strategy dedicated towards developing payment for environmental services (see http://go.worldbank.org/51KUO12O50). In addition, one of the major donor programs of the World Bank, the Global Environment Facility (GEF), has developed a Scientific and Technical Advisory Panel (STAP) to provide guidance on the evaluation of environmental service projects that are seeking funding, since these types of projects are becoming increasingly popular for achieving development and conservation objectives (see http://stapgef.unep.org/resources/ sg/PES). This attention from the development community is a clear link between ongoing conservation efforts and efforts to help enhance global development.

Beyond multilaterals, governments are also paying more attention to the benefits of nature to people. This attention may be best illustrated by the relatively recent formation of an entire office in the U.S. government's Department of Agriculture that is dedicated to catalyzing new markets for ecosystem services (see www.fs.fed.us/ecosystemservices/OEM/index.shtml). This Office of Environmental Markets was formed in 2008 and aims to help support farm bill programs in the U.S., which can have significant benefits for service provision throughout the country.

There is also a growing and rapidly evolving focus on ecosystem services by conservation nonprofits. Organizations such as Conservation International, TNC, and WWF are changing their missions and/or developing projects that focus on integrating people and nature. New partnerships, notably, the Natural Capital Project—a partnership between Stanford University, TNC, and WWF—are being created to provide tools (see www.naturalcapitalproject.org/InVEST.html) to map the flow of various services in geographies around the world and to help more effectively include natural capital in decision-making.

Finally, there is recognition across government, particularly recently by the German Federal Ministry and the European Commission, with other partners, of the importance of making sure we include the true costs of lost biodiversity and associated ecosystem services on our future. In an effort to fully understand the benefits nature has for people's well-being, this group commissioned the TEEB report to sharpen awareness, help facilitate creation of cost-effective policies, and help make better informed decisions

Over the next decade, measuring the impact of ecosystem service–based approaches both on people's well-being and also on nature is critical. Understanding which activities and actions yield a particular outcome and at what cost will provide us with a more evidence-based suite of options for making informed choices in our daily lives.

Critical Thinking

1. Outline briefly how people benefit from nature.
2. What are ecosystem services?
3. If ecosystems did not provide benefits for humans, would we care about maintaining them?

4. How do the benefits of ecosystem services for people living in wealthy, industrialized countries differ from people living in rural, impoverished regions of the world?

5. What is the relationship between a "habitat" and an "ecosystem"? Can you commit habitat degradation and still maintain ecosystem stability? Explain.

6. Identify in this article three key terms, concepts, or principles that are used in your textbook (environmental science, economics, sociology, history, geography, etc.) or employed in the discipline you are currently studying. (Note: The terms, concepts, or principles may be implicit, explicit, implied, or inferred.)

UNIT 8

Quest for Power

Unit Selections

Learning Outcomes

After reading this Unit, you should be able to:

- Identify the key issues surrounding our oil energy dependence and offer insights regarding possible future scenarios of environmental issue #8: our quest for power (energy).
- Discuss the world's latest "infatuations" with alternative energy sources.
- Outline limitations and challenges of these alternatives.
- Identify seven myths about alternative energy.
- Define "peak oil" and asses the validity of its argument.
- Offer some insight into why the demand for oil is changing.
- Describe how the "geopolitical" landscape of oil production and consumption changes.
- Summarize the changes Americans made in 2008 regarding their gasoline consumption behavior and compare what behavior changes remain today.

Student Website

www.mhhe.com/cls

Internet References

Alliance for Global Sustainability (AGS)
http://globalsustainability.org

Alternative Energy Institute (AEI)
www.altenergy.org

Arctic Climate Impact Assessment
www.acia.uaf.edu

Department of Energy—Energy Efficiency and Renewable Energy
www.eere.energy.gov

Energy and the Environment: Resources for a Networked World
http://zebu.uregon.edu/energy.html

Fuel Economy (Department of Energy)
www.fueleconomy.gov

Global Climate Change
http://climate.jpl.nasa.gov

Institute for Global Communication/EcoNet
www.igc.org

Nuclear Power Introduction
http://library.thinkquest.org/17658/pdfs/nucintro.pdf

U.S. Department of Energy
www.energy.gov

I have no doubt that we will be successful in harnessing the sun's energy. . . . If sunbeams were weapons of war, we would have had solar energy centuries ago.

—*Sir George Porter,* quoted in The Observer,
26 August 1973

Top Ten Environmental Issue 8 is *energy* (the capacity to do work). Or, is the issue really *power* (possession of control, authority, or influence over others)? With the coming of the twenty-first century, the two terms have become increasingly confused, or perhaps more correctly, synonymous. If you don't have oil, what do you have? Not much. But maybe that is a good thing for when the oil runs out. You'll have little to lose. Who's afraid of "peak oil"? The peasant intensive subsistent farmers of Southeast Asia? The industrialized world? If oil equals power, then petroleum-based societies could be rendered impotent if they cannot find energy alternatives, power alternatives to petroleum. After all, consider the United States with only 5 percent of the world's population but consuming 25 percent of its energy resources. Or consider the "industrialized" world with 25 percent of the world's population but consuming a whopping 80 percent of the earth's resources. That's beyond energy: that is sheer power.

Although "seeking clean energy alternatives" sounds environmentally correct, a cursory review of the geopolitical economic history of oil makes it quite evident that it's been about the *power* it affords the beholder. The energy of oil is nice as well, to make stuff, raise standards of living, fuel material production, and energize consumption. But the power it affords those who have the means to access it, to produce it, to consume it is intoxicating. The energy of oil has allowed peoples to control their environment and modify it to their needs; to hold ultimate authority over deciding who gets what, thus constructing a planet of haves and have-nots; and to leverage influence over global development policy choices, to name only a few of its powerful benefits. However, **Unit 8** suggests the resource landscape of earth's first and foremost source of energy is changing, and so too are its traditional geopolitical patterns of power. New energy sources, new energy source places, new energy technologies, new energy players, and new energy environmental degradations threatening the continued well-being of the human race are all converging to write the next chapter in the human races' quest for power.

Professor Vaclav Smil's Article 27, "Global Energy: The Latest Infatuations," opens this unit with a review of some of the energy hopes and dreams most likely to appear in our next chapter on energy. Much infatuation seems to be centered around transportation, energy/power transmission transitions, and dealing with the excessive current surplus of our carbon emissions. Should we be asking: Are these *global* infatuations shared by all peoples in all nations? Or are these the hopes and dreams of an *industrialized/ing* world scrambling to not only find new energy but also to maintain old power? Setting aside these questions for the moment, the author points out that although most proposed energy alternatives are reasonable, they are wrought with economic limitations, technical challenges, and exaggerated expectations. Time is of the essence, specifically the cultural,

© BananaStock/PunchStock

social, and political transition times needed to implement changes. But Professor Smil sees more fundamental, more worrisome problems emerging from our questions. Most of humanity of the "developing economies" see a *need to increase* energy consumption a great deal more to experience healthy lives and a modicum of prosperity. On the contrary, affluent nations *need to reduce* their excessive energy use. In conclusion, the author asks, "Could the world's richest countries go wrong by striving for moderation of their energy use?" Probably not until energy no longer equals power.

Author Michael Grunwald agrees with Smil's conclusion and expands on Smil's infatuations theme. Article 28 explores the "seven myths" about energy alternatives and argues that there will be no magic ticket for an alternative to oil. Proposed alternatives and technological fixes still have a lot to prove in, again, very little time. What may be the most effective saving grace, according to the author, is for the industrialized world to "unplug a few digital picture frames, substitute teleconferencing for some business travel, and take it easy on the air conditioner." We need to not only accept we may have to do less, with less, but also we need to encourage the developing world to "Do as we're doing, not as we did."

That oil is a finite energy resource and will eventually, some day, run out is a simple fact. When it will run out, however, is a highly contentious discourse. Article 29, "Half a Tank: The Impending Arrival of Peak Oil," reviews the "peak oil" theory and argues that it may be close at hand. However, Mark Floegel's article is included here because of his effort to convince the reader to consider a world without oil, as well as his referencing the importance of the need for "relocalization" as a way to ready ourselves for that world without oil.

Oil supplies, much like "global warming," present a pretty familiar picture today. But according to Article 30, "It's Still the One," the changing patterns of global oil "demand," global alternatives, global traders, global consumers, and the global geopolitical economics of oil are creating a new energy landscape nothing short of *terra incognita*. The article underscores the importance of appreciating not only the patterns and linkages embedded in the environmental issue of the quest for power, but also that the "energy issue" is more than just about fueling the world.

Finally, Unit 8 concludes with a short survey, Article 31, "Gas Costs Squeeze Daily Life: Survey Reveals How High Prices Have Pushed Us into New Routines." The article is included here to provide the reader with an opportunity to critically reflect on one of the primary components involved in the environmental issue of *energy.* And the component is us, Americans. Note the article was published in 2008, and we have seen significant advances in automobile technologies, but we might consider some questions that the article raises. For example, "What of our fuel consumption behavior has changed in 2012?" "What changes in our American transportation infrastructures have changed?" "How have higher gas prices pushed the other 95 percent (the United States represents 5 percent) of the world's population into new routines?" And finally, "What does this survey say about the other article topics presented in this unit?"

Critical thinking about environmental issues can oftentimes be uncomfortable, as can dieting and physical fitness. But the results can oftentimes save our life. Ultimately, we will have to address several questions in any global discourse about our quest for power. What really is the energy task at hand? What is our energy alternatives endgame? Do we need solar-powered cars or solar-powered minitractors? What exactly do we intend to do with this new energy—make more and more stuff? Or make a better and better world for all humanity? These are very different agendas.

Global Energy

The Latest Infatuations

In energy matters, what goes around, comes around—but perhaps should go away.

VACLAV SMIL

To follow global energy affairs is to have a never-ending encounter with new infatuations. Fifty years ago media ignored crude oil (a barrel went for little more than a dollar). Instead the western utilities were preoccupied with the annual double-digit growth of electricity demand that was to last indefinitely, and many of them decided that only large-scale development of nuclear fission, to be eventually transformed into a widespread adoption of fast breeder reactors, could secure electricity's future. Two decades later, in the midst of the second energy "crisis" (1979–1981, precipitated by Khomeini's takeover of Iran), rising crude oil prices became the world's prime existential concern, growth of electricity demand had slumped to low single digits, France was the only nation that was seriously pursuing a nuclear future, and small cars were in vogue.

After world crude oil prices collapsed in 1985 (temporarily below $5 per barrel), American SUVs began their rapid diffusion that culminated in using the Hummer H1, a civilian version of a U.S. military assault vehicle weighing nearly 3.5 tonnes, for trips to grocery stores—and the multinational oil companies were the worst performing class of stocks of the 1990s. The first decade of the 21st century changed all that, with constant fears of an imminent peak of global oil extraction (in some versions amounting to nothing less than lights out for western civilization), catastrophic consequences of fossil fuel–induced global warming and a grand unraveling of the post–WW II world order.

All of this has prompted incessant calls for the world to innovate its way into a brighter energy future, a quest that has engendered serial infatuations with new, supposedly perfect solutions: Driving was to be transformed first by biofuels, then by fuel cells and hydrogen, then by hybrid cars, and now it is the electrics (Volt, Tesla, Nissan) and their promoters (Shai Agassi, Elon Musk, Carlos Ghosn) that command media attention; electricity generation was to be decarbonized either by a nuclear renaissance or by ubiquitous wind turbines (even Boone Pickens, a veteran Texas oilman, succumbed to that call of the wind), while others foresaw a comfortable future for fossil fuels once their visions of mass carbon capture and sequestration (CCS) were put in practice. And if everything fails, then geoengineering—manipulating the Earth's climate with shades in space, mist-spewing ships or high-altitude flights disgorging sulfur compounds—will save us by cooling the warming planet.

This all brings to mind Lemuel Gulliver's visit to the grand academy of Lagado: No fewer than 500 projects were going on there at once, always with anticipation of an imminent success, much as the inventor who "has been eight years upon a project for extracting sunbeams out of cucumbers" believed that "in eight years more, he should be able to supply the governor's gardens with sunshine, at a reasonable rate"—but also always with complaints about stock being low and entreaties to "give . . . something as an encouragement to ingenuity." Admittedly, ideas for new energy salvations do not currently top 500, but their spatial extent puts Lagado's inventors to shame: Passionately advocated solutions range from extracting work from that meager 20-Kelvin difference between the surface and deep waters in tropical seas (OTEC: ocean thermal energy conversion) to Moon-based solar photovoltaics with electricity beamed to the Earth by microwaves and received by giant antennas.

And continuous hopes for success (at a low price) in eight more years are as fervent now as they were in the fictional 18th century Lagado. There has been an endless procession of such claims on behalf of inexpensive, market-conquering solutions, be they fuel cells or cellulosic ethanol, fast breeder reactors or tethered wind turbines. And energy research can never get enough money to satisfy its promoters: In 2010 the U.S. President's council of advisors recommended raising the total for U.S. energy research to $16 billion a year; that is actually

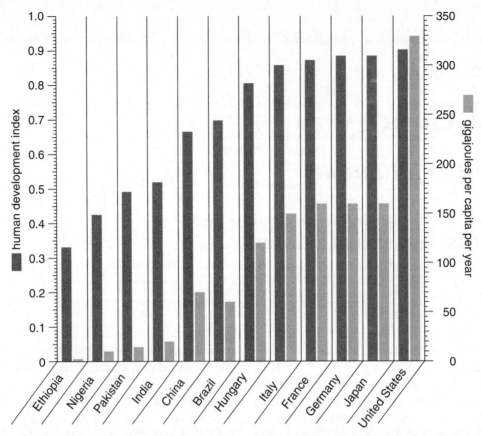

Figure 1 At very low and low per capita consumption levels, higher use of energy is clearly tied to rising index of human development, but once energy per capita reaches about 150 gigajoules per year, the correlation breaks down. More is not better.

too little considering the magnitude of the challenge—but too much when taking into account the astonishing unwillingness to adopt many readily available and highly effective existing fixes in the first place.

Enough to Go Around?

Although all this might be dismissed as an inevitable result of the desirably far-flung (and hence inherently inefficient) search for solutions, as an expected bias of promoters devoted to their singular ideas and unavoidably jockeying for limited funds, I see more fundamental, and hence much more worrisome, problems. Global energy perspective makes two things clear: Most of humanity needs to consume a great deal more energy in order to experience reasonably healthy lives and to enjoy at least a modicum of prosperity; in contrast, affluent nations in general, and the United States and Canada in particular, should reduce their excessive energy use. While the first conclusion seems obvious, many find the second one wrong or outright objectionable.

In 2009 I wrote that, in order to retain its global role and its economic stature, the United States should

provide a globally appealing example of a policy that would simultaneously promote its capacity to innovate, strengthen its economy by putting it on sounder fiscal

foundations, and help to improve Earth's environment. Its excessively high per-capita energy use has done the very opposite, and it has been a bad bargain because its consumption overindulgence has created an enormous economic drain on the country's increasingly limited financial resources without making the nation more safe and without delivering a quality of life superior to that of other affluent nations.

I knew that this would be considered a nonstarter in the U.S. energy policy debate: Any calls for restraint or reduction of North American energy use are still met with rejection (if not derision)—but I see that quest to be more desirable than ever. The United States and Canada are the only two major economies whose average annual per capita energy use surpasses 300 gigajoules (an equivalent of nearly 8 tonnes, or more than 50 barrels, of crude oil). This is twice the average in the richest European Union (E.U.) economies (as well as in Japan)—but, obviously, Pittsburghers or Angelenos are not twice as rich, twice as healthy, twice as educated, twice as secure or twice as happy as inhabitants of Bordeaux or Berlin. And even a multiple adjustment of national per capita rates for differences in climate, typical travel distances and economic structure leaves most of the U.S.–E.U. gap intact: This is not surprising once it is realized that Berlin has more degree heating days than Washington D.C., that red peppers travel the same distance in

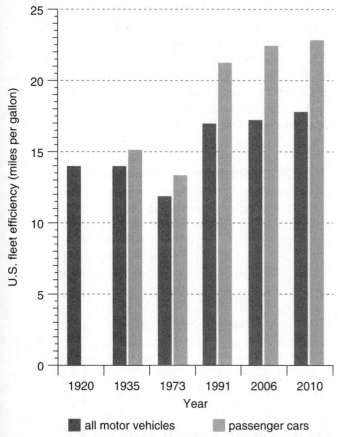

Figure 2 Although the efficiency of internal combustion engines has increased substantially in the past 90 years (particularly when the adoption of diesel-powered cars is taken into account), the average performance of motor vehicles in the Unites States has improved only from about 14 miles per gallon to about 18 mpg.

refrigerated trucks from Andalusia to Helsinki as they do from California's Central Valley to Illinois, and that German exports of energy-intensive machinery and transport-equipment products surpass, even in absolute terms, U.S. sales.

Moreover, those who insist on the necessity and desirability of further growth of America's per capita energy use perhaps do not realize that, for a variety of reasons, a plateau has been reached already and that (again for many reasons) any upward departures are highly unlikely. In 2010 U.S. energy consumption averaged about 330 gigajoules (GJ) per capita, nearly 4 percent lower than in 1970, and even the 2007 (pre-crisis) rate of 355 gigajoules per capita was below the 1980 mean of 359 GJ. This means that the U.S. per capita consumption of primary energy has remained essentially flat for more than one generation (as has British energy use). How much lower it could have been can be illustrated by focusing on a key consumption sector, passenger transport.

Planes, Trains and Automobiles

After 1985 the United States froze any further improvements in its corporate automobile fuel efficiency (CAFE),

encouraged a massive diffusion of exceptionally inefficient SUVs and, at the same time, failed to follow the rest of [the] modernizing world in building fast train links. For 40 years the average performance of the U.S. car fleet ran against the universal trend of improving efficiencies: By 1974 it was lower (at 13.4 miles per gallon [mpg]) than during the mid-1930s! Then the CAFE standards had doubled the efficiency of new passenger cars by 1985, but with those standards subsequently frozen and with the influx of SUVs, vans and light trucks, the average performance of the entire (two-axle, four-wheel) car fleet was less than 26 mpg in 2006 or no better than in 1986—while a combination of continued CAFE upgrades, diffusion of new ultra low-emission diesels (inherently at least 25–30 percent more efficient than gasoline-powered cars) and an early introduction of hybrid drives could have raised it easily to more than 35 or even 40 mpg, massively cutting the U.S. crude oil imports for which the country paid $1.5 trillion during the first decade of the 21st century.

And the argument that its large territory and low population density prevents the United States from joining a growing list of countries with rapid trains (traveling 250–300 kilometers per hour or more) is wrong. The northeastern megalopolis (Boston–Washington) contains more than 50 million people with average population density of about 360 per square kilometer and with nearly a dozen major cities arrayed along a relatively narrow and less than 700-kilometer long coastal corridor. Why is that region less suited to a rapid rail link than France, the pioneer of European rapid rail transport, with a population of 65 million and nationwide density of only about 120 people per square kilometer whose *trains à grande vitesse* must radiate from its capital in order to reach the farthest domestic destinations more than 900 kilometers away? Apparently, Americans prefer painful trips to airports, TSA searches and delayed shuttle flights to going from downtown to downtown at 300 kilometers per hour.

In a rational world animated by rewarding long-term policies, not only the United States and Canada but also the European Union should be boasting about gradual reductions in per capita energy use. In contrast, modernizing countries of Asia, Latin America and, most of all, Africa lag so far behind that even if they were to rely on the most advanced conversions they would still need to at least quadruple (in India's case, starting from about 20 GJ per capita in 2010) their per capita supply of primary energy or increase their use by more than an order of magnitude—Ethiopia now consumes modern energies at a rate of less than 2 GJ per capita—before getting to the threshold of a decent living standard for most of their people and before reducing their huge internal economic disparities.

China has traveled further, and faster, along this road than any other modernizing nation. In 1976 (the year of Mao Zedong's death) its average per capita energy consumption was less than 20 GJ per capita, in 1990 (after the first decade of Deng Xiaoping's modernization) it was still below 25 GJ, and

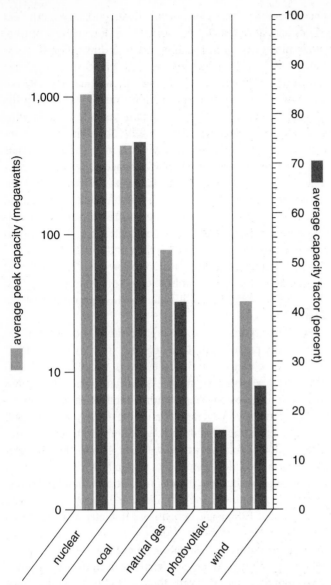

Figure 3 Generation of electricity by wind turbines and photovoltaic (PV) cells differs in two fundamental ways from thermal electricity production. First, as shown in the left column, average capacities of photovoltaic and wind farms are smaller than those of nuclear, coal and even natural gas-powered generators. (Note the logarithmic scale.) Second, the percentage of time that the generators can work at full capacity (load factor) is much lower. (The capacity factor of gas-fired generators is constrained not by their ability to stay online but by their frequent use as intermittent sources to meet demand peaks.) Moreover, differences in capacity factors will always remain large. In 2009 the load factor averaged 74 percent for U.S. coal-fired stations, and the nuclear ones reached 92 percent, whereas wind turbines managed only about 25 percent. (All plots show the U.S. averages in 2009.)

a decade later it had just surpassed 30 GJ per capita. By 2005 the rate had approached 55 GJ and in 2010 it reached 70 or as much as some poorer E.U. countries were consuming during the 1970s. Although China has become a major importer of crude oil (now the world's second largest, surpassed only by the United States) and it will soon be importing large volumes

of liquefied natural gas and has pursued a large-scale program of developing its huge hydrogenation potential, most of its consumption gains have come from an unprecedented expansion of coal extraction. While the U.S. annual coal output is yet to reach one billion tonnes, China's raw coal extraction rose by one billion tonnes in just four years between 2001 and 2005 and by nearly another billion tonnes by 2010 to reach the annual output of 3 billion tonnes.

China's (and, to a lesser degree, India's) coal surge and a strong overall energy demand in Asia and the Middle East have been the main reason for recent rises of CO_2 emissions: China became the world's largest emitter in 2006, and (after a small, economic crisis-induced, decline of 1.3 percent in 2009) the global total of fossil fuel-derived CO_2 emissions set another record in 2010, surpassing 32 billion tonnes a year (with China responsible for about 24 percent). When potential energy consumption increases needed by low-income countries are considered together with an obvious lack of any meaningful progress in reducing the emissions through internationally binding agreements (see the sequential failures of Kyōtō, Bali, Copenhagen and Cancún gatherings), it is hardly surprising that technical fixes appear to be, more than ever, the best solution to minimize future rise of tropospheric temperatures.

Renewable Renaissance?

Unfortunately, this has led to exaggerated expectations rather than to realistic appraisals. This is true even after excluding what might be termed zealous sectarian infatuations with those renewable conversions whose limited, exceedingly diffuse or hard-to-capture resources (be they jet stream winds or ocean waves) prevent them from becoming meaningful economic players during the next few decades. Promoters of new renewable energy conversions that now appear to have the best prospects to make significant near-term contributions—modern biofuels (ethanol and biodiesel) and wind and solar electricity generation—do not give sufficient weight to important physical realities concerning the global shift away from fossil fuels: to the scale of the required transformation, to its likely duration, to the unit capacities of new convertors, and to enormous infrastructural requirements resulting from the inherently low power densities with which we can harvest renewable energy flows and to their immutable stochasticity.

The scale of the required transition is immense. Ours remains an overwhelmingly fossil-fueled civilization: In 2009 it derived 88 percent of its modern energies (leaving traditional biomass fuels, wood and crop residues aside) from oil, coal and natural gas whose global market shares are now surprisingly close at, respectively, 35, 29 and 24 percent. Annual combustion of these fuels has now reached 10 billion tonnes of oil equivalent or about 420 exajoules (420×10^{18} joules). This is an annual fossil fuel flux nearly 20 times larger than at the beginning of the 20th century, when the epochal transition from biomass fuels had just passed its pivotal point (coal and oil began to account for more than half of the global energy supply sometime during the late 1890s).

Energy transitions—shifts from a dominant source (or a combination of sources) of energy to a new supply arrangement, or from a dominant prime mover to a new converter—are inherently prolonged affairs whose duration is measured in decades or generations, not in years. The latest shift of worldwide energy supply, from coal and oil to natural gas, illustrates how the gradual pace of transitions is dictated by the necessity to secure sufficient resources, to develop requisite infrastructures and to achieve competitive costs: It took natural gas about 60 years since the beginning of its commercial extraction (in the early 1870s) to reach 5 percent of the global energy market, and then another 55 years to account for 25 percent of all primary energy supply. Time spans for the United States, the pioneer of natural gas use, were shorter but still considerable: 53 years to reach 5 percent, another 31 years to get to 25 percent.

Displacing even just a third of today's fossil fuel consumption by renewable energy conversions will be an immensely challenging task; how far it has to go is attested by the most recent shares claimed by modern biofuels and by wind and photovoltaic electricity generation. In 2010 ethanol and biodiesel supplied only about 0.5 percent of the world's primary energy, wind generated about 2 percent of global electricity and photovoltaics (PV) produced less than 0.05 percent. Contrast this with assorted mandated or wished-for targets: 18 percent of Germany's total energy and 35 percent of electricity from renewable flows by 2020, 10 percent of U.S. electricity from PV by 2025 and 30 percent from wind by 2030 and 15 percent, perhaps even 20 percent, of China's energy from renewables by 2020.

Unit sizes of new converters will not make the transition any easier. Ratings of 500–800 megawatts (MW) are the norm for coal-fired turbogenerators and large gas turbines have capacities of 200–300 MW, whereas typical ratings of large wind turbines are two orders of magnitude smaller, between 2 and 4 MW, and the world's largest PV plant needed more than a million panels for its 80 MW of peak capacity. Moreover, differences in capacity factors will always remain large. In 2009 the load factor averaged 74 percent for U.S. coal-fired stations and the nuclear ones reached 92 percent, whereas wind turbines managed only about 25 percent—and in the European Union their mean load factor was less than 21 percent between 2003 and 2007, while the largest PV plant in sunny Spain has an annual capacity factor of only 16 percent.

As I write this a pronounced high pressure cell brings deep freeze, and calm lasting for days, to the usually windy heart of North America: If Manitoba or North Dakota relied heavily on wind generation (fortunately, Manitoba gets all electricity from flowing water and exports it south), either would need many days of large imports—yet the mid-continent has

no high-capacity east-west transmission lines. Rising shares of both wind and PV generation will thus need considerable construction of new long-distance high-voltage lines, both to connect the windiest and the sunniest places to major consumption centers and also to assure uninterrupted supply when relying on only partially predictable energy flows. As the distances involved are on truly continental scales—be they from the windy Great Plains to the East Coast or, as the European plans call for, from the reliably sunny Sahara to cloudy Germany (Desertec plan)—those expensive new super grids cannot be completed in a matter of years. And the people who fantasize about imminent benefits of new smart grids should remember that the 2009 report card on the American infrastructure gives the existing U.S. grid a near failing grade of D+.

And no substantial contribution can be expected from the only well-tested non-fossil electricity generation technique that has achieved significant market penetration: Nuclear fission now generates about 13 percent of global electricity, with national shares at 75 percent in France and about 20 percent in the United States. Nuclear engineers have been searching for superior (efficient, safe and inexpensive) reactor designs ever since it became clear that the first generation of reactors was not the best choice for the second, larger, wave of nuclear expansion. Alvin Weinberg published a paper on inherently safe reactors of the second nuclear era already in 1984, at the time of his death (in 2003) Edward Teller worked on a design of a thorium-fueled underground power plant, and Lowell Wood argues the benefits of his traveling-wave breeder reactor fueled with depleted uranium whose huge U.S. stockpile now amounts to about 700,000 tonnes.

But since 2005, construction began annually on only about a dozen new reactors worldwide, most of them in China where nuclear generation supplies only about 2 percent of all electricity, and in early 2011 there were no signs of any western nuclear renaissance. Except for the completion of the Tennessee Valley Authority's Watts Bar Unit 2 (abandoned in 1988, scheduled to go on line in 2012), there was no construction underway in the United States, and the completion and cost overruns of Europe's supposed new showcase units, Finnish Olkiluoto and French Flamanville, were resembling the U.S. nuclear industry horror stories of the 1980s. Then, in March 2011, an earthquake and tsunami struck Japan, leading to Fukushima's loss of coolant, destruction of reactor buildings in explosions and radiation leaks; regardless of the eventual outcome of this catastrophe, these events will cast a long suppressing shadow on the future of nuclear electricity.

Technical Fixes to the Rescue?

New energy conversions are thus highly unlikely to reduce CO_2 emissions fast enough to prevent the rise of atmospheric concentrations above 450 parts per million (ppm).

(They were nearly 390 ppm by the end of 2010). This realization has led to enthusiastic exploration of many possibilities available for carbon capture and sequestration—and to claims that would guarantee, even if they were only half true, futures free of any carbon worries. For example, a soil scientist claims that by 2100 biochar sequestration (essentially converting the world's crop residues, mainly cereal straws, into charcoal incorporated into soils) could store more carbon than the world emits from the combustion of all fossil fuels.

Most of these suggestions have been in the realm of theoretical musings: Notable examples include hiding CO_2 within and below the basalt layers of India's Deccan (no matter that those rocks are already much weathered and fractured), or in permeable undersea basalts of the Juan de Fuca tectonic plate off Seattle (but first we would have to pipe the emissions from Pennsylvania, Ohio and Tennessee coal-fired power plants to the Pacific Northwest)—or using exposed peridotites in the Omani desert to absorb CO_2 by accelerated carbonization (just imagine all those CO_2-laden megatankers from China and Europe converging on Oman with their refrigerated cargo).

One of these unorthodox ideas has been actually tried on a small scale. During the (so far) largest experiment with iron enrichment of the surface ocean (intended to stimulate phytoplankton growth and sequester carbon in the cells sinking to the abyss) an Indo-German expedition fertilized 300 square kilometers of the southwestern Atlantic in March and April 2009—but the resulting phytoplankton bloom was devoured by amphipods (tiny shrimp-like zooplankton). That is why the best chances for CCS are in a combination of well-established engineering practices: Scrubbing CO_2 with aqueous amine has been done commercially since the 1930s, piping the gas and using it in enhanced oil recovery is done routinely in many U.S. oilfields, and a pipeline construction effort matching the extension of U.S. natural gas pipelines during the 1960s or 1970s could put in place plenty of links between large stationary CO_2 sources and the best sedimentary formations used to sequester the gas.

But the scale of the effort needed for any substantial reduction of emissions, its safety considerations, public acceptance of permanent underground storage that might leak a gas toxic in high concentrations, and capital and operation costs of the continuous removal and burial of billions of tonnes of compressed gas combine to guarantee very slow progress. In order to explain the extent of the requisite effort I have been using a revealing comparison. Let us assume that we commit initially to sequestering just 20 percent of all CO_2 emitted from fossil fuel combustion in 2010, or about a third of all releases from large stationary sources. After compressing the gas to a density similar to that of crude oil (800 kilograms per cubic meter) it would occupy about 8 billion cubic meters—meanwhile, global crude oil extraction in 2010 amounted to about 4 billion tonnes or (with

average density of 850 kilograms per cubic meter) roughly 4.7 billion cubic meters.

This means that in order to sequester just a fifth of current CO_2 emissions we would have to create an entirely new worldwide absorption-gathering-compression-transportation-storage industry whose annual throughput would have to be about 70 percent larger than the annual volume now handled by the global crude oil industry whose immense infrastructure of wells, pipelines, compressor stations and storages took generations to build. Technically possible—but not within a timeframe that would prevent CO_2 from rising above 450 ppm. And remember not only that this would contain just 20 percent of today's CO_2 emissions but also this crucial difference: The oil industry has invested in its enormous infrastructure in order to make a profit, to sell its product on an energy-hungry market (at around $100 per barrel and 7.2 barrels per tonne that comes to about $700 per tonne)—but (one way or another) the taxpayers of rich countries would have to pay for huge capital costs and significant operating burdens of any massive CCS.

And if CCS will not scale up fast enough or it will be too expensive we are now offered the ultimate counter-weapon by resorting to geoengineering schemes. One would assume that a favorite intervention—a deliberate and prolonged (decades? centuries?) dispensation of millions of tonnes of sulfur gases into the upper atmosphere in order to create temperature-reducing aerosols—would raise many concerns at any time, but I would add just one obvious question: How would the Muslim radicals view the fleets of American stratotankers constantly spraying sulfuric droplets on their lands and on their mosques?

These are uncertain times, economically, politically and socially. The need for new departures seems obvious, but effective actions have failed to keep pace with the urgency of needed changes—particularly so in affluent democracies of North America, Europe and Japan as they contemplate their overdrawn accounts, faltering economies, aging populations and ebbing global influence. In this sense the search for new energy modalities is part of a much broader change whose outcome will determine the fortunes of the world's leading economies and of the entire global civilization for generations to come. None of us can foresee the eventual contours of new energy arrangements—but could the world's richest countries go wrong by striving for moderation of their energy use?

Bibliography

British Petroleum. 2010. BP Statistical Review of World Energy. www.bp.com/liveassets/bp_internet/globalbp/globalbp_uk_english/reports_and_publications/statistical_energy_review_2008/STAGING/local_assets/2010_downloads/statistical_review_of_world_energy_full_report_2010.pdf.

International Energy Agency. 2010. Key World Energy Statistics. www.iea.org/textbase/nppdf/free/2010/key_stats_2010.pdf.

Jackson, R. B., and J. Salzman. 2010. Pursuing geoengineering for atmospheric restoration. *Issues in Science and Technology* 26:67–76.

Metz, B., et al., eds. 2005. *Carbon Dioxide Capture and Storage.* Cambridge: Cambridge University Press.

Sivak, M., and O. Tsimhoni. 2009. Fuel efficiency of vehicles on US roads: 1923-2006. *Energy Policy* 37:3168–3170.

Smil, V. 2008. *Energy in Nature and Society: General Energetics of Complex Systems.* Cambridge, MA: The MIT Press.

Smil, V. 2009. U.S. energy policy: The need for radical departures. *Issues in Science and Technology* 25:47–50.

Smil, V. 2010. *Energy Transitions: History, Requirements, Prospects.* Santa Barbara, CA: Praeger.

Smil, V. 2010. *Energy Myths and Realities: Bringing Science to the Energy Policy Debate.* Washington, DC: American Enterprise Institute.

U.S. Energy Information Administration. 2010. International Energy Outlook 2010. www.eia.doe.gov/oiaf/ieo/pdf/0484 (2010).pdf.

Critical Thinking

1. In one sentence, describe the point that the author is trying to make in his article.

2. What does the author consider as the two fundamental things we need to be "clear" on when talking global energy? Do you agree? Why or why not?

3. List the energy alternatives discussed, and note each alternative's challenges.

4. Although discussed implicitly in the article, (1) what do we use energy for now, and (2) what will we use alternative energies for in the future?

5. Locate on your text's world map the regional populations this article relates to.

6. Identify in this article, three key terms, concepts, or principles that are used in your textbook (environmental science, economics, sociology, history, geography, etc,) or employed in the discipline you are currently studying. (Note: The terms, concepts, or principles may be implicit, explicit, implied, or inferred.)

VACLAV SMIL is a Distinguished Professor at the University of Manitoba. His interdisciplinary research deals with interactions of energy, environment, food, economy, population and technical advances. He is the author of 30 books on these topics; the latest ones (in 2010) include *Energy Myths and Realities, Energy Transitions* and *Two Prime Movers of Globalization.* E-mail: vsmil@cc.umanitoba.ca. Website: www.vaclavsmil.com.

From *American Scientist*, May/June 2011, pp. 212, 214–219. Copyright © 2011 by American Scientist, magazine of Sigma Xi, The Scientific Research Society. Reprinted by permission.

Seven Myths about Alternative Energy

As the world looks around anxiously for an alternative to oil, energy sources such as biofuels, solar, and nuclear seem like they could be the magic ticket. They're not.

MICHAEL GRUNWALD

What Comes Next?
Imagining the Post-Oil World

Nothing is as fraught with myths, misperceptions, and outright flights of fancy as the conversation about oil's successors. We asked two authors—award-winning environmental journalist Michael Grunwald and energy consultant David J. Rothkopf—to take aim at some of these myths, and look over the horizon to see which technologies might win the day and which ones could cause unexpected new problems. If fossil fuels are indeed saying their very long goodbye, then their would-be replacements still have a lot to prove.

1. "We Need to Do Everything Possible to Promote Alternative Energy"

Not exactly. It's certainly clear that fossil fuels are mangling the climate and that the status quo is unsustainable. There is now a broad scientific consensus that the world needs to reduce green-house gas emissions more than 25 percent by 2020—and more than 80 percent by 2050. Even if the planet didn't depend on it, breaking our addictions to oil and coal would also reduce global reliance on petrothugs and vulnerability to energy-price spikes.

But though the world should do everything sensible to promote alternative energy, there's no point trying to do everything possible. There are financial, political, and technical pressures as well as time constraints that will force tough choices; solutions will need to achieve the biggest emissions reductions for the least money in the shortest time. Hydrogen cars, cold fusion, and other speculative technologies might sound cool, but they could divert valuable resources from ideas that are already achievable and cost-effective. It's nice that someone managed to run his car on liposuction leftovers, but that doesn't mean he needs to be subsidized.

Reasonable people can disagree whether governments should try to pick energy winners and losers. But why not at least agree that governments shouldn't pick losers to be winners? Unfortunately,

that's exactly what is happening. The world is rushing to promote alternative fuel sources that will actually accelerate global warming, not to mention an alternative power source that could cripple efforts to stop global warming.

We can still choose a truly alternative path. But we'd better hurry.

2. "Renewable Fuels Are the Cure for Our Addiction to Oil"

Unfortunately not. "Renewable fuels" sound great in theory, and agricultural lobbyists have persuaded European countries and the United States to enact remarkably ambitious biofuels mandates to promote farm-grown alternatives to gasoline. But so far in the real world, the cures—mostly ethanol derived from corn in the United States or biodiesel derived from palm oil, soybeans, and rapeseed in Europe—have been significantly worse than the disease.

Researchers used to agree that farm-grown fuels would cut emissions because they all made a shockingly basic error. They gave fuel crops credit for soaking up carbon while growing, but it never occurred to them that fuel crops might displace vegetation that soaked up even more carbon. It was as if they assumed that biofuels would only be grown in parking lots. Needless to say, that hasn't been the case; Indonesia, for example, destroyed so many of its lush forests and peat lands to grow palm oil for the European biodiesel market that it ranks third rather than 21st among the world's top carbon emitters.

In 2007, researchers finally began accounting for deforestation and other land-use changes created by biofuels. One study found that it would take more than 400 years of biodiesel use to "pay back" the carbon emitted by directly clearing peat for palm oil. Indirect damage can be equally devastating because on a hungry planet, food crops that get diverted to fuel usually end up getting replaced somewhere. For example, ethanol profits are prompting U.S. soybean farmers to switch to corn, so Brazilian soybean farmers are expanding into cattle pastures to pick up the slack and Brazilian ranchers are invading the Amazon rain forest, which is why another study pegged corn ethanol's payback period at 167 years. It's simple economics: The mandates increase demand for grain, which boosts prices, which makes it lucrative to ravage the wilderness.

Deforestation accounts for 20 percent of global emissions, so unless the world can eliminate emissions from all other sources—cars, coal, factories, cows—it needs to back off forests. That means limiting agriculture's footprint, a daunting task as the world's population grows—and an impossible task if vast expanses of cropland are converted to grow middling amounts of fuel. Even if the United States switched its entire grain crop to ethanol, it would only replace one fifth of U.S. gasoline consumption.

This is not just a climate disaster. The grain it takes to fill an SUV tank with ethanol could feed a hungry person for a year; biofuel mandates are exerting constant upward pressure on global food prices and have contributed to food riots in dozens of poorer countries. Still, the United States has quintupled its ethanol production in a decade and plans to quintuple its biofuel production again in the next decade. This will mean more money for well-subsidized grain farmers, but also more malnutrition, more deforestation, and more emissions. European leaders have paid a bit more attention to the alarming critiques of biofuels— including one by a British agency that was originally established to promote biofuels—but they have shown no more inclination to throw cold water on this $100 billion global industry.

3. "If Today's Biofuels Aren't the Answer, Tomorrow's Biofuels Will Be"

Doubtful. The latest U.S. rules, while continuing lavish support for corn ethanol, include enormous new mandates to jump-start "second-generation" biofuels such as cellulosic ethanol derived from switchgrass. In theory, they would be less destructive than corn ethanol, which relies on tractors, petroleum-based fertilizers, and distilleries that emit way too much carbon. Even first-generation ethanol derived from sugar cane—which already provides half of Brazil's transportation fuel—is considerably greener than corn ethanol. But recent studies suggest that any biofuels requiring good agricultural land would still be worse than gasoline for global warming. Less of a disaster than corn ethanol is still a disaster.

Back in the theoretical world, biofuels derived from algae, trash, agricultural waste, or other sources could help because they require no land or at least unspecific "degraded lands," but they always seem to be "several" years away from large-scale commercial development. And some scientists remain hopeful that fast-growing perennial grasses such as miscanthus can convert sunlight into energy efficiently enough to overcome the land-use dilemmas—someday. But for today, farmland happens to be very good at producing the food we need to feed us and storing the carbon we need to save us, and not so good at generating fuel. In fact, new studies suggest that if we really want to convert biomass into energy, we're better off turning it into electricity.

Then what should we use in our cars and trucks? In the short term . . . gasoline. We just need to use less of it.

Instead of counterproductive biofuel mandates and ethanol subsidies, governments need fuel-efficiency mandates to help the world's 1 billion drivers guzzle less gas, plus subsidies for mass transit, bike paths, rail lines, telecommuting, carpooling, and other activities to get those drivers out of their cars. Policymakers also need to eliminate subsidies for roads to nowhere, mandates that require excess parking and limit dense development in urban areas, and other sprawl-inducing policies. None of this is as enticing as inventing a magical new fuel, but it's doable, and it would cut emissions.

In the medium term, the world needs plug-in electric cars, the only plausible answer to humanity's oil addiction that isn't decades away. But electricity is already the source of even more emissions than oil. So we'll need an answer to humanity's coal addiction, too.

4. "Nuclear Power Is the Cure for Our Addiction to Coal"

Nope. Atomic energy is emissions free, so a slew of politicians and even some environmentalists have embraced it as a clean alternative to coal and natural gas that can generate power when there's no sun or wind. In the United States, which already gets nearly 20 percent of its electricity from nuclear plants, utilities are thinking about new reactors for the first time since the Three Mile Island meltdown three decades ago—despite global concerns about nuclear proliferation, local concerns about accidents or terrorist attacks, and the lack of a disposal site for the radioactive waste. France gets nearly 80 percent of its electricity from nukes, and Russia, China, and India are now gearing up for nuclear renaissances of their own.

But nuclear power cannot fix the climate crisis. The first reason is timing: The West needs major cuts in emissions within a decade, and the first new U.S. reactor is only scheduled for 2017—unless it gets delayed, like every U.S. reactor before it. Elsewhere in the developed world, most of the talk about a nuclear revival has remained just talk; there is no Western country with more than one nuclear plant under construction, and scores of existing plants will be scheduled for decommissioning in the coming decades, so there's no way nuclear could make even a tiny dent in electricity emissions before 2020.

The bigger problem is cost. Nuke plants are supposed to be expensive to build but cheap to operate. Unfortunately, they're turning out to be really, really expensive to build; their cost estimates have quadrupled in less than a decade. Energy guru Amory Lovins has calculated that new nukes will cost nearly three times as much as wind—and that was before their construction costs exploded for a variety of reasons, including the global credit crunch, the atrophying of the nuclear labor force, and a supplier squeeze symbolized by a Japanese company's worldwide monopoly on steel-forging for reactors. A new reactor in Finland that was supposed to showcase the global renaissance is already way behind schedule and way, way over budget. This is why plans for new plants were recently shelved in Canada and several U.S. states, why Moody's just warned utilities they'll risk ratings downgrades if they seek new reactors, and why

149

renewables attracted $71 billion in worldwide private capital in 2007—while nukes attracted zero.

It's also why U.S. nuclear utilities are turning to politicians to supplement their existing loan guarantees, tax breaks, direct subsidies, and other cradle-to-grave government goodies with new public largesse. Reactors don't make much sense to build unless someone else is paying; that's why the strongest push for nukes is coming from countries where power is publicly funded. For all the talk of sanctions, if the world really wants to cripple the Iranian economy, maybe the mullahs should just be allowed to pursue nuclear energy.

Unlike biofuels, nukes don't worsen warming. But a nuclear expansion—like the recent plan by U.S. Republicans who want 100 new plants by 2030—would cost trillions of dollars for relatively modest gains in the relatively distant future.

Nuclear lobbyists do have one powerful argument: If coal is too dirty and nukes are too costly, how are we going to produce our juice? Wind is terrific, and it's on the rise, adding nearly half of new U.S. power last year and expanding its global capacity by a third in 2007. But after increasing its worldwide wattage tenfold in a decade—China is now the leading producer, and Europe is embracing wind as well—it still produces less than 2 percent of the world's electricity. Solar and geothermal are similarly wonderful and inexhaustible technologies, but they're still global rounding errors. The average U.S. household now has 26 plug-in devices, and the rest of the world is racing to catch up; the U.S. Department of Energy expects global electricity consumption to rise 77 percent by 2030. How can we meet that demand without a massive nuclear revival?

Wind is terrific, but it produces less than 2 percent of the world's electricity.

We can't. So we're going to have to prove the Department of Energy wrong.

5. "There Is No Silver Bullet to the Energy Crisis"

Probably not. But some bullets are a lot better than others; we ought to give them our best shot before we commit to evidently inferior bullets. And one renewable energy resource is the cleanest, cheapest, and most abundant of them all. It doesn't induce deforestation or require elaborate security. It doesn't depend on the weather. And it won't take years to build or bring to market; it's already universally available.

It's called "efficiency." It means wasting less energy—or more precisely, using less energy to get your beer just as cold, your shower just as hot, and your factory just as productive. It's not about some austerity scold harassing you to take cooler showers, turn off lights, turn down thermostats, drive less, fly less, buy less stuff, eat less meat, ditch your McMansion, and otherwise change your behavior to save energy. Doing less with less is called conservation. Efficiency is about doing more or

the same with less; it doesn't require much effort or sacrifice. Yet more efficient appliances, lighting, factories, and buildings, as well as vehicles, could wipe out one fifth to one third of the world's energy consumption without any real deprivation.

Efficiency isn't sexy, and the idea that we could use less energy without much trouble hangs uneasily with today's more-is-better culture. But the best way to ensure new power plants don't bankrupt us, empower petrodictators, or imperil the planet is not to build them in the first place. "Negawatts" saved by efficiency initiatives generally cost 1 to 5 cents per kilowatt-hour versus projections ranging from 12 to 30 cents per kilowatt-hour from new nukes. That's because Americans in particular and human beings in general waste amazing amounts of energy. U.S. electricity plants fritter away enough to power Japan, and American water heaters, industrial motors, and buildings are as ridiculously inefficient as American cars. Only 4 percent of the energy used to power a typical incandescent bulb produces light; the rest is wasted. China is expected to build more square feet of real estate in the next 15 years than the United States has built in its entire history, and it has no green building codes or green building experience.

But we already know that efficiency mandates can work wonders because they've already reduced U.S. energy consumption levels from astronomical to merely high. For example, thanks to federal rules, modern American refrigerators use three times less energy than 1970s models, even though they're larger and more high-tech.

The biggest obstacles to efficiency are the perverse incentives that face most utilities; they make more money when they sell more power and have to build new generating plants. But in California and the Pacific Northwest, utility profits have been decoupled from electricity sales, so utilities can help customers save energy without harming shareholders. As a result, in that part of the country, per capita power use has been flat for three decades—while skyrocketing 50 percent in the rest of the United States. If utilities around the world could make money by helping their customers use less power, the U.S. Department of Energy wouldn't be releasing such scary numbers.

6. "We Need a Technological Revolution to Save the World"

Maybe. In the long term, it's hard to imagine how (without major advances) we can reduce emissions 80 percent by 2050 while the global population increases and the developing world develops. So a clean-tech Apollo program modeled on the Manhattan Project makes sense. And we do need carbon pricing to send a message to market makers and innovators to promote low-carbon activities; Europe's cap-and-trade scheme seems to be working well after a rocky start. The private capital already pouring into renewables might someday produce a cheap solar panel or a synthetic fuel or a superpowerful battery or a truly clean coal plant. At some point, after we've milked efficiency for all the negawatts and negabarrels we can, we might need something new.

But we already have all the technology we need to start reducing emissions by reducing consumption. Even if we only hold electricity demand flat, we can subtract a coal-fired megawatt every time we add a wind-powered megawatt. And with a smarter grid, green building codes, and strict efficiency standards for everything from light bulbs to plasma TVs to server farms, we can do better than flat. Al Gore has a reasonably plausible plan for zero-emissions power by 2020; he envisions an ambitious 28 percent decrease in demand through efficiency, plus some ambitious increases in supply from wind, solar, and geothermal energy. But we don't even have to reduce our fossil fuel use to zero to reach our 2020 targets. We just have to use less.

If somebody comes up with a better idea by 2020, great! For now, we should focus on the solutions that get the best emissions bang for the buck.

7. "Ultimately, We'll Need to Change Our Behaviors to Save the World"

Probably. These days, it's politically incorrect to suggest that going green will require even the slightest adjustment to our way of life, but let's face it: Jimmy Carter was right. It wouldn't kill you to turn down the heat and put on a sweater. Efficiency is a miracle drug, but conservation is even better; a Prius saves gas, but a Prius sitting in the driveway while you ride your bike uses no gas. Even energy-efficient dryers use more power than clotheslines.

More with less will be a great start, but to get to 80 percent less emissions, the developed world might occasionally have to do less with less. We might have to unplug a few digital picture frames, substitute teleconferencing for some business travel, and take it easy on the air conditioner. If that's an inconvenient truth, well, it's less inconvenient than trillions of dollars' worth of new reactors, perpetual dependence on hostile petrostates, or a fricasseed planet.

After all, the developing world is entitled to develop. Its people are understandably eager to eat more meat, drive more cars, and live in nicer houses. It doesn't seem fair for the developed world to say: Do as we say, not as we did. But if the developing world follows the developed world's wasteful path to prosperity, the Earth we all share won't be able to accommodate us. So we're going to have to change our ways. Then we can at least say: Do as we're doing, not as we did.

Critical Thinking

1. List the seven myths, and compare each one to the discussion regarding alternative energy in Article 27. Point out similarities and dissimilarities of the two authors' views.

2. In Myths 1, 6, and 7, the author says "we." Who is he referring to?

3. On your text world map, locate current energy suppliers and energy consumers.

4. Why might using new alternative energy sources to simply maintain the consumption patterns of the industrialized world get us nowhere in the end? Offer some insights.

5. Identify in this article three key terms, concepts, or principles that are used in your textbook (environmental science, economics, sociology, history, geography, etc.) or employed in the discipline you are currently studying. (Note: The terms, concepts, or principles may be implicit, explicit, implied, or inferred.)

MICHAEL GRUNWALD, a senior correspondent at *Time* magazine, is an award-winning environmental journalist and author of *The Swamp: The Everglades, Florida, and the Politics of Paradise*.

Half a Tank

The Impending Arrival of Peak Oil

MARK FLOEGEL

imple lessons are not necessarily easy to learn. For example: oil is a non-renewable and limited resource. Students learn this in grade school, but people in the United States got their first real instruction during the 1973 oil embargo, when Middle Eastern nations restricted the flow of petroleum in response to U.S. military support for Israel. A second lesson was delivered during the 1979 Iranian revolution, when that country's internal turmoil temporarily disrupted its oil output.

Sudden hikes in the price of gas and the resulting recession in the early 1980s made it clear the world had begun to change. Fuel-efficient cars, many of Japanese design, appeared on roadways supposedly owned by what were then called "the Big Three auto makers." Congress imposed corporate average fuel efficiency standards in 1975, northern residents dialed back thermostats in winter and President Jimmy Carter installed solar panels on the White House roof.

While the world energy landscape is more complicated than "oil is non-renewable," it's not that much more complicated. The other key factor is also a staple of grade-school economics: the law of supply and demand. Mideast political crises passed, which increased supply; demand was reduced by domestic fuel-efficiency programs. The price of oil dropped, the recession ended and Ronald Reagan yanked the solar panels off the White House roof. (They wound up heating water for a dining hall at Unity College in Maine.)

Oil prices began ripping the roof off again in summer 2005, and that roller coaster is still at full velocity. If the 1970s oil crises were a glimpse of the future, the future may be here.

Peak Oil

The life of an oil field's production can be plotted on a curve. The first oil comes from the ground easily. Many oil fields don't even need to be pumped initially; the pressure of the field will force the oil to the surface through any hole poked into it. The cost of obtaining the oil is almost zero. Later, pump jacks need to be installed or water is pumped into the field to keep pressure up. As time passes, the field produces less oil and the cost per barrel rises until further production is unfeasible. Production curves for these fields are shaped like parabolas; production rises steeply, peaks and falls as quickly as it rose.

In 1956, when the United States produced 2.6 billion barrels of oil, M. King Hubbert, a Shell Oil petro-geologist, extrapolated a typical field-production curve to assess crude oil production for the United States as a whole. Putting down his slide rule, he predicted that the production of U.S. crude would peak in the early 1970s. It was a bold prediction, since many of his colleagues thought new discoveries and improved recovery techniques would keep oil yields growing long after that date.

U.S. oil production did peak in 1970, with 3.5 billion barrels pumped. Production then began to fall, but because production and consumption trends are only clear in hindsight, the peaking of U.S. production was not an accepted fact for several years after. In 2005, according to the federal Energy Information Administration (EIA), domestic production was 1.8 billion barrels, about as much as 1947.

Given the state of petroleum geology and economics, predicting the peak (and eventual end) of global oil production should be a matter of a few simple equations. It is in the interest of oil companies and oil-producing nations, however, to keep hidden the true state of oil reserves. On one hand, oil producers want to foster confidence in their supplies and discourage competing sources of energy. On the other hand, the same interests want to deliver the message that oil is sufficiently scarce to justify continuous price hikes. A nation or corporation can hide how much it has on hand, but how much oil it brings to market is a public fact. Sooner or later, the price of oil will rise higher than before and the usual suspects of the oil world won't increase production, because they can't. Then it will be clear that the world has passed the peak.

This arena of doubt is the stage for "peak oil" theorists, who believe the global peak is at hand, or perhaps has already occurred. In August 2004, Ken Deffeyes, geologist, Hubbert acolyte and professor emeritus at Princeton University, told National Public Radio's Steve Inskeep that global oil production would peak around Thanksgiving, 2005. (He later amended that to mid-December 2005.) Deffeyes was wrong, apparently. The EIA reports daily oil production in the third quarter of 2006 was higher (by 170,000 barrels) than the highest quarter of 2005.

A stouter buttress of peak oil theory is "Peaking of World Oil Production: Impacts, Mitigation and Risk Management," commonly called "the Hirsch report" after its principle author, Robert Hirsch. Commissioned from defense contractor SAIC by the Department of Energy, the report polled a dozen experts—Deffeyes among them—on global oil's expected peak. Predictions ran from almost immediately to beyond the horizon.

The thrust of the Hirsch report is that an oil peak will occur at some point. The sooner society recognizes this peak and energy diversification is realized, the less severe the economic effects. The report's authors estimate 20 years of preparation are needed to fully mitigate the effects of an oil peak. Of the 12 experts polled, only one—energy economist M. C. Lynch—predicted the peak is more than 20 years in the future.

Cambridge Energy Research Associates, which has recently estimated global total recoverable oil at 4.8 trillion barrels—more than twice the long-accepted number of two trillion barrels—nonetheless made a moderate prediction for the Hirsch report, saying the peak would come "after 2020."

Since the February 2005 release of the Hirsch report, two events have underlined suspicions of an imminent oil peak. In November 2005, the Kuwaiti oil ministry announced that the Burgan oil field—the second largest in the world—had peaked at 1.7 million barrels per day. This announcement came years earlier—and hundreds of thousands of barrels lower—than expected by the oil industry.

A month later, the New York Times reported Saudi Arabia was backing away from its traditional role as the world's swing oil producer. (A swing producer is one with substantial supplies of a given commodity on hand. Swing producers stabilize prices by ramping up production in times of scarcity and scaling back in times of glut.) This news came not only on the heels of the Burgan peak announcement, but shortly after energy investment banker Matthew Simmons published *Twilight in the Desert,* a book accusing the Saudis of having far fewer oil reserves than the kingdom claimed. Simmons sees Saudi reluctance to continue its swing producer role as a tacit admission that his suspicions are correct.

Destroying Demand

That's a sketch of what is known—and not known—about the supply side. Demand is constantly increasing. Despite the introduction of conservation and efficiency measures in some parts, rich nations continue to use more oil each year. Developing nations are increasing oil use at a much higher rate. From 2000 to 2005, oil consumption in the United States and Japan grew by 1.5 percent each year; in China, oil consumption grew by 8.5 percent per year.

Even if world oil production has not yet reached its peak, but demand grows faster than supply increases can keep up, the result will be an economic state that mimics post-peak oil. The Hirsch report contains an augury of this. One table displays additional oil reserves as a function of increasing demand. It shows excess capacity began to decline in 1987. In 2002, that excess oil production, the world's "oil cushion," averaged over five million barrels per day. By 2006, excess production was down to one million barrels per day, according to the EIA.

The news gets worse. The easy-to-access, desirable oil was accessed first. "Light and sweet" are the adjectives used to describe the best crude oil. The terms mean the oil is relatively free of impurities and easy to refine into gasoline, kerosene, jet fuel and other petroleum products. Light, sweet crude oil was often found in subterranean pools not too far below the surface of the ground and was easy to pump out, often rising to the surface under its own pressure.

Even those who believe the global oil peak is many years away agree easily accessed, light, sweet crude is substantially gone. Indeed, there are a number of ways in which society can continue with a "business-almost-as-usual" scenario for some time, but the costs, both economic and environmental, will be steep. Rising prices means oil companies are able to afford injecting steam into old wells to increase their output; how long that increased output can be maintained remains to be seen. Much of the petroleum remaining underground tends to be heavy (containing petroleum waxes, which make the oil harder to pump) and sour (meaning high in sulfur and other contaminants). What is not heavy, sour crude oil may be trapped in shale—which must be fractured through hydraulic pressure in the mineral formation, an added cost—or embedded in "tar sands."

Since 2003, the price of oil has made it economically feasible to mine oil from tar sands. If one counts tar sand oil, Canada has the second-largest reserve of oil in the world, after Saudi Arabia's. It covers an area the size of Maryland and Virginia combined. Tar sand strip mines are already being operated near Fort McMurray, Alberta, where oversized machines scoop the sand for processing. The refining process is nothing John D. Rockefeller would recognize. To obtain one barrel of oil, four tons of soil are steamed in two to four barrels of fresh water (removed from local rivers), using substantial amounts of natural gas to heat the process. The environmental result is a devastated landscape, horribly polluted water and added emissions of greenhouses gases into the atmosphere—even before the oil is used for transportation or heating.

Supply-and-demand theory holds that if the price for a commodity rises too high, consumers find a way to "destroy demand," by reducing their consumption of the commodity, either by doing without the commodity or finding ways to use less of it.

As the price of gas rises from three to four to five dollars per gallon, sales of gas-stingy vehicles such as the Toyota Prius will rise, as will demand for a supply of "plug-in hybrids." Plug-ins can be charged with electricity and only resort to gasoline engines when their electrical charges have been spent. Hybrid technologies (although not plug-in hybrids) are currently on the market for vehicles from compact passenger sedans to SUVs to delivery trucks to 18-wheel semi-trailers. Mass transit, light and inter-city rail will likely surge.

The Replacements

As the price of oil soars, a market opens for potential "replacement fuels." Vegetable-based ethanol has received much attention, because it can be mixed with gasoline and burns cleaner than oil-based fuels. Ethanol in the United States is made from corn, which makes it popular in politically important farm states like Iowa.

Unfortunately, corn ethanol is not particularly efficient as a fuel, so gas mileage suffers. Also, it is unclear how economically competitive corn ethanol will be if the price of petroleum-based pesticides (used to grow the corn) rises in a post-peak scenario. More efficient ethanol can be made from soybeans, sugar cane and palm oil, but like corn, all those crops either remove acreage from food production for fuel production or result in forests being cut to make room for the crops, two undesirable environmental consequences.

Hydrogen is an efficient, clean medium for transferring energy for use in vehicles, but almost all the hydrogen on Earth is bound up in complex molecules; substantial amounts of energy are needed to render hydrogen into a usable state. Some have advocated using nuclear power to accomplish this. The environmental and safety drawbacks of nuclear power are clear.

Replacing the energy used to get around on the ground, however, will be easy, compared to finding new ways to fly through the air. If no alternative to petroleum-derived

jet fuel emerges, consumers will likely destroy demand by finding ways to do without flying.

Airlines have already instituted every imaginable fuel-conservation technique—"winglets" on wing tips increase efficiency 3 or 4 percent, ovens have been stripped from galleys, fewer cans of soda are brought on board to save weight. Airlines carried more passengers and cargo in 2005 than in 2000 and did it with 400 million fewer gallons of fuel. Airlines have consolidated routes and cut jobs to save money. The six major airlines cut 167,000 jobs between 2001 and 2006.

With all that efficiency, airfares still rise, because airlines remain at the mercy of fuel costs. As the price of tickets rise, fewer families will fly for vacations and business travelers will replace face-to-face meetings with video conferences. Fewer fliers will probably mean more airline closures, mergers and possibly another federal bailout. It will ultimately mean air travel will be an increasingly rare experience for the average person.

A World without Oil

When the peak comes, the effects on people's lives will be profound, because so much of the economy is subsidized by cheap oil. The post-war "green revolution" in agriculture was based on petroleum-based pesticides. Industrial agriculture, dependent on economies of scale, is also dependent on gas-powered farm machinery and mega-farms miles from markets. Fresh fruit in every season, courtesy of growers in Chile and Argentina, relies on low-cost fuel for freighters, as do bargain wines from Australia.

Cheap-oil subsidies reach their apotheosis in Fiji Water. Drawn from an aquifer in Fiji, the water is packaged in cheap, lightweight bottles made of petroleum-derived plastic. The bottles are shipped around the globe to first-world consumers, almost all of whom have sources of drinking water within a few miles of their homes. It is hard to see Fiji Water surviving peak oil.

But much more significant economic activities will be threatened or transformed as well. Besides Fiji Water bottles, cheap, light, disposable plastic packaging further reduces the cost of food and goods. Petroleum-based plastics account for a substantial portion of electronics gear, auto interiors, plumbing, siding, window frames, flooring, furniture, cookware and tools. After the peak, those products will all cost much more, or be made instead from glass, cloth, wood and metal (or potentially non-petroleum-based plastics). Those omnipresent plastic-shelled electronic gizmos are mostly assembled in low-wage nations in East and South Asia, then floated to big-box stores on a wave of cheap oil.

Post-peak life will extract many costs for consumers by raising the price of almost everything, but at the same time, jobs will be created for local production of things that are now grown or made far away. "Relocalization" is what peak oilers call it.

The economics of peak oil are as simple as those of supply and demand. The debate is about when the peak will arrive and how ready the world will be when it does. There is little debate about the one timeline the Hirsch report makes clear: it will take 20 years to adequately prepare society for the passing of the global oil peak.

Industrialized nations of the Northern Hemisphere have lived off cheap oil's subsidy since the end of the Second World War. It will take two decades to find and perfect the production of alternate fuels, to invent new or relearn old ways of feeding ourselves, to find new or old materials for building tools and essential goods.

For consumers, the law of supply and demand is about choices: how much are we willing to pay? Peak oil theory presents society with a choice. If peak oil theorists are wrong and the peak is well into the future, say 50 years, what advantage is lost by beginning to prepare now? If peak oil theorists are right, even moderately right, and the global peak is, say seven years away, what is risked by failing to prepare?

Critical Thinking

1. Explain the idea of "peak oil."
2. What data/evidence is used to support the idea of peak oil? Who really knows how much oil reserves there are in the world?
3. The article was published in 2007. What has changed? What has not changed?
4. The article states, "When peak comes, the effects on peoples' lives will be profound. . . ." What "people" is he referring to?
5. What makes the idea of "peak oil" an environmental issue? What impacts will radically lower supplies of oil have on the environment?
6. Identify in this article three key terms, concepts, or principles that are used in your textbook (environmental science, economics, sociology, history, geography, etc.) or employed in the discipline you are currently studying. (Note: The terms, concepts, or principles may be implicit, explicit, implied, or inferred.)

MARK FLOEGEL is a freelance writer who lives in Vermont. He posts weekly commentaries at www.markfloegel.org.

From *Multinational Monitor*, January/February 2007. Copyright © 2007 by Essential Information. Reprinted by permission.

It's Still the One

Oil's very future is now being seriously questioned, debated, and challenged. The author of an acclaimed history explains why, just as we need more oil than ever, it is changing faster than we can keep up with.

DANIEL YERGIN

On a still afternoon under a hot Oklahoma sun, neither a cloud nor an ounce of "volatility" was in sight. Anything but. All one saw were the somnolent tanks filled with oil, hundreds of them, spread over the rolling hills, some brand-new, some more than 70 years old, and some holding, inside their silver or rust-orange skins, more than half a million barrels of oil each.

This is Cushing, Oklahoma, the gathering point for the light, sweet crude oil known as West Texas Intermediate—or just WTI. It is the oil whose price you hear announced every day, as in "WTI closed today at. . . ." Cushing proclaims itself, as the sign says when you ride into town, the "pipeline crossroads of the world." Through it passes the network of pipes that carry oil from Texas and Oklahoma and New Mexico, from Louisiana and the Gulf Coast, and from Canada too, into Cushing's tanks, where buyers take title before moving the oil onward to refineries where it is turned into gasoline, jet fuel, diesel, home heating oil, and all the other products that people actually use.

But that is not what makes Cushing so significant. After all, there are other places in the world through which much more oil flows. Cushing plays a unique role in the new global oil industry because WTI is the preeminent benchmark against which other oils are priced. Every day, billions of "paper barrels" of light, sweet crude are traded on the floor of the New York Mercantile Exchange in lower Manhattan and, in ever increasing volumes, at electron speed around the world, an astonishing virtual commerce that no matter how massive in scale, still connects back somehow to a barrel of oil in Cushing changing owners.

That frenetic daily trading has helped turn oil into something new—not only a physical commodity critical to the security and economic viability of nations but also a financial asset, part of that great instantaneous exchange of stocks, bonds, currencies, and everything else that makes up the world's financial portfolio. Today, the daily trade in those "paper barrels"—crude oil futures—is more than 10 times the world's daily consumption of physical barrels of oil. Add in the trades that take place on other exchanges or outside them entirely, and the ratio may be as much as 30 times greater. And though the oil may flow steadily in and out of Cushing at a stately 4 miles per hour, the global oil market is anything but stable.

That's why, as I sat down to work on a new edition of *The Prize* and considered what had changed since the early 1990s, when I wrote this history of the world's most valuable, and misunderstood, commodity, the word "volatility" kept springing to mind. How could it not? Indeed, when people are talking about volatility, they are often thinking oil. On July 11, 2008, WTI hit $147.27. Exactly a year later, it was $59.87. In between, in December, it fell as low as $32.40. (And don't forget a little more than a decade ago, when it was as low as $10 a barrel and consumers were supposedly going to swim forever in a sea of cheap oil.)

These wild swings don't just affect the "hedgers" (oil producers, airlines, heating oil dealers, etc.) and the "speculators," the financial players. They show up in the changing prices at the gasoline station. They stir political passions and feed consumers' suspicions. Volatility also makes it more difficult to plan future energy investments, whether in oil and gas or in renewable and alternative fuels. And it can have a cataclysmic impact on the world economy. After all, Detroit was knocked flat on its back by what happened at the gasoline pump in 2007 and 2008 even before the credit crisis. The enormous impact of these swings is why British Prime Minister Gordon Brown and French President Nicolas Sarkozy were recently moved to call for a global solution to "destructive volatility." But, they were forced to add, "There are no easy solutions."

This volatility is part of the new age of oil. For though Cushing looks pretty much the same as it did when *The Prize* came out, the world of oil looks very different. Some talk today about "the end of oil." If so, others reply, we are entering its very long goodbye. One characteristic of this new age is that oil has developed a split personality—as a physical commodity but also now as a financial asset. Three other defining characteristics of this new age are the globalization of the demand for oil, a vast shift from even a decade ago; the rise of climate change as a political factor shaping decisions on how we will use oil, and how much of it, in the future; and the drive for new technologies that could dramatically affect oil along with the rest of the energy portfolio.

The cast of characters in the oil business has also grown and changed. Some oil companies have become "supermajors," such as ExxonMobil and Chevron, while others, such as Amoco and ARCO, have just disappeared. "Big oil" no longer means the traditional international oil companies, their logos instantly recognizable from corner gas stations, but rather much larger state-owned companies, which, along with governments, today control more than 80 percent of the world's oil reserves. Fifteen of the world's 20 largest oil companies are now state-owned.

The cast of oil traders has also much expanded. Today's global oil game now includes pension funds, institutional money managers, endowments, and hedge funds, as well as individual investors and day traders. The managers at the pension funds and the university endowments see themselves as engaged in "asset allocation," hedging risks and diversifying to protect retirees' incomes and faculty salaries. But, technically, they too are part of the massive growth in the ranks of the new oil speculators.

With all these changes, the very future of this most vital commodity is now being seriously questioned, debated, and challenged, even as the world will need more of it than ever before. Both the U.S. Department of Energy and the International Energy Agency project that, even accounting for gains in efficiency, global energy use will increase almost 50 percent from 2006 to 2030—and that oil will continue to provide 30 percent or more of the world's energy in 2030.

But will it?

$147.27—Closing price per barrel of oil on July 11, 2008. Exactly one year later, it had fallen to $59.87.

From the beginning, oil has been a global industry, going back to 1861 when the first cargo of kerosene was sent from Pennsylvania—the Saudi Arabia of 19th-century oil—to Britain. (The potential crew was so fearful that the kerosene would catch fire that they had to be gotten drunk to shanghai them on board.) But that is globalization of supply, a familiar story. What is decisively new is the globalization of demand.

For decades, most of the market—and the markets that mattered the most—were in North America, Western Europe, and Japan. That's also where the growth was. At the time of the first Gulf War in 1991, China was still an oil exporter.

But now, the growth is in China, India, other emerging markets, and the Middle East. Between 2000 and 2007, the world's daily oil demand increased by 9.4 million barrels. Almost 85 percent of that growth was in emerging markets. There were many reasons that prices soared all the way to $147.27 last year, ranging from geopolitics to a weak dollar to the impact of financial markets and speculation (in all its manifold meanings). But the starting point was the fundamentals—the surge in oil demand driven by powerful economic growth in emerging markets. This shift may be even more powerful than people recognize: So far this year, more new cars have been sold in China than in the United States. When economic recovery

The Capital of Oil

Within three years of its discovery in 1912, the Cushing field in Oklahoma was producing almost 20 percent of U.S. oil. Two years later, it was supplying a substantial part of the fuel used by the U.S. Army in Europe during World War I. The Cushing area was so prolific that it became known as "the Queen of the Oil Fields," and Cushing became one of those classic wild oil boomtowns of the early 20th century. "Any man with red blood gets oil fever" was the diagnosis of one reporter who visited the area during those days. Production grew so fast around Cushing that pipelines had to be hurriedly built and storage tanks quickly thrown up to hold surplus supplies. By the time production began to decline, a great deal of infrastructure was in place, and Cushing turned into a key oil hub, its network of pipelines used to bring in supplies from elsewhere in Oklahoma and West Texas. Those supplies were stored in the tanks at Cushing before being put into other pipelines and shipped to refineries. When the New York Mercantile Exchange—the NYMEX—started to trade oil futures in 1983, it needed a physical delivery point. Cushing, its boom days long gone, but with its network of pipelines and tank farms and its central location, was the obvious answer. As much as 1.1 million barrels per day pass in and out of Cushing—equivalent to about 6 percent of U.S. oil consumption. But prices for much of the world's crude oil are set against the benchmark of the West Texas Intermediate crude oil—also known as "domestic sweet"—sitting in those 300 or so tanks in Cushing, making this sedate Oklahoma town not only an oil hub but one of the hubs of the world economy.

—Daniel Yergin

takes hold again, what happens to oil demand in such emerging countries will be crucial.

The math is clear: More consumers mean more demand, which means more supplies are needed. But what about the politics? There the forecasts are murkier, feeding a new scenario for international tension—a competition, even a clash, between China and the United States over "scarce" oil resources. This scenario even comes with a well-known historical model—the rivalry between Britain and "rising" Germany that ended in the disaster of World War I.

This scenario, though compelling reading, does not really accord with the way that the world oil market works. The Chinese are definitely new players, willing and able to pay top dollar to gain access to existing and new oil sources and, lately, also making loans to oil-producing countries to ensure future supplies. With more than $2 trillion in foreign reserves, China certainly has the wherewithal to be in the lending business.

But the global petroleum industry is not a go-it-alone business. Because of the risk and costs of large-scale development, companies tend to work in consortia with other companies.

Oil-exporting countries seek to diversify the countries and companies they work with. Inevitably, any country in China's position—whose demand had grown from 2.5 million barrels per day to 8 million in a decade and a half—would be worrying about supplies. Such an increase, however, is not a forecast of inevitable strife; it is a message about economic growth and rising standards of living. It would be much more worrying if, in the face of rising demand, Chinese companies were not investing in production both inside China (the source of half of its supply) and outside its borders.

There are potential flash points in this new world of oil. But they will not come from standard commercial competition. Rather, they arise when oil (along with natural gas) gets caught up in larger foreign-policy issues—most notably today, the potentially explosive crisis over the nuclear ambitions of oil- and gas-rich Iran.

Yet, despite all the talk of an "oil clash" scenario, there seems to be less overall concern than a few years ago and much more discussion about "energy dialogue." The Chinese themselves appear more confident about their increasingly important place in this globalized oil market. Although the risks are still there, the Chinese—and the Indians right alongside them—have the same stake as other consumers in an adequately supplied world market that is part of the larger global economy. Disruption of that economy, as the last year has so vividly demonstrated, does not serve their purposes. Why would the Chinese want to get into a confrontation over oil with the United States when the U.S. export market is so central to their economic growth and when the two countries are so financially interdependent?

Oil is not even the most important energy issue between China and the United States. It is coal. The two countries have the world's largest coal resources, and they are the world's biggest consumers of it. In a carbon-constrained world, they share a strong common interest in finding technological solutions for the emissions released when coal is burned.

And that leads directly to the second defining feature of the new age of oil: climate change. Global warming was already on the agenda when *The Prize* came out. It was back in 1992 that 154 countries signed the Rio Convention, pledging to dramatically reduce CO_2 concentrations in the atmosphere. But only in recent years has climate change really gained traction as a political issue—in Europe early in this decade, in the United States around 2005. Whatever the outcome of December's U.N. climate change conference in Copenhagen, carbon regulation is now part of the future of oil. And that means a continuing drive to reduce oil demand.

How does that get done? How does the world at once meet both the challenge of climate change and the challenge of economic growth—steady expansion in the industrial countries and more dramatic growth in China, India, and other emerging markets as tens of millions of their citizens rise from poverty and buy appliances and cars?

The answer has to be in another defining change—an emphasis on technology to a degree never before seen. The energy business has always been a technology business. After all, the men who figured out in 1859, exactly 150 years ago, how to drill that first oil well—Colonel Drake and his New Haven, Conn., investors—would, in today's lingo, be described as a group of disruptive technology entrepreneurs and venture capitalists. Again and again, in researching oil's history, I was struck by how seemingly insurmountable barriers and obstacles were overcome by technological progress, often unanticipated.

9.4 million—Number of barrels by which the world's daily oil demand rose from 2000 to 2007, with 85 percent coming from the developing world.

But the focus today on technology—all across the energy spectrum—is of unprecedented intensity. In the mid-1990s, I chaired a task force for the U.S. Department of Energy on "strategic energy R&D." Our panel worked very hard for a year and a half and produced what many considered a very worthy report. But there was not all that much follow-through. The Gulf War was over, and the energy problem looked like it had been "solved."

Today, by contrast, the interest in energy technology is enormous. And it will only be further stoked by the substantial increases that are ahead in government support for energy R&D. Much of that spending and effort is aimed at finding alternatives to oil. Yet the challenge is not merely to find alternatives; it is to find alternatives that can be competitive at the massive scale required.

What will those alternatives be? The electric car, which is the hottest energy topic today? Advanced biofuels? Solar systems? New building designs? Massive investment in wind? The evolving smart grid, which can integrate electric cars with the electricity industry? Something else that is hardly on the radar screen yet? Or perhaps a revolution in the internal combustion engine, making it two to three times as efficient as the ones in cars today?

We can make educated guesses. But, in truth, we don't know, and we won't know until we do know. For now, it is clear that the much higher levels of support for innovation—along with considerable government incentives and subsidies—will inevitably drive technological change and thus redraw the curve in the future demand for oil.

Indeed, the biggest surprises might come on the demand side, through conservation and improved energy efficiency. The United States is twice as energy efficient as it was in the 1970s. Perhaps we will see a doubling once again. Certainly, energy efficiency has never before received the intense focus and support that it does today.

Just because we have entered this new age of high-velocity change does not mean this story is about the imminent end of oil. Consider the "peak oil" thesis—shorthand for the presumption that the world has reached the high point of production and is headed for a downward slope. Historically, peak-oil thinking gains attention during times when markets are tight and prices are rising, stoking fears of a permanent shortage. In 2007 and 2008, the belief system built around peak oil helped drive prices to $147.27. (It was actually the fifth time that the world had supposedly "run out" of oil. The first such episode was in the 1880s; the last instance before this most recent time was in the 1970s.)

However, careful examination of the world's resource base—including my own firm's analysis of more than 800 of the largest oil fields—indicates that the resource endowment of the planet is sufficient to keep up with demand for decades to come. That, of course, does not mean that the oil will actually make it to consumers. Any number of "aboveground" risks and obstacles can stand in the way, from government policies that restrict access to tax systems to civil conflict to geopolitics to rising costs of exploration and production to uncertainties about demand. As has been the case for decades and decades, the shifting relations between producing and consuming countries, between traditional oil companies and state-owned oil companies, will do much to determine what resources are developed, and when, and thus to define the future of the industry.

There are two further caveats. Many of the new projects will be bigger, more complex, and more expensive. In the 1990s, a "megaproject" might have cost $500 million to $1 billion. Today, the price tag is more like $5 billion to $10 billion. And an increasing part of the new petroleum will come in the form of so-called "unconventional oil"—from ultradeep waters, Canadian oil sands, and the liquids that are produced with natural gas.

But through all these changes, one constant of the oil market is that it is not constant. The changing balance of supply and demand—shaped by economics, politics, technologies, consumer tastes, and accidents of all sorts—will continue to move prices. Economic recovery, expectations thereof, the pent-up demand for "demand," a shift into oil as a "financial asset"—some combination of these could certainly send oil prices up again, even with the current surplus in the market. Yet, the quest for stability is also a constant for oil, whether in reaction to the boom-and-bust world of northwest Pennsylvania in the late 19th century, the 10-cents-a-barrel world of Texas oil in the 1930s, or the $147.27 barrel of West Texas Intermediate in July 2008.

Certainly, the roller-coaster ride of oil prices over the last couple of years, as oil markets and financial markets have become more integrated, has made volatility a central pre-occupation for policymakers who do not want to see their economies whipsawed by huge price swings. Yet without the flexibility and liquidity of markets, there is no effective way to balance supply and demand, no way for consumers and producers to hedge their risks. Nor is there a way to send signals to these consumers and producers about how much oil to use and how much money to invest—or signals to would-be innovators about tomorrow's opportunities.

One part of the solution is not only enhancement of the already considerable regulation of the financial markets where oil is traded, but also greater transparency and better understanding of who the players are in the rapidly expanding financial oil markets. But regulatory changes cannot eliminate market cycles or repeal the laws of supply and demand in the world's largest organized commodity market. Those cycles may not be much in evidence amid the quiet tanks and rolling hills at Cushing. But they are inescapably part of the global landscape of the new world of oil.

Critical Thinking

1. How are the global patterns of oil supply, production, trade, and consumption changing? (Refer to a world map).

2. In what ways might your response to question #1 affect the concerns regarding population, human resources, health, and environment expressed in Unit 1, Article 1?

3. The author believes that we have a planetary oil resource endowment right now sufficient to keep up with demand for decades to come. How does that evidence compare to the "peak oil" idea?

4. What does such an assertion that we have enough oil reserves to "meet decades of demand" imply about efforts to find alternative energies?

5. How might oil be less about energy and more about global political power?

6. Identify in this article three key terms, concepts, or principles that are used in your textbook (environmental science, economics, sociology, history, geography, etc.) or employed in the discipline you are currently studying. (Note: The terms, concepts, or principles may be implicit, explicit, implied, or inferred.)

DANIEL YERGIN received a Pulitzer Prize for *The Prize: The Epic Quest for Oil, Money and Power,* published in an updated edition this year. He is chairman of IHS Cambridge Energy Research Associates.

Reprinted in entirety by McGraw-Hill with permission from *Foreign Policy*, September/October 2009, pp. 90, 92–95. www.foreignpolicy.com. © 2009 Washingtonpost.Newsweek Interactive, LLC.

Gas Costs Squeeze Daily Life

Survey Reveals How High Prices Have Pushed Us into New Routines

Judy Keen and Paul Overberg

Record-high gas prices are prompting Americans to drive less for the first time in nearly three decades, squeezing family budgets and causing major shifts in driving habits, federal data and a USA TODAY/Gallup Poll show.

As prices near—or in some places top—$4 a gallon, most Americans say they are cutting back on other household spending, seriously considering buying more fuel-efficient cars and consolidating their daily errands to save fuel.

Americans worry that steep gas costs are here to stay: eight in 10 say they doubt today's high prices are temporary, the poll finds. It's the first time such a large majority sees pricey gas as a long-term problem.

The $4 mark, compounded by a sagging economy, could be a tipping point that spurs people to make permanent lifestyle changes to reduce dependence on foreign oil and help the environment, says Steve Reich, a program director at the Center for Urban Transportation Research at the University of South Florida.

"This is a more significant shift in behavior than I've seen through other fluctuations in gasoline prices," he says. "People are starting to understand that this resource . . . is not something to be taken for granted or wasted."

The average price of a gallon of gas nationwide is $3.65—the highest ever, adjusted for inflation. California's average: $3.90 a gallon. The federal Energy Information Administration (EIA) expects a $3.66 per-gallon average this summer.

The pinch is reshaping the way Americans use their cars:

- February was the fourth consecutive month in which miles driven in the USA fell, an analysis of Federal Highway Administration data show.

There hasn't been a similar decline since 1979, when shortages created long lines at pumps. In the 12 months ending in February, the latest month for which data are available, miles driven fell 0.4% from a year earlier. The last drop of that scale was in 1980–81.

The decline, while small, is significant because the U.S. population and number of households, drivers and vehicles grow by 1% to 2% a year. A gallon of gas has gone up 59 cents since February, suggesting the trend seems likely to continue. The EIA expects demand for gas to shrink 0.4% this summer from 2007 and fall 0.3% for the year. It would be the first dip in annual consumption since 1991.

- In 2004 and 2005, about one-third of Americans said they cut spending because of rising gas prices. In the new poll, 60 say they are trimming other expenses. Half of households with incomes below $20,000 say they face severe hardships because of soaring gas prices. Three-fourths of households making $75,000 or more also are changing how they use their cars.

Dawn Morris, a consultant in Dover, Del., is blunt about how gas prices are affecting her family.

"It's killing us," she says. She and her husband often stay home on weekends, and when she balances her checkbook, "every third line it says gas: $20, $30, $50."

- Americans' efforts to conserve gas are evident across the USA. At Don Jacobs Used Cars in Lexington, Ky., salesman Tony Morphis says customers are dumping gas guzzlers and ask first about gas mileage when they shop for replacements. Sonya Jensen, owner of Cat's Paw Marina in St. Augustine, Fla., says some boat owners are considering selling their watercraft. At Cycle Cave in Albuquerque, Hervey Hawk says customers are "dragging 30- to 40-year-old bikes out of the garage" and having them fixed so they can pedal to work.

- In the poll, eight in 10 Americans say they use the most fuel-efficient car they own whenever possible. Three-fourths hunt for the cheapest gas available. Six in 10 share rides with friends or neighbors.

Three-fourths say they are getting tune-ups, turning off the air conditioning or driving slower to improve mileage.

Slower speeds might help save lives, says Dennis Hughes, safety chief for the Wisconsin Transportation Department. There have been fewer driving deaths each month since October compared with a year earlier. A harsh winter and record gas prices "conspired to keep a lot of people off the road, or at least to slow down," he says.

Most of those polled expect things to get worse: 54% say they expect gas prices to reach $6 a gallon in the next five years.

For now, they are rethinking the ways they get around, where they buy a home and what they do for fun.

Critical Thinking

1. Gas costs squeeze daily life for whom? Calculate an estimate of the total world population whose lives are "squeezed" by gas prices.

2. At the time of the article (2008), gas was on average $3.65 per gallon. In October 2011, it was $3.46. What is it now? What consumer demand changes do you see? List them.

3. In the article, 54 percent of the people polled in (2008) said they expected to see gas prices reach $6/gallon.

Do you think in two years, this prediction will be correct? Explain.

4. What kind of environmental consequences may result from extremely high gas prices? List several positive and as well as negative impacts.

5. Identify in this article three key terms, concepts, or principles that are used in your textbook (environmental science, economics, sociology, history, geography, etc.) or employed in the discipline you are currently studying. (Note: The terms, concepts, or principles may be implicit, explicit, implied, or inferred.)

UNIT 9

A Global Environmental Ethic?

Unit Selections

Learning Outcomes

After reading this Unit, you should be able to:

- Describe some of the key components of the discourse on environmental ethics and articulate the challenges, obstacles, and realities of environmental issue #9: a global environmental ethic?
- Discuss the diverse range of ideas regarding "sustainable development".
- Outline connections between human attitudes toward sustainable development and behavior.
- Describe the global variables that exist regarding public support for sustainable development.
- Compare and contrast the differences between human-centered and life-centered ethics.
- Explain how the two ideologies in the previous point can result in building different relationships between humans and the environment.
- Identify assumptions underlying any nature ethics.
- Describe the connection between the patterns of poverty and the pattern of environmental degradation.
- Provide examples of "environmental justice."
- Assess to what extent religion can play a role in helping humans establish a sustainable relationship with Earth.
- Describe how different world religions view ideas of ecological balance.
- Offer insights on the challenges of different religions sharing/promoting a universal environmental ethic.
- Articulate what this question means: "What is the endgame of a global environmental ethic?"

Student Website

www.mhhe.com/cls

Internet References

The Earth Institute at Columbia University
www.earth.columbia.edu

Earth Pledge Foundation
www.earthpledge.org

Energy Justice Network
www.energyjustice.net/peak

EnviroLink
http://envirolink.org

Grass-roots.org
www.grass-roots.org

Poverty Mapping
www.povertymap.net

Resources
www.etown.edu/vl

A human being is part of the whole, called by us "Universe," a part limited in time and space. He experiences himself, his thoughts and feelings as something separated from the rest—a kind of optical delusion of his consciousness. This delusion is a kind of prison for us, restricting us to our personal desires and to affection for a few persons nearest to us. Our task must be to free ourselves from this prison by widening our circle of compassion to embrace all living creatures and the whole [of] nature in its beauty.

—*Albert Einstein,* 1950

Sustainable development. Respect for nature. Environmental justice. God–Man–Nature. These are all aspects of "environmental ethics." But should environmental *science* and environmental *ethics* be rubbing elbows? Do ethics have a place in the field of environmental science? Are terms like *intrinsic* and *inherent value, justice, right/wrong/good/bad, respect, stewardship,* or even *environmentalism* appropriate in the vocabulary of science? Does asking if something is "appropriate" have a place in science? Regardless of our philosophical, scientific, religious, or ethical leanings, these and many more questions have and will continue to be a part of environmental science. After all, science, and particularly environmental science, is in the end an effort to understand the workings of life, the universe, the natural world, ourselves. But does it end there, just the understanding? Or does it have something penultimately to do with increasing the probabilities of the survival of a species—us? Is the task of environmental science to help us *discover* a sustainable relationship with Earth, or to *dictate* one?

What does living in harmony with nature mean for human survival? What does living in disharmony with nature—degrading the environment, destroying ecosystems, consuming all our resources, polluting our atmosphere—mean for human survival? Why does adopting a biocentric or lifecentric ethic reduce the risk of human extinction? Why the political and environmental drumbeat to live sustainably with the Earth? Is it because the Earth really doesn't need humans, but we know that we couldn't make it a single day without our wonderful planet? Is this the fundamental reasoning behind our efforts to promulgate a global, environmental ethic? Self-preservation? Although we will never know, Albert Einstein may have kept his thoughts following the preceding quote to himself. Given his shy and quiet nature, there is a good chance he did not want to incite global panic by telling us what would happen to the human race should we not at least consider his scientific observation of humans being just part of a much larger whole, and not *the* whole. There is a bumper sticker that says, "Save the Planet, Kill Yourself." Ethically disconcerting? Perhaps. Religiously offensive? Most likely. Arguable? Very difficult. Would Albert have chuckled? A good chance.

Top Ten Environmental Issue 9: A Global Environmental Ethic? is explored in **Unit 9** because our responses to and resolution of environmental problems will be heavily influenced by how we define such an ethic. Although humanity has many environmental ethics reflecting myriad worldviews of our various cultures, is it

© PBNJ Productions/Getty Images

possible to create a *global* environmental ethic? Is one needed? Scientific logic would posit that doing so could significantly increase the probability for global cooperation, ensuring survival of the human species well into the future. But how should this global environmental ethic read? And more important, how will it be encouraged (enforced)? That, of course, given our diversity of cultures, religions, worldviews, human personalities, and development goals, to name only a few considerations, will make the task challenging. The four articles selected for this unit are intended to help the reader not only appreciate the complexity of such a task as creating a global environmental ethic, but to also reflect critically on some of the components we consider important in defining our relationship (ethics) with the environment: sustainability, respect for nature, environmental justice, Man, Mother Earth, and God.

There are two aspects of any ethic: *attitudes* (or values, opinions, etc.) and *behavior,* and Article 32, the first in this unit, "Do Global Attitudes and Behaviors Support Sustainable Development?" seems to substantiate this premise. Using available survey data, the authors explore peoples' attitudes about one of the primary components of a global environmental ethic—*sustainability.* They point out that although there appears to be no global-scale data identifying attitudes or preferences for a specific end-state of economic development (which would seem to be of the utmost importance for global leaders to know), people around the world do demonstrate a vague consensus regarding the *sensibility* of the idea of sustainability. But they conclude that the evidence suggests that attitudes and expressed values about sustainability do not necessarily translate into actual "sustainability" behaviors. And furthermore, attitudes/values that vary between societies, and even within societies, suggest further the problematic nature of constructing a *global* ethic.

Paul Taylor, in Article 33, "The Ethics for Respect for Nature," proposes what he believes is a new approach to viewing our relationship with nature (an environmental ethic): a life-centered system of ethics. Here, the author believes that only by recognizing and respecting the "inherent worth" of all living organisms can we deconstruct our attitudes of human superiority over nature. Taylor feels we must recognize that the well-being of all of nature, as well as human well-being, is something to be realized "as an end in itself." Of course, global environmental ethics complications arise when humans must assess the "inherent worth" of the mosquito that carries malaria, the Asian carp invasion, the proliferation of disease-carrying rats, or any plant or animal, for that matter, that threatens the well-being of the human species.

Article 34, "Environmental Justice for All," explores another "moral" aspect of the global environmental ethic challenge: how to address human poverty and its association with environmental degradation. The author believes today's environmental justice proponents are changing tactics. They are moving away from "reactive" responses to issues of environmental–social degradation to being "proactive." In a sense, it is an environmental ethic movement that is moving away from saving souls (victims of environmental degradation) to converting would-be sinners (agents of environmental degradation) into environmentally moral beings. In other words, some new advocates for environmental justice believe that by encouraging economic development with an environmental conscience (ethic), we can begin to change the cyclic association of environmental degradation–poverty–degradation. If such an approach proves to be effective in the urban American scene, could the next step be to go global?

Finally, Article 35, "Life, Religion and Everything," has been selected to conclude **Unit 9** for two reasons: (1) religion plays such a prominent role in any discussion of human ethics, and especially when the ethics' discourse involves defining or redefining our spiritual relationship with Earth; and (2) the presentation of the term *sacred,* which may provide the most fundamental, nonpartisan, and universally agreed-upon word among all peoples, religions, languages, cultures, and philosophies used to describe a human's relationship to "place." Excerpted from Webster's Dictionary—*Sacred: dedicated or set apart for the service or worship of deity; worthy of religious veneration, holy; entitled to reverence or respect; of or relating to religion; unassailable, inviolable; highly valued and important.* Such a term may not only help us "broker a better relationship with God, man science and the natural world," as the biologist Rupert Sheldrake believes, but combining the idea of *sacred* with the phenomenon of *environment* may just be the magic key that unlocks the door to our discovering that elusive global environmental ethic.

Do Global Attitudes and Behaviors Support Sustainable Development?

ANTHONY A. LEISEROWITZ, ROBERT W. KATES, AND THOMAS M. PARRIS

Many advocates of sustainable development recognize that a transition to global sustainability—meeting human needs and reducing hunger and poverty while maintaining the life-support systems of the planet—will require changes in human values, attitudes, and behaviors.[1] A previous article in *Environment* described some of the values used to define or support sustainable development as well as key goals, indicators, and practices.[2] Drawing on the few multinational and quasi-global-scale surveys that have been conducted,[3] this article synthesizes and reviews what is currently known about global attitudes and behavior that will either support or discourage a global sustainability transition.[4] (Table 1 provides details about these surveys.)

None of these surveys measured public attitudes toward "sustainable development" as a holistic concept. There is, however, a diverse range of empirical data related to many of the subcomponents of sustainable development: development and environment; the driving forces of population, affluence/poverty/consumerism, technology, and entitlement programs; and the gap between attitudes and behavior.

Development

Concerns for environment and development merged in the early concept of sustainable development, but the meaning of these terms has evolved over time. For example, global economic development is widely viewed as a central priority of sustainable development, but development has come to mean human and social development as well.

Economic Development

The desire for economic development is often assumed to be universal, transcending all cultural and national contexts. Although the surveys in Table 1 have no global-scale data on public attitudes toward economic development per se, this assumption appears to be supported by 91 percent of respondents from 35 developing countries, the United States, and Germany, who said that it is very important (75 percent) or somewhat important (16 percent) to live in a country where there is economic

prosperity.[5] What level of affluence is desired, how that economic prosperity is to be achieved, and how economic wealth should ideally be distributed within and between nations, however, are much more contentious questions. Unfortunately, there does not appear to be any global-scale survey research that has tried to identify public attitudes or preferences for particular levels or end-states of economic development (for example, infinite growth versus steady-state economies) and only limited or tangential data on the ideal distribution of wealth (see the section on affluence below).

Table 1 Multinational Surveys

One-time Surveys

Name	Year(s)	Number of Countries
Pew Global Attitudes Project	2002	43
Eurobarometer	2002	15
International Social Science Program	2000	25
Health of the Planet	1992	24

Repeated Surveys

Name	Year(s)	Number of Countries
GlobeScan International Environmental Monitor	1997–2003	34
World Values Survey	1981–2002	79
Demographic and Health Surveys	1986–2002	17
Organisation for Economic Co-operation and Development	1990–2002	22

Note. Before November 2003, GlobeScan, Inc. was known as Environics International. Surveys before this time bear the older name.

Source: For more detail about these surveys and the countries sampled, see Appendix A in A. Leiserowitz, R. W. Kates, and T. M. Parris, *Sustainability Values, Attitudes and Behaviors: A Review of Multi-national and Global Trends,* CID Working Paper No. 113 (Cambridge, MA: Science, Environment and Development Group, Center for International Development, Harvard University, 2004), www.cid.harvard.edu/cidwp/113.htm.

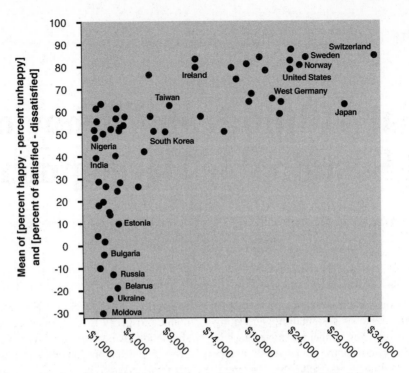

Figure 1 Subjective Well-being by Level of Economic Development.

Note. The subjective well-being index reflects the average of the percentage in each country who describe themselves as "very happy" or "happy" minus the percentage who describe themselves as "not very happy" or "unhappy"; and the percentage placing themselves in the 7–10 range, minus the percentage placing themselves in the 1–4 range, on a 10-point scale on which 1 indicates that one is strongly dissatisfied with one's life as a whole, and 10 indicates that one is highly satisfied with one's life as a whole.

Source: R. Inglehart, "Globalization and Postmodern Values," *Washington Quarterly* 23, no. 1 (1999): 215–228. Subjective well-being data from the 1990 and 1996 World Values Surveys. GNP per capita for 1993 data from *World Bank, World Development Report, 1995* (New York: Oxford University Press, 1995).

Data from the World Values Survey suggest that economic development leads to greater perceived happiness as countries make the transition from subsistence to advanced industrial economies. But above a certain level of gross national product (GNP) per capita—approximately $14,000—the relationship between income level and subjective well-being disappears (see Figure 1). This implies that infinite economic growth does not lead to greater human happiness. Additionally, many of the unhappiest countries had, at the time of these surveys, recently experienced significant declines in living standards with the collapse of the Soviet Union. Yet GNP per capita remained higher in these ex-Soviet countries than in developing countries like India and Nigeria.[6] This suggests that relative trends in living standards influence happiness more than absolute levels of affluence, but the relationship between economic development and subjective well-being deserves more research attention.

Human Development

Very limited data is available on public attitudes toward issues of human development, although it can be assumed that there is near-universal support for increased child survival rates, adult life expectancies, and educational opportunities. However, despite the remarkable increases in these indicators of human well-being since World War II,[7] there appears to be a globally pervasive sense that human well-being has been deteriorating in recent years. In 2002, large majorities worldwide said that a variety of conditions had worsened over the previous five years, including the availability of well-paying jobs (58 percent); working conditions (59 percent); the spread of diseases (66 percent); the affordability of health care (60 percent); and the ability of old people to care for themselves in old age (59 percent). Likewise, thinking of their own countries, large majorities worldwide were concerned about the living conditions of the elderly (61 percent) and the sick and disabled (56 percent), while a plurality was concerned about the living conditions of the unemployed (42 percent).[8]

Development Assistance

One important way to promote development is to extend help to poorer countries and people, either through national governments or nongovernmental organizations and charities. There is strong popular support but less official support for development assistance to poor countries. In 1970, the United Nations General Assembly resolved that each economically advanced country would dedicate 0.7 percent of its gross national income (GNI) to official development assistance (ODA) by the middle of the 1970s—a target that has been reaffirmed in many subsequent international agreements.[9] As of 2004, only five countries had achieved this goal (Denmark, Norway, the Netherlands, Luxembourg, and Sweden). Portugal was close to the target at 0.63, yet all other countries ranged from a high of 0.42 percent (France) to lows of 0.16 and 0.15 percent (the United States and Italy respectively). Overall, the average ODA/GNI among the industrialized countries was only 0.25 percent—far below the UN target.[10]

By contrast, in 2002, more than 70 percent of respondents from 21 developed and developing countries said they would support paying 1 percent more in taxes to help the world's poor.[11] Likewise, surveys in the 13 countries of the Organisation for Economic Co-operation and Development's Development Assistance Committee (OECD-DAC) have found that public support for the principle of giving aid to developing countries (81 percent in 2003) has remained high and stable for more than 20 years.[12] Further, 45 percent said that their government's current (1999–2001) level of expenditure on foreign aid was too low, while only 10 percent said foreign aid was too high.[13] There is also little evidence that the public in OECD countries has developed "donor fatigue." Although surveys have found increasing public concerns about corruption, aid diversion, and inefficiency, these surveys also continue to show very high levels of public support for aid.

Public support for development aid is belied, however, by several factors. First, large majorities demonstrate little understanding of development aid, with most unable to identify their national aid agencies and greatly overestimating the percentage of their national budget devoted to development aid. For example, recent polls have found that Americans believed their government spent 24 percent (mean estimate) of the national budget on foreign assistance, while Europeans often estimated their governments spent 5 to 10 percent.[14] In reality, in 2004 the United States spent approximately 0.81 percent and the European Union member countries an average of approximately 0.75 percent of their national budgets on official development assistance, ranging from a low of 0.30 percent (Italy) to a high of 1.66 percent (Luxembourg).[15] Second, development aid is almost always ranked low on lists of national priorities, well below more salient concerns about (for example) unemployment, education, and health care. Third, "the overwhelming support for foreign aid is based upon the perception that it will be spent on remedying humanitarian crises," not used for other development-related issues like Third World debt, trade barriers, or increasing inequality between rich and poor countries—or for geopolitical reasons (for example, U.S. aid to Israel and Egypt).[16] Support for development assistance has thus been characterized as "a mile wide, but an inch deep" with large majorities supporting aid (in principle) and increasing budget allocations but few understanding what development aid encompasses or giving it a high priority.[17]

Environment

Compared to the very limited or nonexistent data on attitudes toward economic and human development and the overall concept of sustainable development, research on global environmental attitudes is somewhat more substantial. Several surveys have measured attitudes regarding the intrinsic value of nature, global environmental concerns, the trade-offs between environmental protection and economic growth, government policies, and individual and household behaviors.

Human-Nature Relationship

Most research has focused on anthropocentric concerns about environmental quality and natural resource use, with less attention to ecocentric concerns about the intrinsic value of nature. In 1967, the historian Lynn White Jr. published a now-famous and controversial article arguing that a Judeo-Christian ethic and attitude of domination, derived from Genesis, was an underlying historical and cultural cause of the modern environmental crisis.[18] Subsequent ecocentric, ecofeminist, and social ecology theorists have also argued that a domination ethic toward people, women, and nature runs deep in Western, patriarchal, and capitalist culture.[19] The 2000 World Values Survey, however, found that 76 percent of respondents across 27 countries said that human beings should "coexist with nature," while only 19 percent said they should "master nature" (see Figure 2). Overwhelming majorities of Europeans, Japanese, and North Americans said that human beings should coexist with nature, ranging from 85 percent in the United States to 96 percent in Japan. By contrast, only in Jordan, Vietnam, Tanzania, and the Philippines did more than 40 percent say that human beings should master nature.[20] In 2002, a national survey of the United States explored environmental values in more depth and found that Americans strongly agreed that nature has intrinsic value and that humans have moral duties and obligations to animals, plants, and non-living nature (such as rocks, water, and air). The survey found that Americans strongly disagreed that "humans have the right to alter nature to satisfy wants and desires" and that "humans are not part of nature" (see Figure 3).[21] This very limited data suggests that large majorities in the United States and worldwide now reject a domination ethic as the basis of the human-nature relationship, at least at an abstract level. This question, however, deserves much more cross-cultural empirical research.

Environmental Concern

In 2000, a survey of 11 developed and 23 developing countries found that 83 percent of all respondents were concerned a fair amount (41 percent) to a great deal (42 percent) about environmental problems. Interestingly, more respondents from developing countries (47 percent) were "a great deal concerned" about the environment than from developed countries (33 percent), ranging from more than 60 percent in Peru, the Philippines, Nigeria, and India to less than 30 percent in the Netherlands, Germany, Japan, and Spain.[22] This survey also asked respondents to rate the seriousness of several environmental problems (see Figure 4). Large majorities worldwide

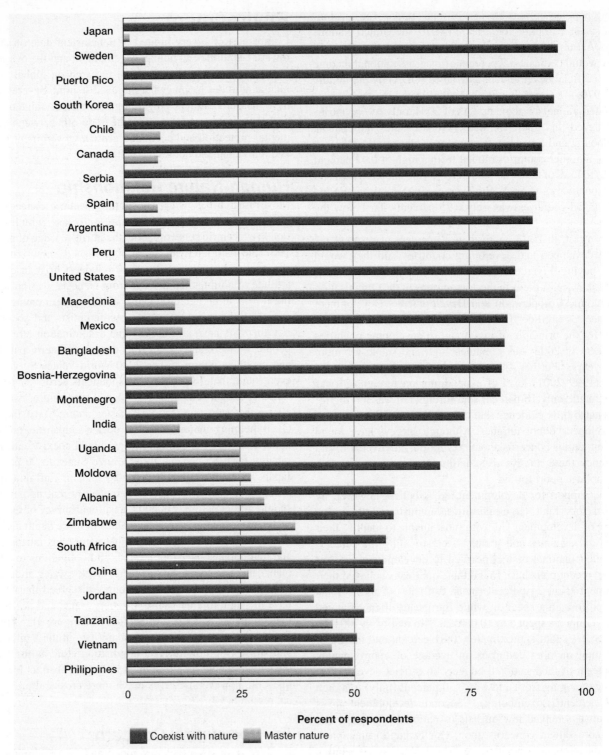

Figure 2 Human-Nature Relationship.

Note. The question asked, "Which statement comes closest to your own views: Human beings should master nature or humans should coexist with nature?"

Source: A. Leiserowitz, 2005. Data from World Values Survey, *The 1999–2002 Values Surveys Integrated Data File 1.0, CD-ROM in R. Inglehart, M. Basanez, J. Diez-Medrano, L. Halman, and R. Luijkx, eds., Human Beliefs and Values: A Cross-Cultural Sourcebook Based on the 1999–2002 Values Surveys, first edition* (Mexico City: Siglo XXI, 2004).

selected the strongest response possible ("very serious") for seven of the eight problems measured. Overall, these results demonstrate very high levels of public concern about a wide range of environmental issues, from local problems like water and air pollution to global problems like ozone depletion and climate change.[23] Further, 52 percent of the global public said

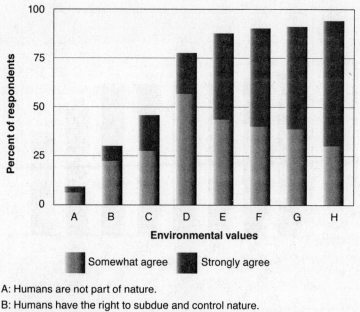

A: Humans are not part of nature.

B: Humans have the right to subdue and control nature.

C: Humankind was created to rule over nature.

D: Humans should adapt to nature rather than modify it to suit them.

E: Humans have moral duties and obligations to non-living nature.

F: Humans have moral duties and obligations to plants and trees.

G: Humans have moral duties and obligations to other animal species.

H: Nature has value within itself regardless of any value humans place on it.

Figure 3   American (U.S.) Environmental Values.

Source: A. Leiserowitz, 2005.

hat if no action is taken, "species loss will seriously affect the planet's ability to sustain life" just 20 years from now.[24]

Environmental Protection versus Economic Growth

n two recent studies, 52 percent of respondents worldwide agreed that "protecting the environment should be given priorty" over "economic growth and creating jobs," while 74 percent of respondents in the G7 countries prioritized environmenal protection over economic growth, even if some jobs were ost.[25] Unfortunately, this now-standard survey question pits the nvironment against economic growth as an either/or dilemma. Rarely do surveys allow respondents to choose an alternative answer, that environmental protection can generate economic growth and create jobs (for example, in new energy system development, tourism, and manufacturing).

ttitudes toward Environmental Policies

n 1995, a large majority (62 percent) worldwide said they would agree to an increase in taxes if the extra money were used to prevent environmental damage," while 33 percent said hey would oppose them.[26] In 2000, there was widespread global support for stronger environmental protection laws and

regulations, with 69 percent saying that, at the time of the survey, their national laws and regulations did not go at all far enough.[27] The 1992 Health of the Planet survey found that a very large majority (78 percent) favored the idea of their own national government "contributing money to an international agency to work on solving global environmental problems." Attitudes toward international agreements in this survey, however, were less favorable. In 1992, 47 percent worldwide agreed that "our nation's environmental problems can be solved without any international agreements," with respondents from low-income countries more likely to strongly agree (23 percent) than individuals from middle-income (17 percent) or high-income (12 percent) countries.[28] In 2001, however, 79 percent of respondents from the G8 countries said that international negotiations and progress on climate change was either "not good enough" (39 percent) or "not acceptable" (40 percent) and needed faster action. Surprisingly, this latter 40 percent supported giving the United Nations "the power to impose legally-binding actions on national governments to protect the Earth's climate."[29]

Environmental Behavior

Material consumption is one of the primary means by which environmental values and attitudes get translated into behavior. (For attitudes toward consumption per se, see the following section on affluence, poverty, and consumerism.)

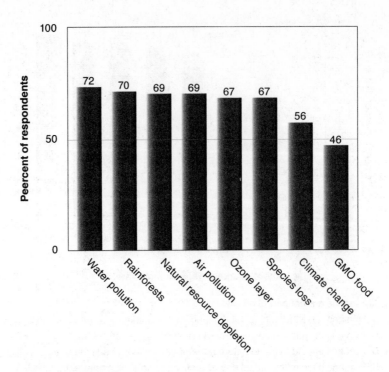

Figure 4 Percent of Global Public Calling Environmental Issues a "Very Serious Problem."

Source: A. Leiserowitz, 2005. Data from Environics International (Globe Scan), *Environics International Environmental Monitor Survey Dataset* (Kingston, Canada: Environics international, 2000), http://jeff-lab.queensu.ca/poadata/info/iem/iemlist.shtml (accessed 5 October 2004).

In 2002, Environics International (GlobeScan) found that 36 percent of respondents from 20 developed and developing countries stated that they had avoided a product or brand for environmental reasons, while 27 percent had refused packaging, and 25 percent had gathered environmental information.[30] Recycling was highly popular, with 6 in 10 people setting aside garbage for reuse, recycling, or safe disposal. These rates, however, reached 91 percent in North America versus only 36–38 percent in Latin America, Eastern Europe, and Central Asia,[31] which may be the result of structural barriers in these societies (for example, inadequate infrastructures, regulations, or markets). There is less survey data regarding international attitudes toward energy consumption, but among Europeans, large majorities said they had reduced or intended to reduce their use of heating, air conditioning, lighting, and domestic electrical appliances.[32]

In 1995, 46 percent of respondents worldwide reported having chosen products thought to be better for the environment, 50 percent of respondents said they had tried to reduce their own water consumption, and 48 percent reported that in the 12 months prior to the survey, they reused or recycled something rather than throwing it away. There was a clear distinction between richer and poorer societies: 67 percent of respondents from high-income countries reported that they had chosen "green" products, while only 30 percent had done so in low-income countries. Likewise, 75 percent of respondents from high-income countries said that they had reused or recycled

something, while only 27 percent in low-income countries said this.[33] However, the latter results contradict the observation of researchers who have noted that many people in developing countries reuse things as part of everyday life (for example, converting oil barrels into water containers) and that million eke out an existence by reusing and recycling items from land fills and garbage dumps.[34] This disparity could be the result c inadequate survey representation of the very poor, who are th most likely to reuse and recycle as part of survival, or, alterna tively, different cultural interpretations of the concepts "reuse and "recycle."

In 2002, 44 percent of respondents in high-income countrie were very willing to pay 10 percent more for an environmentall friendly car, compared to 41 percent from low-income countrie and 29 percent from middle-income countries.[35] These finding clearly mark the emergence of a global market for more energy efficient and less-polluting automobiles. However, while man people appear willing to spend more to buy an environmentall friendly car, most do not appear willing to pay more for gaso line to reduce air pollution. The same 2002 survey found tha among high-income countries, only 28 percent of respondent were very willing to pay 10 percent more for gasoline if th money was used to reduce air pollution, compared to 23 percen in medium-income countries and 36 percent in low-incom countries.[36] People appear to generally oppose higher gasolin prices, although public attitudes are probably affected, at lea in part, by the prices extant at the time of a given survey, th rationale given for the tax, and how the income from the tax will be spent.

Despite the generally pro-environment attitudes and behaviors outlined above, the worldwide public is much less likely to engage in political action for the environment. In 1995, only 13 percent of worldwide respondents reported having donated to an environmental organization, attended a meeting, or signed a petition for the environment in the prior 12 months, with more doing so in high-income countries than in low-income countries.[37] Finally, in 2000, only 10 percent worldwide reported having written a letter or made a telephone call to express their concern about an environmental issue in the past year, 18 percent had based a vote on green issues, and 11 percent belonged to or supported an environmental group.[38]

Drivers of Development and Environment

Many analyses of the human impact on life-support systems focus on three driving forces: population, affluence or income, and technology—the so-called I = PAT identity.[39] In other words, environmental impact is considered a function of these three drivers. In a similar example, carbon dioxide (CO_2) emissions from the energy sector are often considered a function of population, affluence (gross domestic product (GDP) per capita), energy intensity (units of energy per GDP), and technology (CO_2 emissions per unit of energy).[40] While useful, most analysts also recognize that these variables are not fundamental driving forces in and of themselves and are not independent from one another.[41] A similar approach has also been applied to human development (D = PAE), in which development is considered a function of population, affluence, and entitlements and equity.[42] What follows is a review of empirical trends in attitudes and behavior related to population, affluence, technology, and equity and entitlements.

Population

Global population continues to grow, but the rate of growth continues to decline almost everywhere. Recurrent Demographic and Health Surveys (DHS) have found that the ideal number of children desired is declining worldwide. Globally, attitudes toward family planning and contraception are very positive, with 67 percent worldwide and large majorities in 38 out of 40 countries agreeing that birth control and family planning have been a change for the better.[43] Worldwide, these positive attitudes toward family planning are reflected in the behavior of more than 62 percent of married women of reproductive age who are currently using contraception. Within the developing world, the United Nations reports that from 1990 to 2000, contraceptive use among married women in Asia increased from 52 percent to 66 percent, in Latin American and the Caribbean from 57 percent to 69 percent, but in Africa from only 15 percent to 25 percent.[44] Notwithstanding these positive attitudes toward contraception, in 1997, approximately 20 percent to 25 percent of births in the developing world were unwanted, indicating that access to or the use of contraceptives remains limited in some areas.[45]

DHS surveys have found that ideal family size remains significantly larger in western and middle Africa (5.2) than elsewhere in the developing world (2.9).[46] They also found that support for family planning is much lower in sub-Saharan Africa (44 percent) than in the rest of the developing world (74 percent).[47] Consistent with these attitudes, sub-Saharan Africa exhibits lower percentages of married women using birth control as well as lower rates of growth in contraceptive use than the rest of the developing world.[48]

Affluence, Poverty, and Consumerism

Aggregate affluence and related consumption have risen dramatically worldwide with GDP per capita (purchasing-power parity, constant 1995 international dollars) more than doubling between 1975 and 2002.[49] However, the rising tide has not lifted all boats. Worldwide in 2001, more than 1.1 billion people lived on less than $1 per day, and 2.7 billion people lived on less than $2 per day—with little overall change from 1990. However, the World Bank projects these numbers to decline dramatically by 2015—to 622 million living on less than $1 per day and 1.9 billion living on less than $2 per day. There are also large regional differences, with sub-Saharan Africa the most notable exception: There, the number of people living on less than $1 per day rose from an estimated 227 million in 1990 to 313 million in 2001 and is projected to increase to 340 million by 2015.[50]

Poverty

Poverty reduction is an essential objective of sustainable development.[51] In 1995, 65 percent of respondents worldwide said that more people were living in poverty than had been 10 years prior. Regarding the root causes of poverty, 63 percent blamed unfair treatment by society, while 26 percent blamed the laziness of the poor themselves. Majorities blamed poverty on the laziness and lack of willpower of the poor only in the United States (61 percent), Puerto Rico (72 percent), Japan (57 percent), China (59 percent), Taiwan (69 percent), and the Philippines (63 percent) (see Figure 5).[52] Worldwide, 68 percent said their own government was doing too little to help people in poverty within their own country, while only 4 percent said their government was doing too much. At the national level, only in the United States (33 percent) and the Philippines (21 percent) did significant proportions say their own government was doing too much to help people in poverty.[53]

Consumerism

Different surveys paint a complicated and contradictory picture of attitudes toward consumption. On the one hand, majorities around the world agree that, at the societal level, material and status-related consumption are threats to human cultures and the environment. Worldwide, 54 percent thought "less emphasis on money and material possessions" would be a good thing, while

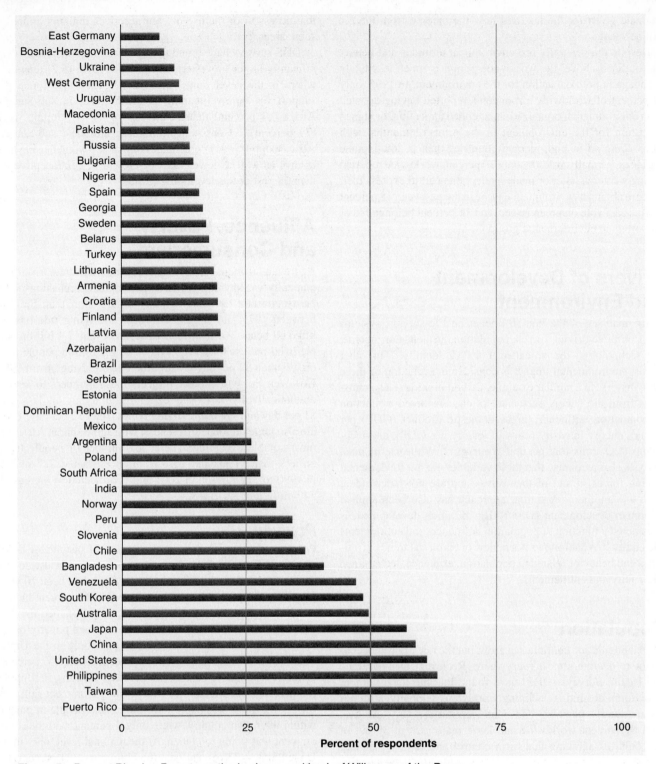

Figure 5 Percent Blaming Poverty on the Laziness and Lack of Willpower of the Poor.

Source: A. Leiserowitz, 2005. Data from R. Inglehart, et al., *World Values Surveys and European Values Surveys, 1981–1984, 1990–1993, and 1995–1997* [computer file], Inter-university Consortium for Political and Social Research (ICPSR) version (Ann Arbor, MI: Institute for Social Research [producer], 2000; Ann Arbor, MI: ICPSR [distributor], 2000).

only 21 percent thought this would be a bad thing.[54] Further, large majorities agreed that gaining more time for leisure activities or family life is their biggest goal in life.[55]

More broadly, in 2002 a global study sponsored by the Pew Research Center for the People & the Press found that 45 percent worldwide saw consumerism and commercialism as a threat to their own culture. Interestingly, more respondents from high-income and upper- middle-income countries (approximately 51 percent) perceived consumerism as a threat than low-middle- and low-income countries (approximately

43 percent).[56] Unfortunately, the Pew study did not ask respondents whether they believed consumerism and commercialism were a threat to the environment. In 1992, however, 41 percent said that consumption of the world's resources by industrialized countries contributed "a great deal" to environmental problems in developing countries."[57]

On the other hand, 65 percent of respondents said that spending money on themselves and their families represents one of life's greatest pleasures. Respondents from low-GDP countries were much more likely to agree (74 percent) than those from high-GDP countries (58 percent), which reflects differences in material needs (see Figure 6).[58]

Likewise, there may be large regional differences in attitudes toward status consumerism. Large majorities of Europeans and North Americans disagreed (78 percent and 76 percent respectively) that other people's admiration for one's possessions is important, while 54 to 59 percent of Latin American, Asian, and Eurasian respondents, and only 19 percent of Africans (Nigeria only), disagreed.[59] There are strong cultural norms against appearing materialistic in many Western societies, despite the high levels of material consumption in these countries relative to the rest of the world. At the same time, status or conspicuous consumption has long been posited as a significant driving force in at least some consumer behavior, especially in affluent societies.[60] While these studies are a useful start, much more research is needed to unpack and explain the roles of values and attitudes in material consumption in different socioeconomic circumstances.

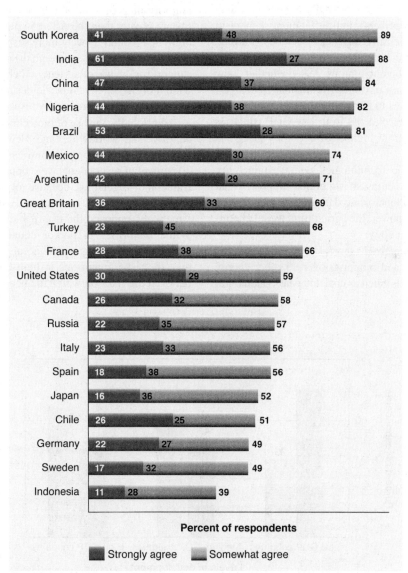

Percent of respondents

■ Strongly agree ■ Somewhat agree

Figure 6 Purchasing for Self and Family Gives Greatest Pleasure ("Strongly" and "Somewhat" agree).

Note. The question was, "To spend money, to buy something new for myself or my family, is one of the greatest pleasures in my life."

Source: Environics International (GlobeScan), *Consumerism: A Special Report* (Toronto: Environics International, 2002).

Science and Technology

Successful deployment of new and more efficient technologies is an important component of most sustainability strategies, even though it is often difficult to assess all the environmental, social, and public health consequences of these technologies in advance. Overall, the global public has very positive attitudes toward science and technology. The 1995 World Values Survey asked respondents, "In the long run, do you think the scientific advances we are making will help or harm mankind?" Worldwide, 56 percent of respondents thought science will help mankind, while 26 percent thought it will harm mankind. Further, 67 percent said an increased emphasis on technological development would be a good thing, while only 9 percent said it would be bad.[61] Likewise, in 2002, GlobeScan found large majorities worldwide believed that the benefits of modern technology outweigh the risks.[62] The support for technology, however, was significantly higher in countries with low GDPs (69 percent) than in high-GDP countries (56 percent), indicating more skepticism among people in technologically advanced societies. Further, this survey found dramatic differences in technological optimism between richer and poorer countries. Asked whether "new technologies will resolve most of our environmental challenges, requiring only minor changes in human thinking and individual behavior," 62 percent of respondents from low-GDP countries agreed, while 55 percent from high-GDP countries disagreed (see Figure 7).

But what about specific technologies with sustainability implications? Do these also enjoy strong public support? What follows is a summary of global-scale data on attitudes toward renewable energy, nuclear power, the agricultural use of chemical pesticides, and biotechnology.

Europeans strongly preferred several renewable energy technologies (solar, wind, and biomass) over all other energy sources, including solid fuels (such as coal and peat), oil, natural gas, nuclear fission, nuclear fusion, and hydroelectric power. Also, Europeans believed that by the year 2050, these energy sources will be best for the environment (67 percent), be the least expensive (40 percent), and will provide the greatest amount of useful energy (27 percent).[63] Further, 37 percent of Europeans and approximately 33 percent of respondents in 16 developed and developing countries were willing to pay 10 percent more for electricity derived from renewable energy sources.[64]

Nuclear power, however, remains highly stigmatized throughout much of the developed world.[65] Among respondents from 18 countries (mostly developed), 62 percent considered nuclear power stations "very dangerous" to "extremely dangerous" for the environment.[66] Whatever its merits or demerits as an alternative energy source, public attitudes about nuclear power continue to constrain its political feasibility.

Regarding the use of chemical pesticides on food crops, a majority of people in poorer countries believed that the benefits are greater than the risks (54 percent), while respondents in high-GDP countries were more suspicious, with only 32 percent believing the benefits outweigh the risks.[67] Since 1998, however, support for the use of agricultural chemicals has dropped worldwide. Further, chemical pesticides are now one of the top food-related concerns expressed by respondents around the world.[68]

Additionally, the use of biotechnology in agriculture remains controversial worldwide, and views on the issue are divided between rich and poor countries. Across the G7 countries, 70 percent of respondents were opposed to scientifically altered fruits and vegetables because of health and environmental concerns,[69] while 62 percent of Europeans and 45 percent of Americans opposed the use of biotechnology in agriculture.[70] While majorities in poorer countries (65 percent) believed the benefits of using biotechnology on food crops are greater than the risks, majorities in high-GDP countries (51 percent) believed the risks outweigh the benefits.[71]

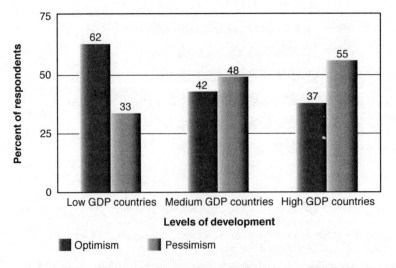

Figure 7 Technological Optimism Regarding Environmental Problems.

Source: A. Leiserowitz, 2005. Data from Environics International (GlobeScan), *International Environmental Monitor* (Toronto: Environics International, 2002), 135.

More broadly, public understanding of biotechnology is still limited, and slight variations in question wordings or framings can have significant impacts on support or opposition. For example, 56 percent worldwide thought that biotechnology will be good for society in the long term, yet 57 percent also agreed that "any attempt to modify the genes of plants or animals is ethically and morally wrong."[72] Particular applications of biotechnology also garnered widely different degrees of support. While 78 percent worldwide favored the use of biotechnology to develop new medicines, only 34 percent supported its use in the development of genetically modified food. Yet, when asked whether they supported the use of biotechnology to produce more nutritious crops, 61 percent agreed.[73]

Income Equity and Entitlements

Equity and entitlements strongly determine the degree to which rising population and affluence affect human development, particularly for the poor. For example, as global population and affluence have grown, income inequality between rich and poor countries has also increased over time, with the notable exceptions of East and Southeast Asia—where incomes are on the rise on a par with (or even faster than) the wealthier nations of the world.[74] Inequality within countries has also grown in many rich and poor countries. Similarly, access to entitlements—the bundle of income, natural resources, familial and social connections, and societal assistance that are key determinants of hunger and poverty[75]—has recently declined with the emergence of market-oriented economies in Eastern and Central Europe, Russia, and China; the rising costs of entitlement programs in the industrialized countries, including access to and quality of health care,

education, housing, and employment; and structural adjustment programs in developing countries that were recommended by the International Monetary Fund. Critically, it appears there is no comparative data on global attitudes toward specific entitlements; however, there is much concern that living conditions for the elderly, unemployed, and the sick and injured are deteriorating, as cited above in the discussion on human development.

In 2002, large majorities said that the gap between rich and poor in their country had gotten wider over the previous 5 years. This was true across geographic regions and levels of economic development, with majorities ranging from 66 percent in Asia, 72 percent in North America, and 88 percent in Eastern Europe (excepting Ukraine) stating that the gap had gotten worse.[76] Nonetheless, 48 percent of respondents from 13 countries preferred a "competitive society, where wealth is distributed according to one's achievement," while 34 percent preferred an "egalitarian society, where the gap between rich and poor is small, regardless of achievement" (see Figure 8).[77]

More broadly, 47 percent of respondents from 72 countries preferred "larger income differences as incentives for individual effort," while 33 percent preferred that "incomes should be made more equal."[78] These results suggest that despite public perceptions of growing economic inequality, many accept it as an important incentive in a more individualistic and competitive economic system. These global results, however, are limited to just a few variables and gloss over many countries that strongly prefer more egalitarian distributions of wealth (such as India). Much more research is needed to understand how important the principles of income equality and equal economic opportunity are considered globally, either as global goals or as means to achieve other sustainability goals.

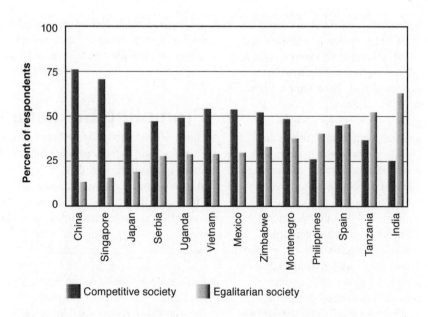

Figure 8 Multinational Preferences for a Competitive Versus Egalitarian Society.

Source: A. Leiserowitz, 2005. Data from World Values Survey. *The 1999–2002 Values Surveys Integrated Data File 1.0,* CD-ROM in R. Inglehart, M. Basanez, J. Diez-Medrano, L. Halman, and R. Luijkx, eds., *Human Beliefs and Values: A Cross-Cultural Sourcebook Based on the 1999–2002 Values Surveys,* first edition (Mexico City: Siglo XX[[]]I, 2004).

Does the Global Public Support Sustainable Development?

Surprisingly, the question of public support for sustainable development has never been asked directly, at least not globally. But two important themes emerge from the multinational data and analysis above. First, in general, the global public supports the main tenets of sustainable development. Second, however, there are many contradictions, including critical gaps between what people say and do—both as individuals and in aggregate. From these themes emerge a third finding: Diverse barriers stand between sustainability attitudes and action.

- *Large majorities worldwide appear to support environmental protection and economic and human development—the three pillars of sustainable development.* They express attitudes and have taken modest actions consonant with support for sustainable development, including support for environmental protection; economic growth; smaller populations; reduced poverty; improved technology; and care and concern for the poor, the marginal, the young, and the aged.

- *Amid the positive attitudes, however, are many contradictions.* Worldwide, all the components of the Human Development Index—life expectancy, adult literacy, and per capita income—have dramatically improved since World War II.[79] Despite the remarkable increases in human well-being, however, there appears to be a globally pervasive sense that human well-being has more recently been deteriorating. Meanwhile, levels of development assistance are consistently overestimated by lay publics, and the use of such aid is misunderstood, albeit strongly supported. Overall, there are very positive attitudes toward science and technology, but the most technologically sophisticated peoples are also the most pessimistic about the ability of technology to solve global problems. Likewise, attitudes toward biotechnology vary widely, depending on how the question is asked.

- Further, there are serious gaps between what people believe and what people do, both as individuals and as polities. Worldwide, the public strongly supports significantly larger levels of development assistance for poor countries, but national governments have yet to translate these attitudes into proportional action. Most people value the environment—for anthropocentric as well as ecocentric reasons—yet many ecological systems around the world continue to degrade, fragment, and lose resilience. Most favor smaller families, family planning, and contraception, but one-fifth to one-quarter of children born are not desired. Majorities are concerned with poverty and think more should be done to alleviate it, but important regions of the world think the poor themselves are to blame, and a majority worldwide accepts large gaps between rich and poor. Most people think that less emphasis on material possessions would be a good thing and that more time for leisure and family should be primary goals, but spending money often provides one of life's greatest pleasures. While many would pay more for fuel-efficient cars, fuel economy has either stagnated or even declined in many countries. Despite widespread public support for renewable energy, it still accounts for only a tiny proportion of global energy production.

- *There are diverse barriers standing between pro-sustainability attitudes and individual and collective behaviors.*[80] These include at least three types of barriers. First are the direction, strength, and priority of particular attitudes. Some sustainability attitudes may be widespread but not strongly or consistently enough relative to other, contradictory attitudes. A second type of barrier between attitudes and behavior relates to individual capabilities. Individuals often lack the time, money, access, literacy, knowledge, skills, power, or perceived efficacy to translate attitudes into action. Finally, a third type of barrier is structural and includes laws; regulations; perverse subsidies; infrastructure; available technology; social norms and expectations; and the broader social, economic, and political context (such as the price of oil, interest rates, special interest groups, and the election cycle).

Thus, each particular sustainability behavior may confront a unique set of barriers between attitudes and behaviors. Further, even the same behavior (such as contraceptive use) may confront different barriers across society, space, and scale—with different attitudes or individual and structural barriers operating in developed versus developing countries, in secular versus religious societies, or at different levels of decisionmaking (for example, individuals versus legislatures). Explaining unsustainable behavior is therefore "dauntingly complex, both in its variety and in the causal influences on it."[81] Yet bridging the gaps between what people believe and what people do will be an essential part of the transition to sustainability.

Promoting Sustainable Behavior

Our limited knowledge about global sustainability values, attitudes, and behaviors does suggest, however, that there are short- and long-term strategies to promote sustainable behavior. We know that socially pervasive values and attitudes are often highly resistant to change. Thus, in the short term, leveraging the values and attitudes already dominant in particular cultures may be more practical than asking people to adopt new value orientations.[82] For example, economic values clearly influence and motivate many human behaviors, especially in the market and cash economies of the developed countries. Incorporating environmental and social "externalities" into prices or accounting for the monetary value of ecosystem services can thus encourage both individual and collective sustainable behavior.[83] Likewise, anthropocentric concerns about the impacts of environmental degradation and exploitative labor conditions on human health and social well-being remain strong motivators for action in both the developed and developing worlds.[84] Additionally, religious values are vital sources of meaning, motivation, and direction for much of the world, and many religions are actively re-evaluating and reinterpreting their traditions in support of sustainability.[85]

In the long term, however, more fundamental changes may be required, such as extending and accelerating the shift from materialist to post-materialist values, from anthropocentric to ecological worldviews, and a redefinition of "the good life."[86] These long-term changes may be driven in part by impersonal forces, like changing economics (globalization) or technologies (for example, mass media and computer networks) or by broadly based social movements, like those that continue to challenge social attitudes about racism, environmental degradation, and human rights. Finally, sustainability science will play a critical role, at multiple scales and using multiple methodologies, as it works to identify and explain the key relationships between sustainability values, attitudes, and behaviors—and to apply this knowledge in support of sustainable development.

Notes

1. For example, see U.S. National Research Council, Policy Division, Board on Sustainable Development, *Our Common Journey: A Transition toward Sustainability* (Washington, DC: National Academy Press, 1999); and P. Raskin et al., *Great Transition: The Promise and Lure of the Times Ahead* (Boston: Stockholm Environment Institute, 2002).

2. R. W. Kates, T. M. Parris, and A. Leiserowitz, "What Is Sustainable Development? Goals, Indicators, Values, and Practice,"*Environment,* April 2005, 8–21.

3. For simplicity, the words "global" and "worldwide" are used throughout this article to refer to survey results. Please note, however, that there has never been a truly representative global survey with either representative samples from every country in the world or in which all human beings worldwide had an equal probability of being selected. Additionally, some developing country results are taken from predominantly urban samples and are thus not fully representative.

4. For more detail about these surveys and the countries sampled, see Appendix A in A. Leiserowitz, R. W. Kates, and T. M. Parris, *Sustainability Values, Attitudes and Behaviors: A Review of Multi-national and Global Trends* (No. CID Working Paper No. 113) (Cambridge, MA: Science, Environment and Development Group, Center for International Development, Harvard University, 2004), www.cid.harvard.edu/cidwp/113.htm.

5. Pew Research Center for the People & the Press, *Views of a Changing World* (Washington, DC: The Pew Research Center for the People & the Press, 2003), T72.

6. See R. Inglehart, "Globalization and Postmodern Values," *Washington Quarterly* 23, no. 1 (1999): 215–28.

7. Leiserowitz, Kates, and Parris, note 4 above, page 8.

8. Pew Research Center for the People & the Press, *The Pew Global Attitudes Project Dataset* (Washington, DC: The Pew Research Center for the People & the Press, 2004).

9. Gross national income (GNI) is "[t]he total market value of goods and services produced during a given period by labor and capital supplied by residents of a country, regardless of where the labor and capital are located. [GNI] differs from GDP primarily by including the capital income that residents earn from investments abroad and excluding the capital income that nonresidents earn from domestic investment." Official development assistance (ODA) is defined as "[t]hose flows to developing countries and multilateral institutions provided by official agencies, including state and local governments, or by their executive agencies, each transaction of which meets the following tests: (a) it is administered with the promotion of the economic development and welfare of developing countries as its main objective; and (b) it is concessional in character and conveys a grant element of at least 25 percent." UN Millennium Project, *The 0.7% Target: An In-Depth Look,* www.unmillenniumproject.org/involved/action07.htm (accessed 24 August 2005). Official development assistance (ODA) does not include aid flows from private voluntary organizations (such as churches, universities, or foundations). For example, it is estimated that in 2000, the United States provided more than $4 billion in private grants for development assistance, versus nearly $10 billion in ODA. U.S. Agency for International Development (USAID), *Foreign Aid in the National Interest* (Washington. DC, 2002), 134.

10. Organisation for Economic Co-operation and Development (OECD), *Official Development Assistance Increases Further—But 2006 Targets Still a Challenge* (Paris: OECD, 2005), www.oecd.org/document/3/0,2340,en_2649_34447_34700611_1_1_1_1,00.html (accessed 30 July 2005).

11. Environics International (GlobeScan), *The World Economic Forum Poll: Global Public Opinion on Globalization* (Toronto: Environics International, 2002), www.globescan.com/brochures/WEF_Poll_Brief.pdf (accessed 5 October 2004), 3. Note that Environics International changed its name to Globe Scan Incorporated in November 2003.

12. OECD, *Public Opinion and the Fight Against Poverty* (Paris: OECD Development Centre, 2003), 17.

13. Ibid, page 19.

14. Program on International Policy Attitudes (PIPA), *Americans on Foreign Aid and World Hunger: A Study of U.S. Public Attitudes* (Washington, DC: PIPA, 2001), www.pipa.org/OnlineReports/BFW (accessed 17 November 2004); and OECD, note 12 above, page 22.

15. See OECD Development Co-operation Directorate, *OECD-DAC Secretariat Simulation of DAC Members' Net ODA Volumes in 2006 and 2010,* www.oecd .org/dataoecd/57/30/35320618.pdf; and Central Intelligence Agency, The World Factbook, www.cia.gov/cia/publications/factbook/.

16. OECD, note 12 above, page 20.

17. I. Smillie and H. Helmich, eds., *Stakeholders: Government-NGO Partnerships for International Development* (London: Earthscan, 1999).

18. L. White Jr., "The Historical Roots of Our Ecologic Crisis," *Science,* l0 March 1967, 1203–07.

19. See C. Merchant, *The Death of Nature: Women, Ecology, and the Scientific Revolution* (1st ed.) (San Francisco: Harper & Row 1980); C. Merchant, *Radical Ecology: The Search for a Livable Worm* (New York: Routledge, 1992); and G. Sessions, *Deep Ecology for the Twenty-First Century* (1st ed.) (New York: Shambhala Press 1995).

20. World Values Survey, *The 1999–2002 Values Surveys Integrated Data File 1.0,* CD-ROM in R. Inglehart, M. Basanez, J. Diez-Medrano, L. Halman, and R. Luijkx, eds., *Human Beliefs and Values: A Cross-Cultural Sourcebook Based on the 1999–2002 Values Surveys,* first edition (Mexico City: Siglo XXI, 2004).

21. These results come from a representative national survey of American Climate change risk perceptions, policy preferences, and behaviors and broader environmental and cultural values. From November 2002 to February 2003, 673 adults (18 and older) completed a mail-out, mail-back questionnaire, for a response rate of 55 percent. The results are weighted to bring them in line with actual population proportions. See A. Leiserowitz, "American Risk Perceptions: Is Climate Change Dangerous?" *Risk Analysis,* in press; and A. Leiserowitz, "Climate Change Risk Perception and Policy Preferences: The Role of Affect, Imagery, and Values," *Climatic Change,* in press.

22. These results support the argument that concerns about the environment are not "a luxury affordable only by those who have enough economic security to pursue quality-of-life goals." See R. E. Dunlap, G. H. Gallup Jr., and A. M. Gallup, "Of Global Concern: Results of the Health of the Planet Survey," *Environment,* November 1993, 7–15, 33–39 (quote at 37); R. E. Dunlap, A. G. Mertig, "Global Concern for the Environment: Is Affluence a Prerequisite?" *Journal of Social Issues* 511, no. 4 (1995): 121–37; S. R. Brechin and W. Kempton, "Global Environmentalism: A Challenge to the Postmaterialism Thesis?" *Social Science Quarterly* 75, no. 2 (1994): 245–69.

23. Environics International (GlobeScan), *Environics International Environmental Monitor Survey Data-set* (Kingston, Canada: Environics International, 2000), http://jeff-lab.queensu.ca/poadata/info/iem/iemlist.shtml (accessed 5 October 2004). These multinational levels of concern and perceived seriousness of environmental problems remained roughly equivalent from 1992 to 2000, averaged across the countries sampled by the 1992 Health of the Planet and the Environics surveys, although some countries saw significant increases in perceived seriousness of environmental problems (India, the Netherlands, the Philippines, and South Korea), while others saw significant decreases (Turkey and Uruguay). See R. E. Dunlap, G. H. Gallup Jr., and A. M. Gallup, *Health of the Planet: Results of a 1992 International Environmental Opinion Survey of Citizens in 24 Nations* (Princeton, N J: The George H. Gallup International Institute, 1993); and R. E. Dunlap, G. H. Gallup Jr., and A. M. Gallup, "Of Global Concern: Results of the Health of the Planet Survey," *Environment,* November 1993, 7–15, 33–39.

24. GlobeScan, *Results of First-Ever Global Poll on Humanity's Relationship with Nature* (Toronto: GlobeScan Incorporated, 2004), www.globescan.com/news_archives/IUCN_PR.html (accessed 30 July 2005).

25. World Values Survey, note 20 above; and Pew Research Center for the People & the Press, *What the World Thinks in 2002* (Washington, DC: The Pew Research Center for the People & the Press, 2002), T-9. The G7 includes Canada, France, Germany, Great Britain, Italy, Japan and the United States. It expanded to the G8 with the addition of Russia in 1998.

26. R. Inglehart, et al., *World Values Surveys and European Values Surveys, 1981–1984, 1990–1993, and 1995–1997* [computer file], Inter-university Consortium for Political and Social Research (ICPSR) version (Ann Arbor, MI: Institute

for Social Research [producer], 2000; Ann Arbor, MI: ICPSR [distributor], 2000).

27. Environics International (GlobeScan), note 23 above.

28. Dunlap, Gallup Jr., and Gallup, *Health of the Planet: Results of a 1992 International Environmental Opinion Survey of Citizens in 24 Nations,* note 23 above.

29. Environics International (GlobeScan), *New Poll Shows G8 Citizens Want Legally-Binding Climate Accord* (Toronto: Environics International, 2001), www.globescan.com/news_archives/IEM_climatechange.pdf (accessed 30 July 2005).

30. Environics International (GlobeScan), *International Environmental Monitor* (Toronto: Environics International, 2002), 44.

31. Ibid., page 49.

32. The European Opinion Research Group, *Eurobarometer: Energy: Issues, Options and Technologies, Science and Society,* EUR 20624 (Brussels: European Commission, 2002), 96–99.

33. Inglehart, note 26 above.

34. C. M. Rogerson, "The Waste Sector and Informal Entrepreneurship in Developing World Cities," *Urban Forum* 12, no. 2 (2001): 247–59.

35. Environics International (GlobeScan), note 30 above, page 63. These results are based on the sub-sample of those who own or have regular use of a car.

36. Environics International (GlobeScan), note 30 above, page 65.

37. Inglehart, note 26 above.

38. Environics International (GlobeScan), note 23 above.

39. P. A. Ehrlich and J. P. Holdren, review of *The Closing Circle,* by Barry Commoner, *Environment,* April 1972, 24, 26–39.

40. Y. Kaya, "Impact of Carbon Dioxide Emission Control on GNP Growth: Interpretation of Proposed Scenarios," paper presented at the Intergovernmental Panel on Climate Change (IPCC) Energy and Industry Subgroup, Response Strategies Working Group, Paris, France, 1990; and R. York, E. Rosa, and T. Dietz, "STIRPAT, IPAT and ImPACT: Analytic Tools for Unpacking the Driving Forces of Environmental Impacts," *Ecological Economics* 46, no. 3 (2003): 351.

41. IPCC, *Emissions Scenarios* (Cambridge: Cambridge University Press, 2000); and E. F. Lambin, et al., "The Causes of Land-Use and Land-Cover Change: Moving Beyond the Myths," *Global Environmental Change: Human and Policy Dimensions* 11, no. 4 (2001):

42. T. M. Parris and R. W. Kates, "Characterizing a Sustainability Transition: Goals, Targets, Trends, and Driving Forces," *Proceedings of the National Academy of Sciences of the United States of America* 100, no. 14 (2003): 6.

43. Pew Research Center for the People & the Press, note 8 above, page T17.

44. United Nations, *Majority of World's Couples Are Using Contraception* (New York: United Nations Population Division, 2001).

45. J. Bongaarts, "Trends in Unwanted Childbearing in the Developing World," *Studies in Family Planning* 28, no. 4 (1997): 267–77.

46. Demographic and Health Surveys (DHS), *STATCompiler* (Calverton, MD: Measure DHS, 2004), www.measuredhs.com (accessed 5 October, 2004).

47. Ibid.

48. U.S. Bureau of the Census, *World Population Profile: 1998,* WP/98 (Washington, DC, 1999), 45.

49. World Bank, *World Development Indicators CD-ROM 2004* [computer file] (Washington, DC: International Bank for Reconstruction and Development (IBRD) [producer], 2004).

50. World Bank, *Global Economic Prospects 2005: Trade, Regionalism, and Development* [computer file] (Washington, DC: IBRD [producer] 2005).

51. For more information on poverty reduction strategies, see T. Banuri, review of *Investing in Development: A Practical Plan to Achieve the Millennium Goals,* by UN Millennium Project, *Environment,* November 2005 (this issue), 37.

52. Inglehart, note 26 above.

53. Inglehart, note 26 above.

54. Inglehart, note 26 above.

55. Environics International (GlobeScan), *Consumerism: A Special Report* (Toronto: Environics International, 2002), 6.

56. Pew Research Center for the People & the Press, note 25 above.

57. Dunlap, Gallup Jr., and Gallup, *Health of the Planet: Results of a 1992 International Environmental Opinion Survey of Citizens in 24 Nations,* note 23 above, page 57.

58. Environics International (GlobeScan), note 55 above, pages 3–4.

59. Environics International (GlobeScan), note 55 above, pages 3–4.

60. T. Veblen, *The Theory of the Leisure Class: An Economic Study of Institutions* (New York: Macmillan 1899).

61. Inglehart, note 26 above.

62. Environics International (GlobeScan), note 30 above, page 133.

63. The European Opinion Research Group, note 32 above, page 70.

64. Environics International (GlobeScan), note 23 above.

65. For example, see J. Flynn, P. Slovic, and H. Kunreuther, *Risk, Media and Stigma: Understanding Public Challenges to Modern Science and Technology* (London: Earthscan, 2001).

66. International Social Science Program, *Environment II,* (No. 3440) Cologne: Zentralarchiv für Empirische Sozialforschung, Universitaet zu Koeln (Central Archive for Empirical Social Research, University of Cologne), 2000, 114.

67. Environics International (GlobeScan), note 30 above, page 139.

68. Environics International (GlobeScan), note 30 above, page 141.

69. Pew Research Center for the People & the Press, note 25 above, page T20.

70. Chicago Council on Foreign Relations (CCFR), *Worldviews 2002* (Chicago: CCFR, 2002), 26.

71. Environics International (GlobeScan), note 30 above, page 163.

72. Environics International (GlobeScan), note 30 above, page 156–57.

73. Environics International (GlobeScan), note 30 above, page 57.

74. W. J. Baumol, R. R. Nelson, and E. N. Wolff, *Convergence of Productivity: Cross-National Studies and Historical Evidence* (New York: Oxford University Press, 1994).

75. A. K. Sen, *Poverty and Famines: An Essay on Entitlement and Deprivation* (Oxford: Oxford University Press, 1981).

76. Pew Research Center for the People & the Press, note 5 above, page 37.

77. World Values Survey, note 20 above.

78. World Values Survey, note 20 above.

79. The human development index (HDI) measures a country's average achievements in three basic aspects of human development: longevity, knowledge, and a decent standard of living. Longevity is measured by life expectancy at birth; knowledge is measured with the adult literacy rate and the combined primary, secondary, and tertiary gross enrollment ratio; and standard of living is measured by gross domestic product per capita (purchase-power parity US$). The UN Development Programme (UNDP) has used the HDI for its annual reports since 1993. UNDP, *Questions About the Human Development Index (HDI),* www.undp.org/hdr2003/faq.html#21 (accessed 25 August 2005).

80. See, for example, J. Blake, "Overcoming the 'Value-Action Gap' in Environmental Policy: Tensions Between National Policy and Local Experience," *Local Environment* 4, no. 3 (1999): 257–78; A. Kollmuss and J. Agyeman, "Mind the Gap: Why Do People Act Environmentally and What Are the Barriers to Pro-Environmental Behavior?" *Environmental Education Research* 8, no. 3 (2002): 239–60; and E C. Stem, "Toward a Coherent Theory of Environmentally Significant Behavior," *Journal of Social Issues* 56, no. 3 (2000): 407–24.

81. Stern, ibid., page 421.

82. See, for example, P. W. Schultz and L. Zelezny, "Reframing Environmental Messages to Be Congruent with American Values," *Human Ecology Review* 10, no. 2 (2003): 126–36.

83. Millennium Ecosystem Assessment, *Ecosystems and Human Well-Being: Synthesis* (Washington, DC: Island Press, 2005).

84. Dunlap, Gallup Jr., and Gallup, *Health of the Planet: Results of a 1992 International Environmental Opinion Survey of Citizens in 24 Nations,* note 23 above, page 36.

85. See *The Harvard Forum on Religion and Ecology,* http://environment.harvard.edu/religion/main.html; R. S. Gottlieb, *This Sacred Earth: Religion, Nature, Environment* (New York: Routledge, 1996); and G. Gardner, *Worldwatch Paper #164: Invoking the Spirit: Religion and Spirituality in the Quest for a Sustainable World* (Washington, DC: Worldwatch Institute, 2002).

86. R. Inglehart, *Modernization and Postmodernization: Cultural, Economic and Political Change in 43 Societies* (Princeton: Princeton University Press, 1997); T. O'Riordan, "Frameworks for Choice: Core Beliefs and the Environment," *Environment,* October 1995, 4–9, 25–29; and E Raskin and Global Scenario Group, *Great Transition: The Promise and Lure of the Times Ahead* (Boston: Stockholm Environment Institute, 2002).

Critical Thinking

1. Construct a one-page outline of this article and try to write a one-sentence conclusion that expresses the central message of the article.

2. Review Table 1 in the article and make a list of critical questions regarding the information provided. (For example, what countries; how many people in survey.)

3. The author points out that "amid the positive attitudes, however, are many contradictions." What are those contradictions? What role do you think place, culture, and environmental health play in these contradictions?

4. How do you think positive attitudes about sustainable development can be transformed into behaviors?

5. Describe what differences may exist regarding "sustainable development" ideas between impoverished societies and wealthy societies.

6. Identify in this article three key terms, concepts, or principles that are used in your textbook (environmental science, economics, sociology, history, geography, etc.) or employed in the discipline you are currently studying. (Note: The terms, concepts, or principles may be implicit, explicit, implied, or inferred.)

ANTHONY A. LEISEROWITZ is a research scientist at Decision Research and an adjunct professor of environmental studies at the University of Oregon, Eugene. He is also a principal investigator at the Center for Research on Environmental Decisions at Columbia University. Leiserowitz may be reached at (541) 485-2400 or by email at ecotone@uoregon.edu. ROBERT W. KATES is an independent scholar based in Trenton, Maine, and a professor emeritus at Brown University, where he served as director of the Feinstein World Hunger Program. He is also a former vice-chair of the Board of Sustainable Development of the U.S National Academy's National Research Council. In 1991, Kates was awarded the National Medal of Science for his work on hunger, environment, and natural hazards. He is an executive editor of *Environment* and may be contacted at rkates@acadia.net. THOMAS M. PARRIS is a research scientist at and director of the New England office of ISCIENCES, LLC. He is a contributing editor of *Environment*. Parris may be reached at parris@isciences.com. The authors retain copyright.

Article originally appeared in *Environment*, November 2005, pp. 23–38. Published by Heldref Publications, Washington, DC. Copyright © 2005 by Anthony A. Leiserowitz, Robert W. Kates, and Thomas M. Parris. (Figure 1 © 1999 by Ronald Inglehart). Reprinted by permission.

The Ethics of Respect for Nature

PAUL W. TAYLOR

Human-Centered and Life-Centered Systems of Environmental Ethics

In this paper I show how the taking of a certain ultimate moral attitude toward nature, which I call "respect for nature," has a central place in the foundations of a life-centered system of environmental ethics. I hold that a set of moral norms (both standards of character and rules of conduct) governing human treatment of the natural world is a rationally grounded set if and only if, first, commitment to those norms is a practical entailment of adopting the attitude of respect for nature as an ultimate moral attitude, and second, the adopting of that attitude on the part of all rational agents can itself be justified. When the basic characteristics of the attitude of respect for nature are made clear, it will be seen that a life-centered system of environmental ethics need not be holistic or organicist in its conception of the kinds of entities that are deemed the appropriate objects of moral concern and consideration. Nor does such a system require that the concepts of ecological homeostasis, equilibrium, and integrity provide us with normative principles from which could be derived (with the addition of factual knowledge) our obligations with regard to natural ecosystems. The "balance of nature" is not itself a moral norm, however important may be the role it plays in our general outlook on the natural world that underlies the attitude of respect for nature. I argue that finally it is the good (well-being, welfare) of individual organisms, considered as entities having inherent worth, that determines our moral relations with the Earth's wild communities of life.

In designating the theory to be set forth as life-centered, I intend to contrast it with all anthropocentric views. According to the latter, human actions affecting the natural environment and its nonhuman inhabitants are right (or wrong) by either of two criteria: they have consequences which are favorable (or unfavorable) to human well-being, or they are consistent (or inconsistent) with the system of norms that protect and implement human rights. From this human-centered standpoint it is to humans and only to humans that all duties are ultimately owed. We may have responsibilities *with regard to* the natural ecosystems and biotic communities of our planet, but these responsibilities are in every case based on the contingent fact that our treatment of those ecosystems and communities of life can further the realization of human values and/or human rights. We

have no obligation to promote or protect the good of nonhuman living things, independently of this contingent fact.

A life-centered system of environmental ethics is opposed to human-centered ones precisely on this point. From the perspective of a life-centered theory, we have prima facie moral obligations that are owed to wild plants and animals themselves as members of the Earth's biotic community. We are morally bound (other things being equal) to protect or promote their good for *their* sake. Our duties to respect the integrity of natural ecosystems, to preserve endangered species, and to avoid environmental pollution stem from the fact that these are ways in which we can help make it possible for wild species populations to achieve and maintain a healthy existence in a natural state. Such obligations are due those living things out of recognition of their inherent worth. They are entirely additional to and independent of the obligations we owe to our fellow humans. Although many of the actions that fulfill one set of obligations will also fulfill the other, two different grounds of obligation are involved. Their well-being, as well as human well-being, is something to be realized *as an end in itself.*

If we were to accept a life-centered theory of environmental ethics, a profound reordering of our moral universe would take place. We would begin to look at the whole of the Earth's biosphere in a new light. Our duties with respect to the "world" of nature would be seen as making prima facie claims upon us to be balanced against our duties with respect to the "world" of human civilization. We could no longer simply take the human point of view and consider the effects of our actions exclusively from the perspective of our own good.

The Good of a Being and the Concept of Inherent Worth

What would justify acceptance of a life-centered system of ethical principles? In order to answer this it is first necessary to make clear the fundamental moral attitude that underlies and makes intelligible the commitment to live by such a system. It is then necessary to examine the considerations that would justify any rational agent's adopting that moral attitude.

Two concepts are essential to the taking of a moral attitude of the sort in question. A being which does not "have" these concepts, that is, which is unable to grasp their meaning and conditions of applicability, cannot be said to have the attitude as part of its moral outlook. These concepts are, first, that of the good (well-being, welfare) of a living thing, and second, the idea of an entity possessing inherent worth. I examine each concept in turn.

(1) Every organism, species population, and community of life has a good of its own which moral agents can intentionally further or damage by their actions. To say that an entity has a good of its own is simply to say that, without reference to any *other* entity, it can be benefited or harmed. One can act in its overall interest or contrary to its overall interest, and environmental conditions can be good for it (advantageous to it) or bad for it (disadvantageous to it). What is good for an entity is what "does it good" in the sense of enhancing or preserving its life and well-being. What is bad for an entity is something that is detrimental to its life and well-being.

We can think of the good of an individual nonhuman organism as consisting in the full development of its biological powers. Its good is realized to the extent that it is strong and healthy. It possesses whatever capacities it needs for successfully coping with its environment and so preserving its existence throughout the various stages of the normal life cycle of its species. The good of a population or community of such individuals consists in the population or community maintaining itself from generation to generation as a coherent system of genetically and ecologically related organisms whose average good is at an optimum level for the given environment. (Here *average good* means that the degree of realization of the good of *individual organisms* in the population or community is, on average, greater than would be the case under any other ecologically functioning order of interrelations among those species populations in the given ecosystem).

The idea of a being having a good of its own, as I understand it, does not entail that the being must have interests or take an interest in what affects its life for better or for worse. We can act in a being's interest or contrary to its interest without its being interested in what we are doing to it in the sense of wanting or not wanting us to do it. It may, indeed, be wholly unaware that favorable and unfavorable events are taking place in its life. I take it that trees, for example, have no knowledge or desires or feelings. Yet it is undoubtedly the case that trees can be harmed or benefited by our actions. We can crush their roots by running a bulldozer too close to them. We can see to it that they get adequate nourishment and moisture by fertilizing and watering the soil around them. Thus we can help or hinder them in the realization of their good. It is the good of trees themselves that is thereby affected. We can similarly act so as to further the good of an entire tree population of a certain species (say, all the redwood trees in a California valley) or the good of a whole community of plant life in a given wilderness area, just as we can do harm to such a population or community.

When construed in this way, the concept of a being's good is not coextensive with sentience or the capacity for feeling pain. William Frankena has argued for a general theory of environmental ethics in which the ground of a creature's being worthy of moral consideration is its sentience. I have offered some criticisms of this view elsewhere, but the full refutation of such a position, it seems to me,

finally depends on the positive reasons for accepting a life-centered theory of the kind I am defending in this essay.

It should be noted further that I am leaving open the question of whether machines—in particular, those which are not only goal-directed, but also self-regulating—can properly be said to have a good of their own. Since I am concerned only with human treatment of wild organisms, species populations, and communities of life as they occur in our planet's natural ecosystems, it is to those entities along that the concept "having a good of its own" will here be applied. I am not denying that other living things, whose genetic origin and environmental condition have been produced, controlled, and manipulated by humans for human ends, do have a good of their own in the same sense as do wild plants and animals. It is not my purpose in this essay, however, to set out or defend the principles that should guide our conduct with regard to their good. It is only insofar as their production and use by humans have good or ill effects upon natural ecosystems and their wild inhabitants that the ethics of respect for nature comes into play.

(2) The second concept essential to the moral attitude of respect for nature is the idea of inherent worth. We take that attitude toward wild living things (individuals, species populations, or whole biotic communities) when and only when we regard them as entities possessing inherent worth. Indeed, it is only because they are conceived in this way that moral agents can think of themselves as having validly binding duties, obligations, and responsibilities that are *owed* to them as their *due*. I am not at this juncture arguing why they *should* be so regarded; I consider it at length below. But so regarding them is a presupposition of our taking the attitude of respect toward them and accordingly understanding ourselves as bearing certain moral relations to them. This can be shown as follows:

What does it mean to regard an entity that has a good of its own as possessing inherent worth? Two general principles are involved: the principle of moral consideration and the principle of intrinsic value.

According to the principle of moral consideration, wild living things are deserving of the concern and consideration of all moral agents simply in virtue of their being members of the Earth's community of life. From the moral point of view their good must be taken into account whenever it is affected for better or worse by the conduct of rational agents. This holds no matter what species the creature belongs to. The good of each is to be accorded some value and so acknowledged as having some weight in the deliberations of all rational agents. Of course, it may be necessary for such agents to act in ways contrary to the good of this or that particular organism or group or organisms in order to further the good of others, including the good of humans. But the principle of moral consideration prescribes that, with respect to each being an entity having its own good, every individual is deserving of consideration.

The principle of intrinsic value states that, regardless of what kind of entity it is in other respects, if it is a member of the Earth's community of life, the realization of its good is something *intrinsically* valuable. This means that its good is prima facie worthy of being preserved or promoted as an end in itself and for the sake of the entity whose good it is. Insofar as we regard any organism, species population, or life community as an entity having inherent worth, we believe that it must never be treated as if it were a mere object or thing whose entire value lies in being instrumental to the good of some other entity. The well-being of each is judged to have value in and of itself.

Combining these two principles, we can now define what it means for a living thing or group of living things to possess inherent worth. To say that it possesses inherent worth is to say that its good is deserving of the concern and consideration of all moral agents, and that the realization of its good has intrinsic value, to be pursued as an end in itself and for the sake of the entity whose good it is.

The duties owed to wild organisms, species populations, and communities of life in the Earth's natural ecosystems are grounded on their inherent worth. When rational, autonomous agents regard such entities as possessing inherent worth, they place intrinsic value on the realization of their good and so hold themselves responsible for performing actions that will have this effect and for refraining from actions having the contrary effect.

The Attitude of Respect for Nature

Why should moral agents regard wild living things in the natural world as possessing inherent worth? To answer this question we must first take into account the fact that, when rational, autonomous agents subscribe to the principles of moral consideration and intrinsic value and so conceive of wild living things as having that kind of worth, such agents are *adopting a certain ultimate moral attitude toward the natural world*. This is the attitude I call "respect for nature." It parallels the attitude of respect for persons in human ethics. When we adopt the attitude of respect for persons as the proper (fitting, appropriate) attitude to take toward all persons as persons, we consider the fulfillment of the basic interests of each individual to have intrinsic value. We thereby make a moral commitment to live a certain kind of life in relation to other persons. We place ourselves under the direction of a system of standards and rules that we consider validly binding on all moral agents as such.

Similarly, when we adopt the attitude of respect for nature as an ultimate moral attitude we make a commitment to live by certain normative principles. These principles constitute the rules of conduct and standards of character that are to govern our treatment of the natural world. This is, first, an *ultimate* commitment because it is not derived from any higher norm. The attitude of respect for nature is not grounded on some other, more general, or more fundamental attitude. It sets the total framework for our responsibilities toward the natural world. It can be justified, as I show below, but its justification cannot consist in referring to a more general attitude or a more basic normative principle.

Second, the commitment is a *moral* one because it is understood to be a disinterested matter of principle. It is this feature that distinguishes the attitude of respect for nature from the set of feelings and dispositions that comprise the love of nature.

The latter seems from one's personal interest in and response to the natural world. Like the affectionate feelings we have toward certain individual human beings, one's love of nature is nothing more than the particular way one feels about the natural environment and its wild inhabitants. And just as our love for an individual person differs from our respect for all persons as such (whether we happen to love them or not), so love of nature differs from respect for nature. Respect for nature is an attitude we believe all moral agents ought to have simply as moral agents, regardless of whether or not they also love nature. Indeed, we have not truly taken the attitude of respect for nature ourselves unless we believe this. To put it in a Kantian way, to adopt the attitude of respect for nature is to take a stance that one wills it to be a universal law for all rational beings. It is to hold that stance categorically, as being validly applicable to every moral agent without exception, irrespective of whatever personal feelings toward nature such an agent might have or might lack.

Although the attitude of respect for nature is in this sense a disinterested and universalizable attitude, anyone who does adopt it has certain steady, more or less permanent dispositions. These dispositions, which are themselves to be considered disinterested and universalizable, comprise three interlocking sets: dispositions to seek certain ends, dispositions to carry on one's practical reasoning and deliberation in a certain way, and dispositions to have certain feelings. We may accordingly analyze the attitude of respect for nature into the following components. (a) The disposition to aim at, and to take steps to bring about, as final and disinterested ends, the promoting and protecting of the good of organisms, species populations, and life communities in natural ecosystems. (These ends are "final" in not being pursued as means to further ends. They are "disinterested" in being independent of the self-interest of the agent.) (b) The disposition to consider actions that tend to realize those ends to be prima facie obligatory *because* they have that tendency. (c) The disposition to experience positive and negative feelings toward states of affairs in the world *because* they are favorable or unfavorable to the good of organisms, species populations, and life communities in natural ecosystems.

The logical connection between the attitude of respect for nature and the duties of a life-centered system of environmental ethics can now be made clear. Insofar as one sincerely takes that attitude and so has the three sets of dispositions, one will at the same time be disposed to comply with certain rules of duty (such as nonmaleficence and noninterference) and with standards of character (such as fairness and benevolence) that determine the obligations and virtues of moral agents with regard to the Earth's wild living things. We can say that the actions one performs and the character traits one develops in fulfilling these moral requirements are the way one *expresses* or *embodies* the attitude in one's conduct and character. In his famous essay, "Justice as Fairness," John Rawls describes the rules of the duties of human morality (such as fidelity, gratitude, honesty, and justice) as "forms of conduct in which recognition of others as persons is manifested." I hold that the rules of duty governing our treatment of the natural world and its inhabitants are forms of conduct in which the attitude of respect for nature is manifested.

The Justifiability of the Attitude of Respect for Nature

I return to the question posed earlier, which has not yet been answered: why *should* moral agents regard wild living things as possessing inherent worth? I now argue that the only way we can answer this question is by showing how adopting the attitude of respect for nature is justified for all moral agents. Let us suppose that we were able to establish that there are good reasons for adopting the attitude, reasons which are intersubjectively valid for every rational agent. If there are such reasons, they would justify anyone's having the three sets of dispositions mentioned above as constituting what it means to have the attitude. Since these include the disposition to promote or protect the good of wild living things as a disinterested and ultimate end, as well as the disposition to perform actions for the reason that they tend to realize that end, we see that such dispositions commit a person to the principles of moral consideration and intrinsic value. To be disposed to further, as an end in itself, the good of any entity in nature just because it is that kind of entity, is to be disposed to give consideration to *every* such entity and to place intrinsic value on the realization of its good. Insofar as we subscribe to these two principles we regard living things as possessing inherent worth. Subscribing to the principles is what is *means* to so regard them. To justify the attitude of respect for nature, then, is to justify commitment to these principles and thereby to justify regarding wild creatures as possessing inherent worth.

We must keep in mind that inherent worth is not some mysterious sort of objective property belonging to living things that can be discovered by empirical observation or scientific investigation. To ascribe inherent worth to an entity is not to describe it by citing some feature discernible by sense perception or inferable by inductive reasoning. Nor is there a logically necessary connection between the concept of a being having a good of its own and the concept of inherent worth. We do not contradict ourselves by asserting that an entity that has a good of its own lacks inherent worth. In order to show that such an entity "has" inherent worth we must give good reasons for ascribing that kind of value to it (placing that kind of value upon it, conceiving of it to be valuable in that way). Although it is humans (persons, valuers) who must do the valuing, for the ethics of respect for nature, the value so ascribed is not a human value. That is to say, it is not a value derived from considerations regarding human well-being or human rights. It is a value that is ascribed to nonhuman animals and plants themselves, independently of their relationship to what humans judge to be conducive to their own good.

Whatever reasons, then, justify our taking the attitude of respect for nature as defined above are also reasons that show why we *should* regard the living things of the natural world as possessing inherent worth. We saw earlier that, since the attitude is an ultimate one, it cannot be derived from a more fundamental attitude nor shown to be a special case of a more general one. On what sort of grounds, then, can it be established?

The attitude we take toward living things in the natural world depends on the way we look at them, on what kind of beings we conceive them to be, and on how we understand the relations we bear to them. Underlying and supporting our attitude is a certain *belief system* that constitutes a particular world view or outlook on nature and the place of human life in it. To give good reasons for adopting the attitude of respect for nature, then, we must first articulate the belief system which underlies and supports that attitude. If it appears that the belief system is internally coherent and well-ordered, and if, as far as we can now tell, it is consistent with all known scientific truths relevant to our knowledge of the object of the attitude (which in this case includes the whole set of the Earth's natural ecosystems and their communities of life), then there remains the task of indicating why scientifically informed and rational thinkers with a developed capacity of reality awareness can find it acceptable as a way of conceiving of the natural world and our place in it. To the extent we can do this we provide at least a reasonable argument for accepting the belief system and the ultimate moral attitude it supports.

I do not hold that such a belief system can be *proven* to be true, either inductively or deductively. As we shall see, not all of its components can be stated in the form of empirically verifiable propositions. Nor is its internal order governed by purely logical relationships. But the system as a whole, I contend, constitutes a coherent, unified, and rationally acceptable "picture" or "map" of a total world. By examining each of its main components and seeing how they fit together, we obtain a scientifically informed and well-ordered conception of nature and the place of humans in it.

The belief system underlying the attitude of respect for nature I call (for want of a better name) "the biocentric outlook on nature." Since it is not wholly analyzable into empirically confirmable assertions, it should not be thought of as simply a compendium of the biological sciences concerning our planet's ecosystems. It might best be described as a philosophical world view, to distinguish it from a scientific theory or explanatory system. However, one of its major tenets is the great lesson we have learned from the science of ecology: the interdependence of all living things in an organically unified order whose balance and stability are necessary conditions for the realization of the good of its constituent biotic communities.

Before turning to an account of the main components of the biocentric outlook, it is convenient here to set forth the overall structure of my theory of environmental ethics as it has now emerged. The ethics of respect for nature is made up of three basic elements: a belief system, an ultimate moral attitude, and a set of rules of duty and standards of character. These elements are connected with each other in the following manner. The belief system provides a certain outlook on nature which supports and makes intelligible an autonomous agent's adopting, as an ultimate moral attitude, the attitude of respect for nature. It supports and makes intelligible the attitude in the sense that, when an autonomous agent understands its moral relations to the natural world in terms of this outlook, it recognizes the attitude of respect to be the only *suitable* or *fitting* attitude to take toward all wild forms of life in the Earth's biosphere. Living things are now viewed as *the appropriate objects of the attitude of respect* and are accordingly regarded as entities possessing inherent worth. One then places intrinsic value on the promotion and protection of their good. As a consequence of this, one makes a moral commitment to abide by a set of rules of duty and to fulfill (as far as one can by one's own efforts) certain standards of good character. Given one's adoption of the attitude of respect, one makes that moral commitment because one considers those rules and standards to be validly binding on all moral agents. They are seen as embodying forms of conduct and character structures in which the attitude of respect for nature is manifested.

This three-part complex which internally orders the ethics of respect for nature is symmetrical with a theory of human ethics grounded on respect for persons. Such a theory includes, first, a conception of oneself and others as persons, that is, as centers of autonomous choice. Second, there is the attitude of respect for persons as persons. When this is adopted as an ultimate moral attitude it involves the disposition to treat every person as having inherent worth or "human dignity." Every human being, just in virtue of her or his humanity, is understood to be worthy of moral consideration, and intrinsic value is placed on the autonomy and well-being of each. This is what Kant meant by conceiving of persons as ends in themselves. Third, there is an ethical system of duties which are acknowledged to be owed by everyone to everyone. These duties are forms of conduct in which public recognition is given to each individual's inherent worth as a person.

This structural framework for a theory of human ethics is meant to leave open the issue of consequentialism (utilitarianism) versus nonconsequentialism (deontology). That issue concerns the particular kind of system of rules defining the duties of moral agents toward persons. Similarly, I am leaving open in this paper the question of what particular kind of system of rules defines our duties with respect to the natural world.

The Biocentric Outlook on Nature

The biocentric outlook on nature has four main components. (1) Humans are thought of as members of the Earth's community of life, holding that membership on the same terms as apply to all the nonhuman members. (2) The Earth's natural ecosystems as a totality are seen as a complex web of interconnected elements, with the sound biological functioning of each being dependent on the sound biological functioning of the others. (This is the component referred to above as the great lesson that the science of ecology has taught us). (3) Each individual organism is conceived as a teleological center of life, pursuing its own good in its own way. (4) Whether we are concerned with standards of merit or with the concept of inherent worth, the claim that humans by their very nature are superior to other species in a groundless claim and, in the light of elements (1), (2), and (3) above, must be rejected as nothing more than an irrational bias in our own favor.

The conjunction of these four ideas constitutes the biocentric outlook on nature.

Critical Thinking

1. Explain: what is a life-centered system of environmental ethics, and how it is different than an "anthropocentric" outlook?

2. What may be some problems that words like *good, moral, inherent worth,* and *intrinsic value* present when trying to resolve environmental issues?

3. The author states: "Each individual organism is conceived as a teleological center of life, pursuing its own good in its own way." Is this true for organisms, say, that carry diseases that kill humans? Explain.

4. What is the ultimate purpose of our developing an environmental ethic?

5. Would the Earth and nature be better off without humans consuming them? Offer some insights.

6. Would you contribute to the extinction of a species if it meant saving your (children, siblings, parents—pick one)?

7. Identify in this article three key terms, concepts, or principles that are used in your textbook (environmental science, economics, sociology, history, geography, etc.) or employed in the discipline you are currently studying. (Note: The terms, concepts, or principles may be implicit, explicit, implied, or inferred.)

Environmental Justice for All

Leyla Kokmen

Manuel Pastor ran bus tours of Los Angeles a few years back. These weren't the typical sojourns to Disneyland or the MGM studios, though; they were expeditions to some of the city's most environmentally blighted neighborhoods—where railways, truck traffic, and refineries converge, and where people live 200 feet from the freeway.

The goal of the "toxic tours," explains Pastor, a professor of geography and of American studies and ethnicity at the University of Southern California (USC), was to let public officials, policy makers, and donors talk to residents in low-income neighborhoods about the environmental hazards they lived with every day and to literally see, smell, and feel the effects.

"It's a pretty effective forum," says Pastor, who directs USC's Program for Environmental and Regional Equity, noting that a lot of the "tourists" were eager to get back on the bus in a hurry. "When you're in these neighborhoods, your lungs hurt."

Like the tours, Pastor's research into the economic and social issues facing low-income urban communities highlights the environmental disparities that endure in California and across the United States. As stories about global warming, sustainable energy, and climate change make headlines, the fact that some neighborhoods, particularly low-income and minority communities, are disproportionately toxic and poorly regulated has, until recently, been all but ignored.

A new breed of activists and social scientists are starting to capitalize on the moment. In principle they have much in common with the environmental justice movement, which came of age in the late 1970s and early 1980s, when grassroots groups across the country began protesting the presence of landfills and other environmentally hazardous facilities in predominantly poor and minority neighborhoods.

In practice, though, the new leadership is taking a broader-based, more inclusive approach. Instead of fighting a proposed refinery here or an expanded freeway there, all along trying to establish that systematic racism is at work in corporate America, today's environmental justice movement is focusing on proactive responses to the social ills and economic roadblocks that if removed would clear the way to a greener planet.

The new movement assumes that society as a whole benefits by guaranteeing safe jobs, both blue-collar and white-collar, that pay a living wage. That universal health care would both decrease disease and increase awareness about the quality of everyone's air and water. That better public education and easier access to job training, especially in industries that are emerging to address the global energy crisis, could reduce crime, boost self-esteem, and lead to a homegrown economic boon.

That green rights, green justice, and green equality should be the environmental movement's new watchwords.

"This is the new civil rights of the 21st century," proclaims environmental justice activist Majora Carter.

A lifelong resident of Hunts Point in the South Bronx, Carter is executive director of Sustainable South Bronx, an eight-year-old nonprofit created to advance the environmental and economic future of the community. Under the stewardship of Carter, who received a prestigious MacArthur Fellowship in 2005, the organization has managed a number of projects, including a successful grassroots campaign to stop a planned solid waste facility in Hunts Point that would have processed 40 percent of New York City's garbage.

Her neighborhood endures exhaust from some 60,000 truck trips every week and has four power plants and more than a dozen waste facilities. "It's like a cloud," Carter says. "You deal with that, you're making a dent."

The first hurdle Carter and a dozen staff members had to face was making the environment relevant to poor people and people of color who have long felt disenfranchised from mainstream environmentalism, which tends to focus on important but distinctly nonurban issues, such as preserving Arctic wildlife or Brazilian rainforest. For those who are struggling to make ends meet, who have to cobble together adequate health care, education, and job prospects, who feel unsafe on their own streets, these grand ideas seem removed from reality.

That's why the green rights argument is so powerful: It spans public health, community development, and economic growth to make sure that the green revolution isn't just for those who can afford a Prius. It means cleaning up blighted communities like the South Bronx to prevent potential health problems and to provide amenities like parks to play in, clean trails to walk on, and fresh air to breathe. It also means building green industries into the local mix, to provide healthy jobs for residents in desperate need of a livable wage.

Historically, mainstream environmental organizations have been made up mostly of white staffers and have focused more on the ephemeral concept of the environment rather than on the people who are affected. Today, though, as climate change and gas prices dominate public discourse, the concepts driving the new environmental justice movement are starting to catch on. Just recently, for instance, *The New York Times* columnist Thomas Friedman dubbed the promise of public investment in the green economy the "Green New Deal."

Van Jones, whom Friedman celebrated in print last October, is president of the Ella Baker Center for Human Rights in Oakland, California. To help put things in context, Jones briefly sketches the history of environmentalism:

The first wave was conservation, led first by Native Americans who respected and protected the land, then later by Teddy Roosevelt, John Muir, and other Caucasians who sought to preserve green space.

The second wave was regulation, which came in the 1970s and 1980s with the establishment of the Environmental Protection Agency (EPA) and Earth Day. Increased regulation brought a backlash against poor people and people of color, Jones says. White, affluent communities sought to prevent environmental hazards from entering their neighborhoods. This "not-in-my-backyard" attitude spurred a new crop of largely grassroots environmental justice advocates who charged businesses with unfairly targeting low-income and minority communities. "The big challenge was NIMBY-ism," Jones says, noting that more toxins from power plants and landfills were dumped on people of color.

The third wave of environmentalism, Jones says, is happening today. It's a focus on investing in solutions that lead to "eco-equity." And, he notes, it invokes a central question: "How do we get the work, wealth, and health benefits of the green economy to the people who most need those benefits?"

There are a number of reasons why so many environmental hazards end up in the poorest communities.

Property values in neighborhoods with environmental hazards tend to be lower, and that's where poor people—and often poor people of color—can afford to buy or rent a home. Additionally, businesses and municipalities often choose to build power plants in or expand freeways through low-income neighborhoods because the land is cheaper and poor residents have less power and are unlikely to have the time or organizational infrastructure to evaluate or fight development.

"Wealthy neighborhoods are able to resist, and low-income communities of color will find their neighborhoods plowed down and [find themselves] living next to a freeway that spews pollutants next to their schools," USC's Manuel Pastor says.

Moreover, regulatory systems, including the EPA and various local and state zoning and environmental regulatory bodies, allow piecemeal development of toxic facilities. Each new chemical facility goes through an individual permit process, which doesn't always take into account the overall picture in the community. The regulatory system isn't equipped to address potentially dangerous cumulative effects.

In a single neighborhood, Pastor says, you might have toxins that come from five different plants that are regulated by five different authorities. Each plant might not be considered dangerous on its own, but if you throw together all the emissions from those static sources and then add in emissions from moving sources, like diesel-powered trucks, "you've created a toxic soup," he says.

In one study of air quality in the nine-county San Francisco Bay Area, Pastor found that race, even more than income, determined who lived in more toxic communities. That 2007 report, "Still Toxic After All These Years: Air Quality and Environmental Justice in the San Francisco Bay Area," published by the Center for Justice, Tolerance & Community at the University of California at Santa Cruz, explored data from the EPA's Toxic Release Inventory, which reports toxic air emissions from large industrial facilities. The researchers examined race, income, and the likelihood of living near such a facility.

More than 40 percent of African American households earning less than $10,000 a year lived within a mile of a toxic facility, compared to 30 percent of Latino households and fewer than 20 percent of white households.

As income rose, the percentages dropped across the board but were still higher among minorities. Just over 20 percent of African American and Latino households making more than $100,000 a year lived within a mile of a toxic facility, compared to just 10 percent of white households.

The same report finds a connection between race and the risk of cancer or respiratory hazards, which are both associated with environmental air toxics, including emissions both from large industrial facilities and from mobile sources. The researchers looked at data from the National Air Toxics Assessment, which includes estimates of such ambient air toxics as diesel particulate matter, benzene, and lead and mercury compounds. The areas with the highest risk for cancer had the highest proportion of African American and Asian residents, the lowest rate of home ownership, and the highest proportion of people in poverty. The same trends existed for areas with the highest risk for respiratory hazards.

According to the report, "There is a general pattern of environmental inequity in the Bay Area: Densely populated communities of color characterized by relatively low wealth and income and a larger share of immigrants disproportionately bear the hazard and risk burden for the region."

Twenty years ago, environmental and social justice activists probably would have presented the disparities outlined in the 2007 report as evidence of corporations deliberately targeting minority communities with hazardous waste. That's what happened in 1987, when the United Church of Christ released findings from a study that showed toxic waste facilities were more likely to be located near minority communities. At the 1991 People of Color Environmental Leadership Summit, leaders called the disproportionate burden both racist and genocidal.

In their 2007 book *Break Through: From the Death of Environmentalism to the Politics of Possibility,* authors Ted Nordhaus and Michael Shellenberger take issue with this strategy. They argue that some of the research conducted in the name of environmental justice was too narrowly focused and that activists have spent too much time looking for conspiracies of environmental racism and not enough time looking at the multifaceted problems facing poor people and people of color.

"Poor Americans of all races, and poor Americans of color in particular, disproportionately suffer from social ills of every kind," they write. "But toxic waste and air pollution are far from being the most serious threats to their health and well-being. Moreover, the old narratives of intentional discrimination fail to explain or address these disparities. Disproportionate environmental health outcomes can no more be reduced to intentional discrimination than can disproportionate economic and educational outcomes. They are due to a larger and more complex set of historic, economic, and social causes."

Today's environmental justice advocates would no doubt take issue with the finer points of Nordhaus and Shellenberger's criticism—in particular, that institutional racism is a red herring. Activists and researchers are acutely aware that they are facing a multifaceted spectrum of issues, from air pollution to a dire lack of access to regular health care. It's because of that complexity, however, that they are now more geared toward proactively addressing an array of social and political concerns.

"The environmental justice movement grew out of putting out fires in the community and stopping bad things from happening, like a landfill," says Martha Dina Argüello, executive director of Physicians for Social Responsibility—Los Angeles, an organization that connects environmental groups with doctors to promote public health. "The more this work gets done, the more you realize you have to go upstream. We need to stop bad things from happening."

"We can fight pollution and poverty at the same time and with the same solutions and methods," says the Ella Baker Center's Van Jones.

Poor people and people of color have borne all the burden of the polluting industries of today, he says, while getting almost none of the benefit from the shift to the green economy. Jones stresses that he is not an environmental justice activist, but a "social-uplift environmentalist." Instead of concentrating on the presence of pollution and toxins in low-income communities, Jones prefers to focus on building investment in clean, green, healthy industries that can help those communities. Instead of focusing on the burdens, he focuses on empowerment.

With that end in mind, the Ella Baker Center's Green-Collar Jobs Campaign plans to launch the Oakland Green Jobs Corps this spring. The initiative, according to program manager Aaron Lehmer, received $250,000 from the city of Oakland and will give people ages 18 to 35 with barriers to employment (contact with the criminal justice system, long-term unemployment) opportunities and paid internships for training in new energy skills like installing solar panels and making buildings more energy efficient.

The concept has gained national attention. It's the cornerstone of the Green Jobs Act of 2007, which authorizes $125 million annually for "green-collar" job training that could prepare 30,000 people a year for jobs in key trades, such as installing solar panels, weatherizing buildings, and maintaining

wind farms. The act was signed into law in December as part of the Energy Independence and Security Act.

While Jones takes the conversation to a national level, Majora Carter is focusing on empowerment in one community at a time. Her successes at Sustainable South Bronx include the creation of a 10-week program that offers South Bronx and other New York City residents hands-on training in brownfield remediation and ecological restoration. The organization has also raised $30 million for a bicycle and pedestrian greenway along the South Bronx waterfront that will provide both open space and economic development opportunities.

As a result of those achievements, Carter gets calls from organizations across the country. In December she traveled to Kansas City, Missouri, to speak to residents, environmentalists, businesses, and students. She mentions exciting work being done by Chicago's Blacks in Green collective, which aims to mobilize the African American community around environmental issues. Naomi Davis, the collective's founder, told Chicago Public Radio in November that the group plans to develop environmental and economic opportunities—a "green village" with greenways, light re-manufacturing, ecotourism, and energy-efficient affordable housing—in one of Chicago's most blighted areas.

Carter stresses that framing the environmental debate in terms of opportunities will engage the people who need the most help. It's about investing in the green economy, creating jobs, and building spaces that aren't environmentally challenged. It won't be easy, she says. But it's essential to dream big.

"It's about sacrifice," she says, "for something better and bigger than you could have possibly imagined."

Critical Thinking

1. The author argues that "given rights, green justice, and green equality should be the environmental movement's new watchwords." What does that mean?

2. How does the green rights movement engage poor people and people of color?

3. Can the green rights strategy work for all poor people and people of color around the world? Why or why not?

4. Referring to your text's world map, and your own knowledge, locate nations where ensuing environmental justice will be difficult. Why?

5. Identify in this article three key terms, concepts, or principles that are used in your textbook (environmental science, economics, sociology, history, geography, etc.) or employed in the discipline you are currently studying. (Note: The terms, concepts, or principles may be implicit, explicit, implied, or inferred.)

Life, Religion and Everything

Biologist and author Rupert Sheldrake believes that the world's religions have a crucial role in restoring the earth's ecological balance. Laura Sevier meets the man trying to broker a better relationship between God, man, science and the natural world.

LAURA SEVIER

A long, low drone fills the air. We are all chanting the same sound: OOOOOOOHHHHHHHH. Whistling overtones start to ring out above the group sound and then a lone female voice sings out:

Where I sit is holy

Holy is the ground

Forest mountain river listen to my sound

Great spirit circling all around.

The voice then invites everyone to join in and we sing this verse seven or eight times. Then there is silence.

It's not often you attend a talk given by an eminent scientist that begins with a session of Mongolian overtone chanting followed by a Native American Indian song about the holiness of the earth. It's especially surreal given that we're sitting on neat little rows of chairs in a Unitarian Church in Hampstead, in London.

I was there to listen to renowned English biologist Rupert Sheldrake talk about how the world's religions can learn to live with ecological integrity. The chanting, it appears, is the warm-up act, led by Sheldrake's wife, Jill Purce, a music healer.

So far so extraordinary, but then Sheldrake is no ordinary man. A respected scientist from a largely conventional educational background, he's devoted much of the past 17 years of his life to studying the sort of phenomena that most 'serious' scientists dismiss out of hand, such as telepathy, our 'seventh sense'. But religion? Given the current trend for militant atheism within science, I'm amazed. Besides, isn't religion incompatible with science? Not according to Sheldrake, an Anglican Christian. 'One of my main concerns is the opening up of science. Another is exploring the connections between science and spirituality,' he says.

His take on religion—and science—is refreshingly unorthodox precisely because it factors in a crucial new element: nature. 'The thrust of my work is trying to break out of the

mechanistic view of nature as inanimate, dead and machine-like.' In fact his 1991 book *The Rebirth of Nature: the Greening of Science and God* (Inner Traditions Bear & Company, £11.99) was devoted to showing 'how we can once again think of nature as alive'—and sacred.

The Sacred Earth

Our culture seems to have lost touch with any idea of the land as being alive and sacred and anyone who considers it to be so is often branded a tree-hugging hippie and treated with ridicule or suspicion. Land is mostly valued purely in economic terms. Yet no value is attributed to the irreplaceable benefits derived from the normal functioning of the natural world, which assures the stability of our climate, the fertility of our soil, the replenishment of our water.

Our culture seems to have lost any idea of the land being alive and sacred, and values it only in economic terms.

Religion has, until recently, remained pretty quiet on the issue.

As Edward Goldsmith wrote in the *Ecologist* in 2000, mainstream religions have become increasingly 'otherworldly'. They have 'scarcely any interest' in the natural world at all. Traditionally, religion used to play an integral role in linking people to the natural world, imbuing people with the knowledge and values that make caring for it a priority. 'Mainstream religion' Goldsmith wrote 'has failed the earth. It has lost its way, and needs to return to its roots.'

So if the world's religions are to play a part in saving what remains of the natural world, they not only need to return to their roots but also to confront the threat and scale of the global ecological crisis we now face. This means being open to a dialogue with science. 'No religions, when they were growing up, had to deal with our present situation and ecological crisis,' says Sheldrake. 'People thought they could take the earth more or less for granted. Certainly the idea that human beings could transform the climate through their actions was unheard of. This is a new situation for everybody, for religious people and scientists, for traditional cultures and modern scientific ones. We're all in this together.'

Environmental Sin

'Religion and Ecology' is now a subject of serious academic study. The Forum on Religion and Ecology at Yale University, for example, recently explored the ecological dimension of all the major world religions. The ongoing environmental crisis has sparked a 'bringing together' of the world's religions in a series of interreligious meetings and conferences around the world on the theme of 'Religion, Science and the Environment', exploring the response that religious communities can make. These brought together scientists, bishops, rabbis, marine biologists and philosophers in a way that, according to Sheldrake, 'really worked'.

Within many religions, including all branches of Christianity, there's an attempt to recover that sense of connection with nature. 'There's a lot going on,' says Sheldrake, 'even within the group seen as lagging the furthest behind—the American Evangelicals, who are somewhat retrogressive in relation to the environment.'

Some evangelicals who believe in the Rapture and think the world is soon to end have expressed the view that there's no point in attempting to save the environment because it's all going to be discarded like a used tissue.

But a more environmentally friendly view is held by the Evangelical Environmental Network (EEN), a group of individuals and organisations including World Vision, World Relief and the International Bible Society. An Evangelical Declaration on the Care of Creation, its landmark credo published in 1991, begins: 'We believe that biblical faith is essential to the solution of our ecological problems. . . . Because we worship and honour the Creator we seek to cherish and care for the creation. Because we have sinned, we have failed our stewardship of creation. Therefore we repent of the way we have polluted, distorted or destroyed so much of the Creator's work.'

Nature informs us and it is our obligation to read nature as you would a book, to feel it as you would a poem.

It then commits to work for reconciliation of people and the healing of suffering creation.

The belief that environmental destruction is a sin isn't a new concept. The spirituality of native American Indians, for instance, is a land-based one. In this culture, the world is animate, natural things are alive and everything is imbued with spirit.

In the words of John Mohawk, native American chief: 'The natural world is our Bible. We don't have chapters and verses; we have trees and fish and animals . . . The Indian sense of natural law is that nature informs us and it is our obligation to read nature as you would a book, to feel nature as you would a poem, to touch nature as you would yourself, to be part of that and step into its cycles as much as you can.'

Most importantly, environmental destruction is seen as a sin.

Loss of the Sacred

The question is, how did we lose the sacred connection with the natural world? Where did religion and culture go wrong? According to Sheldrake, the break began in the 16th century. Until then there were pagan festivals, such as May Day, that celebrated the seasons and the fertility of the land; there were nature shrines, holy wells and sacred places.

But with the Protestant Reformation in the 16th century there was an attempt by the reformers, who couldn't find anything about these 'pagan' practices in the Bible, to stamp them out. In the 17th century the Puritans brought a further wave of suppression of these things—banning, for example, Maypole dancing (Maypoles being a symbol of male fertility). 'There was a deliberate attempt to get rid of all the things that connected people to the sacredness of the land and it largely succeeded,' says Sheldrake.

Another factor he believes severed our connection is the view of nature as a machine. 'From the time of our remotest ancestors until the 17th century, it was taken for granted that the world of nature was alive, that the universe was alive and that all animals were not only alive but had souls—the word "animal" comes from the word "anima", meaning soul. This was the standard view, even within the Church. Medieval Christianity was based on an animate form of nature—a kind of Christian animism.'

But this model of a living world was replaced by the idea of the universe as a machine, an idea that stems from the philosophy of Rene Descartes. Nature was no more than dead matter and everything was viewed as mechanical, governed by mathematical principles instead of animating souls.

'This mechanistic view of nature,' Sheldrake says, 'is an extremely limiting and alienating one. It forces the whole of our understanding of nature into a machine metaphor— the universe as a machine, animals and plants as machines, you as a machine, the brain as a machine. It's a very man-centred metaphor, as only people make machines. So looking at nature in this way projects one aspect of human activity onto the whole of nature.'

It is this view, he says, that led to our current crisis. 'If you assume that nature is inanimate, then nothing natural has a life, purpose, or value. Natural resources are there to be developed, and the only value placed on them is by market forces and official planners. And if you assume that only humans are conscious, only humans have reason, and therefore only

Hinduism

- The Vedas (ancient Hindu scriptures) describe how the creator god Vishnu made the universe so that every element is interlinked. A disturbance in one part will upset the balance and impact all the other elements.
- Three important principles of Hindu environmentalism are *yajna* (sacrifice), *dhana* (giving) and *tapas* (penance).
- *Yajna* entails that you should sacrifice your needs for the sake of others, for nature, the poor or future generations.
- *Dhana* entails that whatever you consume you must give back.
- *Tapas* commends self-restraint in your lifestyle.
- Mother Earth is personified in the Vedas as the goddess Bhumi, or Prithvi.
- Hindu businessman Balbir Mathur, inspired by his faith, founded Trees for Life (www.treesforlife.org), a non-profit movement that plants fruit trees in developing countries, to provide sustainable and environmentally-friendly livelihoods.

Islam

- Allah has appointed humankind *khalifah* (steward) over the created world.
- This responsibility is called *al-amanah* (the trust) and Man will be held accountable to it at the Day of Judgment.
- The Qur'an warns against disturbing God's natural balance: 'Do no mischief on the earth after it hath been set in order' (7:56).
- Shari'ah (Islamic law) designates *haram* zones, used to contain urban development in protection of natural resources, and *hima,* specific conservation areas.
- The Islamic foundation for ecology and environmental sciences www.ifes.org.uk publishes a newsletter called *Eco Islam* and organised an organic *iftar* (the evening meal during Ramadan) in 2006.
- In 2000, IFEES led an Islamic educational programme on the Muslim-majority island of Misali, in response to the destruction to the aquatic ecosystem by over-fishing and the use of dynamite in coral reefs. The environmental message based on the Qur'an initiated sustainable fishing practices.

Judaism

- The Torah prohibits harming God's earth: 'Do not cut down trees even to prevent ambush, do not foul waters, or burn crops even to cause an enemy's submission' (Devarim 20:19)
- It teaches humility in the face of nature: 'Ask the beasts, and they will teach you; the birds of the sky, and they will tell you; or speak to the earth and it will teach you; the fish of the sea, they will inform you' (Job 12:7–9)
- The Talmudic law *bal tashchit* (do not destroy) was developed by Jewish scholars into a series of specific prohibitions against wasteful actions.
- The Noah Project (www.noahproject.org.uk) is a UK-based Jewish environmental organisation, engaged in hands-on conservation work, and promoting environmental responsibility by emphasising the environmental dimensions of Jewish holidays such as Tu B'Shevat (New Year of the Trees).

Christianity

- Genesis gives a picture of God creating the heavens and earth—and when it was all finished, 'God saw all that he had made, and it was very good.' (1:31) Having made man, he 'put him in the Garden of Eden to work it and take care of it' (2:15).
- Romans 8:19–22 has been interpreted as a message of redemption for the environment, calling on Christians to work towards the time when 'the creation itself will be liberated from its bondage to decay'.
- At the UN Rio Earth Summit in 1992, the World Council of Churches formed a working group on climate change. Their manifesto expresses a concern for justice towards developing countries, who are disproportionately affected by climate change, to future generations and to the world.
- www.christian-ecology.org.uk represents Churches Together in Britain and Ireland. It includes links and a daily prayer guide with references both to the Bible and to scientific and news data. Operation Noah is the climate change campaign.

Buddhism

- Buddhist religious ecology is based on three principles: nature as teacher, as a spiritual force, and as a way of life.
- Buddhists believe that nature can teach us about the interdependence and impermanence of life, and that living near to and in tune with nature gives us spiritual strength.
- Buddha commended frugality, avoiding waste, and non-violence.
- Buddhists believe that man should be in harmonious interaction with nature, not a position of authority.
- Philosopher Dr Simon James, based at Durham University, has studied the Buddhist basis of environmentalism and virtue ethics. A spiritually enlightened individual shows compassion, equanimity and humility—qualities that are intrinsic to an environmentally friendly lifestyle.
- The Zen Environmental Studies Institute (www.mro.org/zesi) in New York runs programmes in nature study and environmental advocacy, informed by Zen Buddhist meditation.

Baha'ism

- Bahá'u'lláh, founder of the Baha'i faith and regarded as a messenger from God, stated 'nature is God's Will and is its expression in and through the contingent world' (the Tablets of Baha'u'llah).
- Baha'is believe that the world reflects God's qualities and attributes and therefore must be cherished.
- The Baha'i Office of the Environment states: 'Baha'u'llah's promise that civilisation will exist on this planet for a minimum of 5,000 centuries makes it unconscionable to ignore the long-term impact of decisions made today. The world community must, therefore, learn to make use of the earth's natural resources . . . in a manner that ensures sustainability into the distant reaches of time.'
- The Barli Rural Development Institute in India was inspired by Baha'i social activism. It has trained hundreds of rural women in conservation strategies such as rainwater harvesting and solar cooking.
- www.onecountry.org is the newsletter of the Bahai international community.

humans have true value, then it's fine to have animals in factory farms and to exploit the world in whatever way you like, and if you do conserve any bit of the earth then you have to conserve it with human ends in mind. Everything is justified in human terms.'

The mechanistic theory has become a kind of religion that is built into the official orthodoxy of economic progress and, through technology's successes, is now triumphant on a global scale. 'So,' says Sheldrake, 'this combination of science, technology, secular humanism and rationalism—all these philosophies that dominate the modern age—open the way for untrammelled exploitation of the earth that is going on everywhere today.'

The Living Universe

It seems like a pretty bleak vision. But there is an alternative: to allow our own experience and intuition to help us see nature and the universe as alive. 'Many people have emotional connections with particular places associated with their childhood, or feel an empathy with animals or plants, or are inspired by the beauty of nature, or experience a mystical sense of unity with the natural world,' Sheldrake says. 'Our private relationship with nature presupposes that nature is alive.'

In other words, we don't need to be told by science, religion or anyone that it is alive, valuable and worthy of respect and reverence. Deep down, we can feel it for ourselves. Many people have urges to get 'back to nature' in some way, to escape the confines of concrete and head for the hills, the sea, a park or even a small patch of grass. These impulses are moving us in the right direction.

Another way forward is through new revolutionary insights within science. 'Science itself is leading us away from this view of nature as a machine towards a much more organic view of living in the world,' says Sheldrake. 'The changes are happening in independent parts of science for different reasons, but all of them are pointing in the same direction: the view of a very organic, creative world.'

The big bang theory gives a new model of the universe that is more like a developing organism, growing spontaneously and forming totally new structures within it. The concept of quantum physics has broken open many of our ideas of the mechanistic universe. The old idea of determinism has given way to indeterminism and chaos theory. The old idea of the earth as dead has given way to Gaia, the living earth. The old idea of the universe as uncreative has given way to the new idea of creative evolution, first in the realm of living things, through Darwin, and now we see that the whole cosmos is in creative evolution. So, if the whole universe is alive, if the universe is like a great organism, then everything within it is best understood as alive.

Encouraging Dialogue

This has opened up new possibilities for a dialogue between science and religion. 'These changing frontiers of science are making it much easier to see that we're all part of, and dependent on, a living earth; and for those of us who follow a religion, to see the living God as the living world,' says Sheldrake. Such insights breathe new meaning into traditional religions, their practices and seasonal festivals.

For example, all religions provide opportunities for giving thanks, both through simple everyday rituals, like saying grace, and also in collective acts of thanksgiving. These expressions of gratitude can help to remind us that we have much to be thankful for. But as Sheldrake points out, 'It's hard to feel a sense of gratitude for an inanimate, mechanical world.'

Helping people see the land as sacred again, Sheldrake maintains, is one of the major roles of religion. 'They all point towards a larger whole: the wholeness of creation and a larger story than our own individual story. All religions tell stories about our place in the world, our relation to other people and to the world in which we live. In that sense all religions relate us to the earth and the heavens.'

Sheldrake thinks we need stories: 'It's part of our nature. Science gives us stories, too—the universe story. So does TV, fiction, books.' And these stories, in his view, unify us in a way that, for instance, some New Age practices (such as personal shrines) don't. While those things have personal value, they don't have the unifying function that a traditional religion does. 'When you go to a Hindu festival or pilgrimage, you see thousands of people coming together, the whole community united by a common story or a celebration of a sacred place.'

The fascinating thing about Rupert Sheldrake is his ability to assimilate ideas from an array of different subjects that are normally kept separate, draw new connections and conclusions and open up new dialogues. He's certainly not afraid to explore new territory or use new metaphors. Thus the big bang is like 'the primal orgasm' or like 'the breaking open of the cosmic egg'.

When talking about the discovery that 95 percent of the universe is 'dark matter' or unknown, he says, 'it is as if science has discovered the cosmic unconscious'. He embraces the idea of 'Mother Nature'—in fact he believes the old intuition of nature as Mother still affects our personal responses to it and conditions our response to the ecological crisis. 'We feel uncomfortable when we recognise that we are polluting our own Mother; it is easier to rephrase the problem in terms of "inadequate waste management".' He sees the green movement as one aspect of 'Mother Nature reasserting herself, whether we like it or not.'

One of the most significant implications of Sheldrake's worldview is that it connects people to the natural world and 'if people feel more connected to the world around them, they might be less likely to accept its destruction,' he says. Reframing our view to encompass a world that is alive also, effectively, puts humans back in our proper place in the scheme of things.

Sheldrake's scientific and philosophical investigation is fuelled by a passionate concern for all of life, and his vision of life expands to the cosmos. If the earth is alive, if the universe is alive, if solar systems are alive, if galaxies are alive, if planets are alive, then causing harm to any of these systems really is a sin; one that we have committed all too willingly for far too long.

For Further Information

Rupert Sheldrake: www.sheldrake.org

Forum on Religion and Ecology: http://environment.harvard.edu/
religion—the ecological dimension of various religions

The Alliance of Religions and Conservation (ARC):
www.arcworld.org

Critical Thinking

1. Describe the thesis of this article in one sentence.

2. Explain the difference between a "mechanistic view of nature" and a "sacred" view of nature.

3. What would most religions say about a human committing suicide to preserve the well-being of nature?

4. Would you say the idea of "sustainability," and in fact the idea of "environmental science," is more "mechanistic" or "spiritual"?

5. "Who" is "nature," and why should nature care about humans? Offer some insights/reflections.

6. Identify in this article three key terms, concepts, or principles that are used in your textbook (environmental science, economics, sociology, history, geography, etc.) or employed in the discipline you are currently studying. (Note: The terms, concepts, or principles may be implicit, explicit, implied, or inferred.)

LAURA SEVIER is a freelance journalist and regular contributor to the *Ecologist*.

UNIT 10
Consuming the Earth

Unit Selections

Learning Outcomes

After reading this Unit, you should be able to:

- Appraise the validity of the argument that human consumption and consumption patterns are the fundamental causes behind all our present-day environmental problems.

- What is the geographical component of the "overconsumption" issue?

- Outline the connections between economic development, carbon emissions, and population growth.

- Elaborate on the question "How much should a person consume?"

- Outline the argument that human consumption patterns contribute to environmental degradation.

- Define "omnivores" and "ecosystem people" and explain the association of these terms to the idea of "resource catchment".

- Make some predictions regarding the future global patterns of "inequalities of consumption".

- Summarize the current state of global consumption and consumerism.

- Expand on the idea that "today's consumption is undermining the environmental resource base".

- Offer insights on linkages between human happiness and the environment.

- Evaluate the validity of the statement, "Our single-minded focus on increasing wealth has succeeded in driving the planet's ecological systems to the brink of failure, even as it's failed to make us happy".

- Discuss the evidence presented to support the argument that bigger is not always better, and that it may in fact be counterproductive to create a sustainable relationship with our environment.

Student Website

www.mhhe.com/cls

Internet References

Alliance for Global Sustainability (AGS)
www.global-sustainability.org

The Dismal Scientist
www.dismal.com

Global Footprint Network
http://footprints@footprintnetwork.org

U.S. Information Agency (USIA)
www.america.gov

World Population and Demographic Data
http://geography.about.com/cs/worldpopulation

We're in a giant car heading towards a brick wall and everyone's arguing over where they're going to sit.

—*David Suzuki* (Canadian Environmentalist, Scientist and Broadcaster b. 1936)

"Oh, how we consume our planet! How ravenous, how insatiable our appetite seems to be!" Such a sentiment seems to be shared by increasing numbers of Earth's scholars, scientists, philosophers, political and religious leaders, indigenous peoples, and global citizens. **Unit 10** is designed to conclude with *Top Ten List of Environmental Issues 10:* Consuming the Earth by connecting full circle to the beginning of this text. Nature is circular, life is circular, karma is circular, and so it seems only fitting that this final unit be circular as well. Unit 10—Humans consume the Earth; Unit 1—Human population continues to grow; Unit 2—Societies and nations continue to "develop"; Unit 3—Humans need food; Unit 4—Humans need water; Unit 5—Humans need energy; Unit 6—Human need for energy to consume food and water is impacting climate; Unit 7—Human need for energy, food, and water is threatening the survival of plant and animal species; Unit 8—Human need for energy, food, and water is degrading ecosystems; Unit 9—Humans (some) see a need (in the midst of Units 1–8) to build a sustainable relationship with Earth ethic; Unit 10—Humans consume the Earth. Full circle.

Fortunately, the growing global focus on our finding a sustainable relationship with Planet Earth appears to suggest that we may now be admitting, reluctantly at times (kicking and screaming at other times), that if we continue our consumption behavior, the Earth may very well consume us—at least our dead and decomposing bodies. And if the evidence is correct—both scientific and based on our personal daily observations—a change in our attitudes is not occurring a moment too soon. Demographers have predicted that Earth will reach a population of 7 billion humans by the end of 2011. But while the birthrate is falling, before this era of population growth levels in 2050, there is a good chance habitat Earth will be home to more than 9 billion humans. Add to this the fact that the human species is seeing more and more of its offspring living to adulthood, and seeing its adults living longer lives, rates of earth resource consumption are destined to rise, as well. That's the *biological* aspect of human demographics.

On the *sociological* side of those demographics, we see the percentage of humans who achieved a decent standard of living is higher now than it has ever been, but conditions of inequality still abound. For instance, data suggests only 2 percent of the world's population own 50 percent of the wealth. On the other hand, the good news about humanity is that the gap between the world's poorest and richest is now filling with a broad middle-income group that had not existed a half century ago. But is that good news? Could it be possible that within such a "silver lining" may lay hidden an ominous cloud? The four articles included in this unit encourage the reader to seriously and critically asses the potential catastrophic impact this twenty-first-century "middle-income group" of the human species may have on the planet should their Earth consumption behavior be anything like that of their twentieth-century ancestors.

© Ingram Publishing

The unit opens with Article 36, "Consumption, Not Population Is Our Main Environmental Threat." This brief essay by Fred Pearce invites the reader to begin to reflect on and consider the possibility that human material consumption behavior, not human population growth, may be our greatest environmental threat. The author argues that by nearly every measure, a small proportion of the world's population consumes the majority of the world's resources and, as such, is responsible for the majority of its pollution (because when we talk about consumption we are really talking about carbon emissions). Who is this "small proportion?" Why are they (we?) eating us out of house and home? More critical, however, is to ask, is this minority of the human population serving as a role model for the majority of its population? Or will the larger population of humans, as they evolve into middle class, instinctually become mega-omnivores like their predecessors?

Article 37, "Consumption and Consumerism," by Anup Shah reports a similar human consumption behavior scenario. Although the global wealth gap (consumption gap) is reducing and more humans are enjoying better "living conditions," it will become even more important for us to be clear regarding the distinction between our defining "better living conditions" as *higher standards of living* (increased material production/consumption) or as *improved quality of life* (ensuring basic sustenance, self-esteem, and freedom).

Consumption, according to the United Nations, is a leading cause of environmental degradation. Nonconsumptive behaviors such as education, democracy, freedom from tyranny, empowerment of women, ensuring human rights, and the like have yet to demonstrate any associative or causal relationship to environmental degradation.

That there are vast inequalities in global consumption is well recognized. Also, the association between material production/consumption in industrialized countries and carbon emissions/climate change is clearly documented. Given this information, the question of any conscientious global citizen would have to be then, "How much should a person consume?" In Article 38, Ramachandra Guha explores such a question, but does not dictate a specific answer. In his analysis, he presents the evidence once again, how the industrialized North lays excessive claim to the "environmental space" of the South, and also supports the fact that developing countries'/emerging economies' peoples (ecosystem people) would also like to enhance their own resource consumption to obtain some of the more positive fruits of the Industrial Revolution. Can everybody possibly enjoy the fruit? Probably. Can everybody in the world own a car?

According to the author, never. Then how much should we consume? Far, far less.

Finally, the unit closes on inviting the reader to perhaps reframe Guha's question earlier to find our answer. Perhaps human beings first need to look deep into our humanity and ask: "What, fundamentally, is the purpose of our consumption? What is our endgame? To eat, to drink, to survive as a species? Of course. But beyond that? To eat, to drink, *and* be merry? Could simple human happiness be what we are seeking? Author Bill McKibben, in Article 39, "Reversal of Fortune," wonders. He believes that our single-minded focus on an unbridled growth (material possession and wealth) ethic is pushing our planet, and us, to profound ecological limits. And, as we continue to degrade our natural capital in our incessant merry-go-round quest for "more," we also continue to fail in grasping the golden rings of what made us happy *originally*—love, peace, joy, laughter, compassion, kindness, family, community, freedom, basic food, and shelter.

It may be a timeworn cliché, but during our evolution as a species, our transformation of the Earth, our ascension to civilization, and our metamorphosis from merely humans to humanity, have we lost sight of the forest for the trees?

Consumption, Not Population Is Our Main Environmental Threat

FRED PEARCE

It's the great taboo, I hear many environmentalists say. Population growth is the driving force behind our wrecking of the planet, but we are afraid to discuss it.

It sounds like a no-brainer. More people must inevitably be bad for the environment, taking more resources and causing more pollution, driving the planet ever farther beyond its carrying capacity. But hold on. This is a terribly convenient argument—"over-consumers" in rich countries can blame "over-breeders" in distant lands for the state of the planet. But what are the facts?

The world's population quadrupled to six billion people during the 20th century. It is still rising and may reach 9 billion by 2050. Yet for at least the past century, rising per-capita incomes have outstripped the rising head count several times over. And while incomes don't translate precisely into increased resource use and pollution, the correlation is distressingly strong.

Moreover, most of the extra consumption has been in rich countries that have long since given up adding substantial numbers to their population.

By almost any measure, a small proportion of the world's people take the majority of the world's resources and produce the majority of its pollution.

Take carbon dioxide emissions—a measure of our impact on climate but also a surrogate for fossil fuel consumption. Stephen Pacala, director of the Princeton Environment Institute, calculates that the world's richest half-billion people—that's about 7 percent of the global population—are responsible for 50 percent of the world's carbon dioxide emissions. Meanwhile the poorest 50 percent are responsible for just 7 percent of emissions.

Although overconsumption has a profound effect on greenhouse gas emissions, the impacts of our high standard of living extend beyond turning up the temperature of the planet. For a wider perspective of humanity's effects on the planet's life support systems, the best available measure is the "ecological footprint," which estimates the area of land required to provide each of us with food, clothing, and other resources, as well as to soak up our pollution. This analysis has its methodological problems, but its comparisons between nations are firm enough to be useful.

They show that sustaining the lifestyle of the average American takes 9.5 hectares, while Australians and Canadians require 7.8 and 7.1 hectares respectively; Britons, 5.3 hectares; Germans, 4.2; and the Japanese, 4.9. The world average is 2.7 hectares. China is still below that figure at 2.1, while India and most of Africa (where the majority of future world population growth will take place) are at or below 1.0.

The United States always gets singled out. But for good reason: It is the world's largest consumer. Americans take the greatest share of most of the world's major commodities: corn, coffee, copper, lead, zinc, aluminum, rubber, oil seeds, oil, and natural gas. For many others, Americans are the largest per-capita consumers. In "super-size-me" land, Americans gobble up more than 120 kilograms of meat a year per person, compared to just 6 kilos in India, for instance.

I do not deny that fast-rising populations can create serious local environmental crises through overgrazing, destructive farming and fishing, and deforestation. My argument here is that viewed at the global scale, it is overconsumption that has been driving humanity's impacts on the planet's vital life-support systems during at least the past century. But what of the future?

We cannot be sure how the global economic downturn will play out. But let us assume that Jeffrey Sachs, in his book *Common Wealth,* is right to predict a 600 percent increase in global economic output by 2050. Most projections put world population then at no more than 40 percent above today's level, so its contribution to future growth in economic activity will be small.

Of course, economic activity is not the same as ecological impact. So let's go back to carbon dioxide emissions. Virtually all of the extra 2 billion or so people expected on this planet in the coming 40 years will be in the poor half of the world. They will raise the population of the poor world from approaching 3.5 billion to about 5.5 billion, making them the poor two-thirds.

Sounds nasty, but based on Pacala's calculations—and if we assume for the purposes of the argument that per-capita emissions in every country stay roughly the same as today—those extra two billion people would raise the share of emissions contributed by the poor world from 7 percent to 11 percent.

Look at it another way. Just five countries are likely to produce most of the emissions and UN fertility projections. He found that an extra child in the United States today will, down the generations, produce an eventual carbon footprint seven times that of an extra Chinese child, 46 times that of a Pakistan child, 55 times that of an Indian child, and 86 times that of a Nigerian child.

Of course those assumptions may not pan out. I have some confidence in the population projections, but per-capita emissions of carbon dioxide will likely rise in poor countries for some time yet, even in optimistic scenarios. But that is an issue of consumption, not population.

In any event, it strikes me as the height of hubris to downgrade the culpability of the rich world's environmental footprint because generations of poor people not yet born might one day get to be as rich and destructive as us. Overpopulation is not driving environmental destruction at the global level; overconsumption is. Every time we talk about too many babies in Africa or India, we are denying that simple fact.

At root this is an ethical issue. Back in 1974, the famous environmental scientist Garret Hardin proposed something he called "lifeboat ethics." In the modern, resource-constrained world, he said, "each rich nation can be seen as a lifeboat full of comparatively rich people. In the ocean outside each lifeboat swim the poor of the world, who would like to get in." But there were, he said, not enough places to go around. If any were let on board, there would be chaos and all would drown. The people in the lifeboat had a duty to their species to be selfish—to keep the poor out.

Hardin's metaphor had a certain ruthless logic. What he omitted to mention was that each of the people in the lifeboat was occupying ten places, whereas the people in the water only wanted one each. I think that changes the argument somewhat.

Critical Thinking

1. Articulate in one sentence the primary argument employed by the author to support his thesis.

2. What implications does the author's argument have with regard to any one of the unit topics presented in this text? Pick one topic and describe some connections/ implications.

3. How does the author connect scientist Garret Hardin's "lifeboat ethics" to the consumption argument?

4. Do you agree/disagree with question #3? Why or why not?

5. Why does such a small portion of the earth's population consume the majority of the earth's resources?

6. Identify in this article, three key terms, concepts, or principles that are used in your textbook (environmental science, economics, sociology, history, geography, etc.) or employed in the discipline you are currently studying. (Note: The terms, concepts, or principles may be implicit, explicit, implied, or inferred.)

FRED PEARCE is a freelance author and journalist based in the UK. He is an environment consultant for New Scientist magazine and author of recent books *When The Rivers Run Dry* and *With Speed and Violence* (Beacon Press).

Consumption and Consumerism

ANUP SHAH

Global inequality in consumption, while reducing, is still high.

Using latest figures available, in 2005, the wealthiest 20% of the world accounted for 76.6% of total private consumption. The poorest fifth just 1.5%.

Breaking that down slightly further, the poorest 10% accounted for just 0.5% and the wealthiest 10% accounted for 59% of all the consumption (Figure 1).

In 1995, the inequality in consumption was wider, but the United Nations also provided some eye-opening statistics (which do not appear available, yet, for the later years) worth noting here:

Today's consumption is undermining the environmental resource base. It is exacerbating inequalities. And the dynamics of the consumption-poverty-inequality-environment nexus are accelerating. If the trends continue without change—not redistributing from high-income to low-income consumers, not shifting from polluting to cleaner goods and production technologies, not promoting goods that empower poor producers, not shifting priority from consumption for conspicuous display to meeting basic needs—today's problems of consumption and human development will worsen.

. . . The real issue is not consumption itself but its patterns and effects.

. . . Inequalities in consumption are stark. Globally, the 20% of the world's people in the highest-income countries account for 86% of total private consumption expenditures—the poorest 20% a minuscule 1.3%. More specifically, the richest fifth:

- Consume 45% of all meat and fish, the poorest fifth 5%
- Consume 58% of total energy, the poorest fifth less than 4%
- Have 74% of all telephone lines, the poorest fifth 1.5%
- Consume 84% of all paper, the poorest fifth 1.1%
- Own 87% of the world's vehicle fleet, the poorest fifth less than 1%

Runaway growth in consumption in the past 50 years is putting strains on the environment never before seen.

— *Human Development Report 1998 Overview,*[1]
United Nations Development Programme (UNDP)

— *Emphasis Added. Figures quoted use data from 1995*

If they were available, it would likely be that the breakdowns shown for the 1995 figures will not be as wide in 2005. However, they are likely to still show wide inequalities in consumption. Furthermore, as a few developing countries continue to develop and help make the numbers show a narrowing gap, there are at least two further issues:

- Generalized figures hide extreme poverty and inequality of consumption on the whole (for example, between 1995 and 2005, the inequality in consumption for the poorest fifth of humanity has hardly changed)
- If emerging nations follow the same path as today's rich countries, their consumption patterns will also be damaging to the environment

And consider the following, reflecting world priorities:

Global Priority	$U.S. Billions
Cosmetics in the United States	8
Ice cream in Europe	11
Perfumes in Europe and the United States	12
Pet foods in Europe and the United States	17
Business entertainment in Japan	35
Cigarettes in Europe	50
Alcoholic drinks in Europe	105
Narcotics drugs in the world	400
Military spending in the world	780

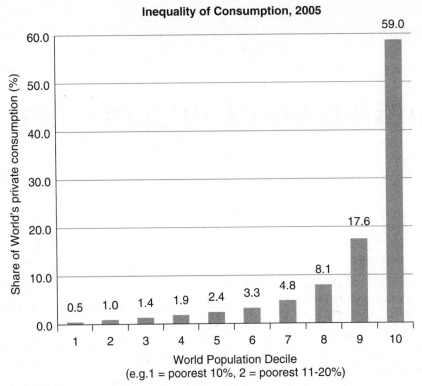

Inequality of Consumption, 2005

Share of World's private consumption (%)

Decile	Value
1	0.5
2	1.0
3	1.4
4	1.9
5	2.4
6	3.3
7	4.8
8	8.1
9	17.6
10	59.0

World Population Decile
(e.g.1 = poorest 10%, 2 = poorest 11-20%)

Source: World Bank Development Indicators 2008

And compare that to what was estimated as *additional* costs to achieve universal access to basic social services in all developing countries:

Global Priority	$U.S. Billions
Basic education for all	6
Water and sanitation for all	9
Reproductive health for all women	12

"Over" population is usually blamed as the major cause of environmental degradation, but the above statistics strongly suggests otherwise. As we will see, consumption patterns today are not to meet everyone's needs. The system that drives these consumption patterns also contributes to inequality of consumption patterns too.

This section of the globalissues.org web site will attempt to provide an introductory look at various aspects of what we consume and how.

- We will see possible "hidden" costs of convenient items to society, the environment and individuals, as well as the relationship with various sociopolitical and economic effects on those who do consume, and those who are unable to consume as much (due to poverty and so on).
- We will look at how some luxuries were turned into necessities in order to increase profits.
- This section goes beyond the "don't buy this product" type of conclusion to the deeper issues and ramifications.

- We will see just a hint at how wasteful all this is on resources, society and capital. The roots of such disparities in consumption are inextricably linked to the roots of poverty. There is such enormous waste in the way we consume that an incredible amount of resources is wasted as well. Furthermore, the processes that lead to such disparities in unequal consumption are themselves wasteful and is structured deep into the system itself. Economic efficiency is for making profits, not necessarily for social good (which is treated as a side effect). The waste in the economic system is, as a result, deep. Eliminating the causes of this type of waste are related to the elimination of poverty and bringing rights to all. Eliminating the waste also allows for further equitable consumption for all, as well as a decent standard of consumption.
- So these issues go beyond just consumption, and this section only begins to highlight the enormous waste in our economy which is not measured as such.
- A further bold conclusion is also made that elimination of so much wasted capital would actually require a reduction of people's workweek. This is because the elimination of such waste means entire industries are halved in size in some cases. So much labor redundancy cannot be tolerated, and hence the answer is therefore to share the remaining productive jobs, which means reducing the workweek!

- We will see therefore, that political causes of poverty are very much related to political issues and roots of consumerism. Hence solutions to things like hunger, environmental degradation, poverty and other problems have many commonalities that would need to be addressed.

Critical Thinking

1. The report states: "The real issue is not consumption itself but its patterns and effects." Refer to the text world map, and sketch on the map what you see are the current "patterns" of consumption.

2. Explain the statement: "Today's consumption is undermining the environmental resource base."

3. What does this report imply about the current use of our fossil fuels? Will future energy alternatives be used as well to maintain consumption?

4. The report argues "There are important issues around consumerism that need to be understood." Select 3 issues and provide a brief answer for each. Try to center your answers around the concept of "environment."

5. Identify in this article three key terms, concepts, or principles that are used in your textbook (environmental science, economics, sociology, history, geography, etc,) or employed in the discipline you are currently studying. (Note: The terms, concepts, or principles may be implicit, explicit, implied, or inferred.)

From *Global Issues*, March 6, 2011. Copyright © 2011 by Anup Shah. Reprinted by permission. www.globalissues.org/issue/235/consumption-and-consumerism.

How Much Should a Person Consume?

Ramachandra Guha

This paper takes as its point of departure an old essay by John Kenneth Galbraith, an essay so ancient and obscure that it might very well have been forgotten even by its prolific author. The essay was written in 1958, the same year that Galbraith published *The Affluent Society,* a book that wryly anatomised the social consequences of the mass-consumption age.

In his book, Galbraith highlighted the "preoccupation with productivity and production" in postwar America and western Europe. The population in these societies had for the most part been adequately housed, clothed and fed; now they expressed a desire for "more elegant cars, more exotic food, more erotic clothing, more elaborate entertainment".[1] When Galbraith termed 1950s America the "affluent society", he meant not only that this was a society most of whose members were hugely prosperous when reckoned against other societies and other times, but also that this was a society so dedicated to affluence that the possession and consumption of material goods became the exclusive standard of individual and collective achievement. He quoted the anthropologist Geoffrey Gorer, who remarked that this was a culture in which "any device or regulation which interferes, or can be conceived as interfering, with [the] supply of more and better things is resisted with unreasoning horror, as the religious resist blasphemy, or the warlike pacifism".[2]

The Unasked Question

The essay I speak of was written months after the book, which made Galbraith's name and reputation. "How Much Should a Country Consume?" is its provocative title, and it can be read as a reflective footnote to *The Affluent Society.* In the book itself, Galbraith had noted the disjunction between "private affluence and public squalor", of how this single-minded pursuit of wealth had diverted attention and resources from the nurturing of true democracy, which he defined as the provision of public infrastructure, the creation of decent schools, parks and hospitals. Now the economist turned his attention, all too fleetingly, to the long-term and global consequences of this collective promotion of consumption, of the "gargantuan and growing appetite" for resources in contemporary America. The American conservation movement, he remarked, had certainly noted the massive exploitation of resources and materials in the postwar period. However, its response was to look for more efficient methods of extraction, or the substitution of one material for another through technological innovation. There was, wrote Galbraith, a marked "selectivity in the conservationist's approach to materials consumption". For

if we are concerned about our great appetite for materials, it is plausible to seek to increase the supply, or decrease waste, to make better use of the stocks that are available, and to develop substitutes. But what of the appetite itself? Surely this is the ultimate source of the problem. If it continues its geometric course, will it not one day have to be restrained? Yet in the literature of the resource problem this is the forbidden question. Over it hangs a nearly total silence. It is as though, in the discussion of the chance for avoiding automobile accidents, we agree not to make any mention of speed![3]

A cultural explanation for this silence had been previously provided by the great Berkeley geographer Carl Sauer. Writing in 1938, Sauer remarked that "the doctrine of a passing frontier of nature replaced by a permanent and sufficiently expanding frontier of technology is a contemporary and characteristic expression of occidental culture, itself a historical–geographical product". This frontier attitude, he went on, "has the recklessness of an optimism that has become habitual, but which is residual from the brave days when north European freebooters overran the world and put it under tribute". Warning that the surge of growth at the expense of nature would not last indefinitely, Sauer—speaking for his fellow Americans—noted wistfully that "we have not yet learned the difference between yield and loot. We do not like to be economic realists".[4]

Galbraith himself identified two major reasons for the silence as regards consumption. One was ideological, the worship of the great god Growth. The principle of Growth (always with a capital G) was a cardinal belief of the American people, which necessarily implied a continuous increase in the production of consumer goods. The second reason was political, the widespread scepticism about the state. For the America of the fifties had witnessed the "resurgence of a notably over-simplified view of economic life which [ascribed] a magical automatism to the price system". Now, Galbraith was himself an unreconstructed New Dealer, who would tackle the problem of overconsumption as he would tackle the problem of underemployment, that is, through purposive state intervention. At the time he wrote, however, free-market economics ruled, and "since consumption could not be discussed without raising the question of an increased role for the state, it was not discussed".[5]

Four years later, Rachel Carson published *Silent Spring,* and the modern American environmental movement gathered pace. One might have expected this new voice of civil society to undertake what the market could not. As it happened, consumption continued to be the great unasked question of the conservation movement. The movement principally focused on two things: the threats to human health posed by pollution, and the threats to wild species and wild habitats posed by economic expansion. The latter concern became, in fact, the defining motif of the movement. The dominance of wilderness protection in American environmentalism has promoted an essentially negativist agenda—the protection of the parks and their animals by freeing them of human habitation and productive activities. As the historian Samuel Hays points out, "natural environments, which formerly had been looked upon as 'useless', waiting only to be developed, now came to be thought of as 'useful' for filling human wants and needs. They played no less a significant role in the advanced consumer society than did such material goods as hi-fi sets or indoor gardens".[6] While saving these islands of biodiversity, environmentalists paid scant attention to what was happening outside them. In the American economy as a whole, the consumption of energy and materials continued to rise.

The growing popular interest in the wild and the beautiful thus not merely accepted the parameters of the affluent society but was wont to see nature itself as merely one more good to be consumed. The uncertain commitment of most nature lovers to a more comprehensive environmental ideology is illustrated by the paradox that they were willing to drive thousands of miles—consuming scarce oil and polluting the atmosphere—to visit national parks and sanctuaries, thus using anti-ecological means to marvel at the beauty of forests, swamps or mountains protected as specimens of a "pristine" and "untouched" nature.[7]

The Real Population Problem

The selectivity of the conservationist approach to consumption was underlined in the works of biologists obsessed with the "population problem". Influential American scientists such as Paul Ehrlich and Garret Hardin identified human population growth as the single most important reason for environmental degradation. This is how Ehrlich began the first chapter of his best-selling book, *The Population Bomb:*

> I have understood the population explosion intellectually for a long time. I came to understand it emotionally one stinking hot night in Delhi a couple of years ago. My wife and daughter and I were returning to our hotel in an ancient taxi. The seats were hopping with fleas. The only functional gear was third. As we crawled through the city, we entered a crowded slum area. The temperature was well over 100, and the air was a haze of dust and smoke. The streets seemed alive with people. People eating, people washing, people sleeping. People visiting, people arguing and screaming. People thrusting their hands through the taxi window, begging. People defecating and urinating. People clinging to buses. People herding animals. People, people, people, people.[8]

Here exploding numbers are blamed for increasing pollution, stinking hot air and even technological obsolescence (that ancient taxi!). During the 1970s and 80s, neo-Malthusian interpretations gained wide currency. Countries such as India and Bangladesh were commonly blamed for causing an environmental crisis. Not surprisingly, activists in these countries have been quick to take offence, pointing out that the West, and especially the United States, consumes, per capita as well as in the aggregate, a far greater proportion of the world's resources. Table 1 gives partial evidence of this. For apart from its overuse of nature's stock (which the table documents), the Western world has also placed an unbearable burden on nature's sink (which the table ignores). Thus, the atmosphere and the oceans can absorb about 13 billion metric tons of carbon dioxide annually. This absorptive capacity, if distributed fairly among all the people of the world, would give each human being the right to emit about 2.3 tons of carbon dioxide per year. At present, an American discharges in excess of 20 tons annually, a German 12 tons, a Japanese 9 tons, an Indian less than one ton. If one looks at the process historically the charges mount, for it is the industrialised countries, led by the United States,

Table 1 US Share of World Consumption of Key Materials, 1995
(figures in millions of metric tons)

Material	World Production	US Consumption	US Consumption as Percentage of World Production
Minerals	7,641	2,410	31.54
Wood products	724	170	23.48
Metals	1,196	132	11.03
Synthetics	252	131	51.98
All materials	9,813	2,843	28.97

Source: Computed from *State of the World 1999* (New York: Worldwatch Institute and W. W. Norton, 1999).

Note: US population is approximately 4.42 percent of total world population.

which have principally been responsible for the build-up of greenhouse gases over the past hundred years.

These figures explain why Third World scholars and activists like to argue that the real "population problem" is in America, since the birth of a child there has an impact on the global environment equivalent to the birth of (say) seventy Indonesian or Indian children. There was a Bangladeshi diplomat who made this case whenever he could, in the United Nations and elsewhere. But after a visit to an American supermarket he was obliged to modify his argument—to state instead that the birth of an American dog (or cat) was the equivalent, ecologically speaking, of the birth of a dozen Bangladeshi children.[9]

As a long-time admirer of American scholarship, I might add my own words of complaint here. Consider the rich and growing academic field of environmental history, which is most highly developed in the United States. Scholars in other parts of the world have taken much inspiration from the works of American exemplars, from their methodological subtlety and fruitful criss-crossing of disciplinary boundaries. For all this, there is a studied insularity among the historians of North America. There were, at last count, more than three hundred professional environmental historians in the United States, and yet not one has seriously studied the global consequences of the consumer society, the impact on land, soil, forests, climate, etc., of the American Way of Life.

One striking example of this territorial blindness is the Gulf War. In that prescient essay of 1958, John Kenneth Galbraith remarked that "it remains a canon of modern diplomacy that any preoccupation with oil should be concealed by calling on our still ample reserves of sanctimony".[10] To be sure, there were Americans who tore the veil of this sanctimonious hypocrisy, who pointed out that it was the US government that had carefully armed and consolidated the dictator it now wished to overthrow. Yet the essentially material imperatives of the

war remained unexamined. It was the left-wing British newspaper, the *Guardian,* which claimed that the Gulf War was carried out to safeguard the American Way of Driving. No American historian, however, has taken to heart the wisdom in that throwaway remark, to reveal in all its starkness the ecological imperialism of the world's sole superpower.

Germany's Greens

I would now like to contrast the American case with the German one. Environmentalists in Germany have been more forthright in their criticisms of the consumer society. "The key to a sustainable development model worldwide," writes Helmut Lippelt, "is the question of whether West European societies really are able to reconstruct their industrial systems in order to permit an ecologically and socially viable way of production and consumption." That Lippelt does not here include the United States or Japan is noteworthy, an expression of his (and his movement's) willingness to take the burden upon themselves. West Europeans should reform themselves, rather than transfer their existing "patterns of high production and high consumption to eastern Europe and the 'Third World' [and thus] destroy the earth".[11]

For the German greens, economic growth in Europe and North America has been made possible only through the economic and ecological exploitation of the Third World. Rudolf Bahro is characteristically blunt: "The present way of life of the most industrially advanced nations," he remarked in 1984, "stands in a global and antagonistic contradiction to the natural conditions of human existence. We are eating up what other nations and future generations need to live on." From this perspective, indeed,

The working class here [in the North] is the richest lower class in the world. And if I look at the problem from the point of view of the whole of humanity, not just from that of Europe, then I must say that the metropolitan working class is the worst exploiting class in history. . . . What made poverty bearable in eighteenth- or nineteenth-century Europe was the prospect of escaping it through exploitation of the periphery. But this is no longer a possibility, and continued industrialism in the Third World will mean poverty for whole generations and hunger for millions.[12]

Bahro was a famous "Fundi", a leader of that section of the German greens which stood in the most uncompromising antagonism to modern society. But even the most hardheaded members of the other, or "Realo", faction acknowledge the unsustainability, on the global plane, of industrial society. The parliamentarian (and now foreign minister) Joschka Fischer, asked by a reporter where he planned to spend his old age, replied: "In the Frankfurt cemetery, although by that time we may pose an environmental hazard with all the poisons, heavy metals and dioxin that we carry around in our bodies." Or as a party document more matter-of-factly put it: "The global spread of industrial economic policies and lifestyles is exhausting the basic ecological health of our planet faster than it can be replenished." This global view, coupled with the stress on accountability, calls for "far-reaching voluntary commitments to restraint by wealthy nations". The industrialised countries, which consume three-quarters of the world's energy and resources, and which contribute the lion's share of "climate-threatening gaseous emissions", must curb their voracious appetite while allowing Southern nations to grow out of poverty. The greens urge the cancellation of all international debt, the banning of trade in products that destroy vulnerable ecosystems, and most radical of all, the freer migration of peoples from poor countries to rich ones.[13]

These elements in the green programme were, of course, forged as an alternative to the policies promoted by the two dominant political parties in Germany, themselves committed to the great god Growth. Since October 1998, the greens find themselves sharing power at the federal level, junior partners, but partners nevertheless, in a coalition dominated by the Social Democrats. Being in power will certainly tame them. They will work only for incremental change, instead of the wholesale restructuring of the consumption and production system some of them previously advocated.

Gandhi

Fifty years before the founding of the German green party, and thirty years before the article by Galbraith alluded to above, an Indian politician pointed to the unsustainability,

at the global level, of the Western model of economic development. "God forbid," he wrote, "that India should ever take to industrialization after the manner of the West. The economic imperialism of a single tiny island kingdom (England) is today keeping the world in chains. If an entire nation of 300 million took to similar economic exploitation, it would strip the world bare like locusts."[14]

The man was Mahatma Gandhi, writing in the weekly journal *Young India* in December 1928. Two years earlier, Gandhi had claimed that to "make India like England and America is to find some other races and places of the earth for exploitation". As it appeared that the Western nations had already "divided all the known races outside Europe for exploitation and there are no new worlds to discover", he pointedly asked: "What can be the fate of India trying to ape the West?"[15]

Gandhi's critique of Western industrialisation has, of course, profound implications for the way we live and relate to the environment today. For him, "the distinguishing characteristic of modern civilisation is an indefinite multiplicity of wants," whereas ancient civilisations were marked by an "imperative restriction upon, and a strict regulating of, these wants".[16] In uncharacteristically intemperate tones, he spoke of his "wholeheartedly detest[ing] this mad desire to destroy distance and time, to increase animal appetites, and [to] go to the ends of the earth in search of their satisfaction. If modern civilization stands for all this, and I have understood it to do so, I call it satanic".[17]

At the level of the individual, Gandhi's code of voluntary simplicity also offered a sustainable alternative to modern lifestyles. One of his best-known aphorisms, that the "world has enough for everybody's need, but not enough for everybody's greed", is in effect an exquisitely phrased one-line environmental ethic. This was an ethic he himself practised, for resource recycling and the minimisation of wants were integral to his life.

Gandhi's arguments have been revived and elaborated by the present generation of Indian environmentalists. Their country is veritably an ecological disaster zone, marked by excessively high rates of deforestation, species loss, land degradation, and air and water pollution. The consequences of this wholesale abuse of nature have chiefly been borne by the poor in the countryside—the peasants, tribespeople, fisherfolk and pastoralists who have seen their resources snatched away or depleted by more powerful economic interests. For in the last few decades, the men who rule India have attempted precisely to "make India like England and America". Without the access to resources and markets enjoyed by those two nations when they began to industrialise, India has had perforce to rely on the exploitation of its own people and environment. The natural resources of the country-side have been increasingly channelled to meet the needs of the urban–industrial sector, the diversion of forests, water, etc., to the elite having accelerated environmental degradation even as it has deprived rural and tribal communities of their traditional rights of access and use. Meanwhile, the modern sector has moved aggressively into the remaining resource frontiers of India—the northeast and the Andaman and Nicobar islands. This bias towards urban–industrial development has resulted only in a one-sided exploitation of the hinterland, thus proving Gandhi's contention that "the blood of the villages is the cement with which the edifice of the cities is built".[18]

The preceding paragraph brutally summarises arguments and evidence provided in a whole array of Indian environmentalist tracts.[19] Simplifying still further, one might say that the key contribution of the Indian environmental movement has been to point to inequalities of consumption *within* a society (or nation). In this respect they have complemented the work of their German counterparts, who have most effectively documented and criticised the inequalities of consumption *between* societies and nations.

Omnivores and Others

The criticisms of these environmentalists are strongly flavoured by morality, by the sheer injustice of one group or country consuming more than its fair share of the earth's resources, by the political imperative of restoring some sense of equality in global or national consumption. I now present an analytical framework that might more dispassionately explain these asymmetries in patterns of consumption.[20] Derived in the first instance from the Indian experience, this model rests on a fundamental opposition between two groups, *omnivores* and *ecosystem people.* These are distinguished above all by the size of their "resource catchment". Thus, omnivores, who include industrialists, rich farmers, state officials and the growing middle class based in the cities (estimated at in excess of one hundred million people), are able to draw upon the natural resources of the whole of India to maintain their lifestyles. Ecosystem people, on the other hand—who would include roughly two-thirds of the rural population, say about four hundred million people—rely for the most part on the resources of their own vicinity, from a catchment of a few dozen square miles at best. Such are the small and marginal farmers in rain-fed tracts, the landless labourers and also the heavily resource-dependent communities of hunter–gatherers, swidden agriculturists, animal herders and woodworking artisans, all stubborn premodern survivals in an increasingly postmodern landscape.

The process of development in independent India has been characterised by a basic asymmetry between omnivores and ecosystem people. A one-sentence definition of development, as it has unfolded over the last fifty years, would be: "Development is the channelling of an ever-increasing volume of natural resources, through the intervention of the state apparatus and at the cost of the state exchequer, to serve the interests of the rural and urban omnivores." Some central features of this process have been:

1. The concentration of political power/decision-making in the hands of omnivores.[21]
2. Hence the use of the state machinery to divert natural resources to islands of omnivore prosperity, especially through subsidies. Wood for paper mills, fertilisers for rich farmers, water and power for urban dwellers are all supplied by the state to omnivores at well below market prices.
3. The culture of subsidies has fostered an indifference among omnivores to the environmental degradation they cause, aided by their ability to pass on its costs to ecosystem people or to society at large.
4. Projects based on the capture of wood, water or minerals—such as eucalyptus plantations, large dams or opencast mining—have tended to dispossess the ecosystem people who previously enjoyed ready access to those resources. This has led to a rising tide of protests by the victims of development—Chipko, Narmada and dozens of other protests that we know collectively as the "Indian environmental movement".
5. But development has also *permanently* displaced large numbers of ecosytem people from their homes. Some twenty million Indians have been uprooted by steel mills, dams and the like; countless others have been forced to move to the cities in search of a legitimate livelihood denied to them in the countryside (sometimes as a direct consequence of environmental degradation).[22] Thus has been created a third class, of *ecological refugees,* living in slums and temporary shelters in the towns and cities of India.

This framework, which divides the Indian population into the three socio-ecological classes of omnivores, ecosystem people and ecological refugees, can help us understand why economic development since 1947 has destroyed nature while failing to remove poverty. The framework synthesises the insights of ecology with sociology, in that it distinguishes social classes by their respective resource catchments, by their cultures and styles of consumption, and also by their widely varying powers to influence state policy.

The framework is analytical as well as value-laden, descriptive and prescriptive. It helps us understand and interpret nature-based conflicts at various spatial levels: from the village community upwards through the district and region and on to the nation. Stemming from the study of the history of modern India, it might also throw light on the dynamics of socio-ecological change in other large developing "Third World" countries such as Brazil and Malaysia, where conflicts between omnivores and eco-system people have also erupted and whose cities are likewise marked by a growing population of "ecological refugees". At a pinch, it might explain asymmetries and inequalities at the global level, too. More than a hundred years ago a famous German radical proclaimed, "Workers of the World, Unite!" But as another German radical[23] recently reminded this writer, the reality of our times is very nearly the reverse: the process of globalisation, whose motto might well be, "Omnivores of the World, Unite!"

Conflicting Fallacies

What, then, is the prospect for the future? Consider two well-known alternatives already prominent in the market place of ideas:

1. *The Fallacy of the Romantic Economist,* which states that everyone can become an omnivore if only we allow the market full play. That is the hope, and the illusion, of globalisation, which promises a universalisation of American styles of consumption. But this is nonsense, for although businessmen and economists resolutely refuse to recognise it, there are clear ecological limits to a global consumer society, to all Indians or Mexicans attaining the lifestyle of an average middle-class North American. Can there be a world with one billion cars, an India with two hundred million cars?

2. *The Fallacy of the Romantic Environmentalist,* which claims that ecosystem people want to remain ecosystem people. This is the anti-modern, anti-Western, anti-science position of some of India's best-known, neo-Gandhian environmentalists.[24] This position is also gaining currency among some sections of Western academia. Anthropologists in particular are almost falling over themselves in writing epitaphs for development, in works that seemingly dismiss the very prospect of directed social change in much of the Third World. It is implied that development is a nasty imposition on the innocent peasant and tribesperson who, left to themselves, would not willingly partake of Enlightenment rationality, modern technology or modern consumer goods.[25] This literature has become so abundant and so influential that it has even been anthologised, in a volume called (what else?) *The Post-Development Reader.*[26]

The editor of this volume is a retired Iranian diplomat now living in the south of France. The authors of those other demolitions of the development project mentioned in footnote 25 are, without exception, tenured professors at well-established American universities. I rather suspect that the objects of their sympathy would cheerfully exchange their own social position for that of their chroniclers. For it is equally a fallacy that ecosystem people want to remain as they are, that they do not want to enhance their own resource consumption, to get some of the benefits of science, development and modernity.

This point can be made more effectively by way of anecdote. Some years ago, a group of Indian scholars and activists gathered in the southern town of Manipal for a national meeting in commemoration of the 125th anniversary of the birth of Mahatma Gandhi. They spoke against the backdrop of a life-size portrait of Gandhi, depicting him clad in the loincloth he wore for the last thirty-three years of his life. Speaker after speaker invoked his mode of dress as symbolising the Mahatma's message. Why did we all not follow his example and give up everything, thus to mingle more definitively with the masses?

Then, on the last evening of the conference, the Dalit (low-caste) poet Devanur Mahadeva got up to speak. He read out a short poem in the Kannada language of southwest India, written not by him but by a Dalit woman of his acquaintance. The poem spoke reverentially of the great Untouchable leader B. R. Ambedkar (1889–1956) and especially of the dark blue suit that Ambedkar invariably wore in the last three decades of *his* life. Why did the Dalit lady focus on Ambedkar's suit, asked Mahadeva? Why, indeed, did the countless statues of Ambedkar put up in Dalit hamlets always have him clad in suit and tie? His answer was deceptively and eloquently simple: if Gandhi wears a loincloth, we all marvel at his *tyaga,* his sacrifice. The scantiness of dress is in this case a marker of what the man has given up. A high-caste, well-born, English-educated lawyer had voluntarily chosen to renounce power and position and live the life of an Indian peasant. That is why we memorialise that loincloth.

However, if Ambedkar had worn a loincloth, that would not occasion wonder or surprise. He is a Dalit, we would say—what else should he wear? Millions of his caste fellows wear nothing else. It is the fact that he has escaped this fate, that his extraordinary personal achievements (a law degree from Lincoln's Inn, a PhD from Columbia University, the drafting of the Constitution of India) have allowed him to escape the fate that society and history had

Table 2 Hierarchies of Resource Consumption

Fuel Used	Mode of Housing	Mode of Transport
Grass	Cave	Feet
Wood	Thatched hut	Bullock cart
Coal	Wooden house	Bicycle
Gas	Stone house	Motor scooter
Electricity	Cement house	Car

allotted him, that is so effectively symbolised in that blue suit. Modernity, not tradition, development, not stagnation, are responsible for this inversion, for this successful and all-too infrequent storming of the upper-caste citadel.

A Blueprint for India

Let me now attempt to represent the story of Dr B. R. Ambedkar's suit in more material terms. Consider the simple hierarchies of fuel, housing and transportation set out in Table 2.

To move down any level of this table is to move towards a more reliable, more efficient, longer-lasting and generally safer mode of consumption. Why, then, would one abjure cheap and safe cooking fuel, for example, or quick and reliable transport, or stable houses that can outlive one monsoon? To prefer gas to dung for your stove, a car to a bullock-cart for your mobility, a wooden home to a straw hut for your family's shelter, is to choose greater comfort, wellbeing and freedom. These are choices that, despite specious talk of cultural difference, must be made available to all humans.

At the same time, to move down these levels is generally to move towards a more intensive and possibly unsustainable use of resources. Unsustainable at the global level, that is, for while a car admittedly expands freedom, there is no possibility whatsoever of every human on Earth being able to possess a car. As things stand, some people consume too much, while others consume far too little. It is these asymmetries that a responsible politics would seek to address. Confining ourselves to India, for instance, one would work to enhance the social power of ecological refugees and ecosystem people, their ability to govern their lives and to gain from the transformation of nature into artefact. This policy would simultaneously force omnivores to internalise the costs of their profligate behaviour. A new, "left–green" development strategy would feature the following five central elements:

1. A move towards a genuinely participatory democracy, with a strengthening of the institutions of local governance (at village, town or district levels) mandated by the Constitution of India but aborted by successive central governments in New Delhi. The experience of the odd state, such as West Bengal and Karnataka, which has experimented successfully with the *panchayat* or self-government system suggests that local control is conducive to the successful management of forests, water, etc.

2. Creation of a process of natural resource use which is open, accessible and accountable. This would include a Freedom of Information Act, so that citizens are fully informed about the intentions of the state and better able to challenge or welcome them, thus making officials more responsive to their public.

3. The use of decentralisation to stop the widespread undervaluing of natural resources. The removal of subsidies and the putting of proper price tags will make resource use more efficient and less destructive of the environment.

4. The encouragement of a shift to private enterprise for producing goods and services, while ensuring that there are no hidden subsidies and that firms properly internalise externalities. There is at present an unfortunate distaste for the market among Indian radicals, whether Gandhian or Marxist. But one cannot turn one's back on the market; the task rather is to tame it. The people and environment of India have already paid an enormous price for allowing state monopolies in sectors such as steel, energy, transport and communications.

5. This kind of development can, however, succeed only if India is a far more equitable society than is the case at present. Three key ways of enhancing the social power of ecological refugees and ecosystem people (in all of which India has conspicuously failed) are land reform, literacy (especially female literacy) and proper health care. These measures would also help bring population growth under control. In the provision of health and education the state might be aided by the voluntary sector, paid for by communities out of public funds.

Remedying Global Inequalities

The charter of sustainable development outlined here[27] applies, of course, only to one country, albeit a large and representative one. Its *raison d'etre* is the persistent and grave inequalities of consumption within the nation. What, then, of inequalities of consumption between nations? This question has been authoritatively addressed n a recent study of the prospects for a "sustainable Germany" sponsored by the Wüppertal Institute for Climate and Ecology.[28] Its fundamental premise is that the North lays excessive claim to the "environmental space" of the South. The way the global economy is currently structured,

> The North gains access to cheap raw materials and hinders access to markets for processed products from those countries; it imposes a system (World Trade Organisation) that favours the strong; it makes use of large areas of land in the South, tolerating soil degradation, damage to regional eco-systems, and disruption of local self-reliance; it exports toxic waste; it claims patent rights to utilisation of biodiversity in tropical regions, etc.[29]

Seen "against the backdrop of a divided world," says the report, "the excessive use of nature and its resources n the North is a principal block to greater justice in the world. . . . A retreat of the rich from overconsumption s thus a necessary first step towards allowing space for nprovement of the lives of an increasing number of people." The problem thus identified, the report goes on to emise, in meticulous detail, how Germany can take the ead in reorienting its economy and society towards a more ustainable path. It begins with an extended treatment of verconsumption, of the excessive use of the global commons by the West over the past two hundred years, of the errestrial consequences of profligate lifestyles—soil erosion, forest depletion, biodiversity loss, air and water pollution. It then outlines a long-term plan for reducing the throughput" of nature in the economy and cutting down n emissions. The report sets targets for substantial cuts y the year 2010 in the consumption of energy (at least 0 percent) and non-renewable raw materials (25 percent), nd in the release of substances such as carbon dioxide 35 percent), sulphur dioxide (80–90 percent), synthetic itrogen fertilisers (100 percent) and agricultural biocides 100 percent).

The policy and technical changes necessary to achieve aese targets are identified as including the elimination of ubsidies for chemical farming, the levying of ecological axes (on gasoline, for example), the adoption of slower nd fuel-efficient cars and the movement of goods by ail instead of road. Some examples of resource conservation in practice are given, such as the replacement of oncrete girders by those made with steel, water conservation and recycling within the city, and a novel contract etween the Munich municipal authorities and organic armers in the countryside. By adopting such measures, ermany would transform itself from a nature-abusing a nature-saving country.

The Wüppertal Institute study is notable for its mix of moral ends with material means, as well as its judicious blending of economic and technical options. More striking still has been its reception. The original German book sold forty thousand copies, an abbreviated version selling an additional hundred thousand copies. It was made into an award-winning television film and discussed by trade unions, political parties, consumer groups, scholars, church congregations and countless lay citizens. In several German towns and regions attempts have begun to put some of its proposals into practice.

Inequalities of consumption thus need to be addressed at both national and international levels. Indeed, the two are interconnected. The Spanish economist Juan Martinez-Alier provides one telling example. In the poorer countries of Asia and Africa, firewood and animal dung are often the only sources of cooking fuel. These are inefficient and polluting, and their collection involves much drudgery. The provision of oil or liquefied petroleum gas for the cooking stoves of Somali or Nepalese peasant women would greatly improve the quality of their lives. This could easily be done, says Martinez-Alier, if the rich were very moderately taxed. He calculates that to replace the fuel used by the world's three thousand million poor people would require about two hundred million tons of oil a year. Now, this is only a quarter of the United States' annual consumption. But the bitter irony is that "oil at $15 a barrel is so cheap that it can be wasted by rich countries, but [is] too expensive to be used as domestic fuel by the poor". The solution is simple: oil consumption in the rich countries should be taxed, while the use of liquefied petroleum gas or kerosene for fuel in the poor countries should be subsidised.[30]

Allowing the poor to ascend just one rung up the hierarchies of resource consumption requires a very moderate sacrifice by the rich. In the present climate, however, any proposal with even the slightest hint of redistribution would be shot down as smacking of "socialism". But this might change, as conflicts over consumption begin to sharpen, as they assuredly shall. Within countries, access to water, land, and forest and mineral resources will be fiercely fought over by contending groups. Between countries, there will be bitter arguments about the "environmental space" occupied by the richer nations.[31] As these divisions become more manifest, the global replicability of Western styles of living will be more directly and persistently challenged. Sometime in the middle decades of the twenty-first century, Galbraith's great unasked question, "How Much Should a Country Consume?", with its corollary, "How Much Should a Person Consume?", will come, finally, to dominate intellectual and political debate.

Notes

1. John Kenneth Galbraith, *The Affluent Society* (London: Hamish Hamilton, 1958), pp. 109–10.

2. Ibid., p. 96.

3. John Kenneth Galbraith, "How Much Should a Country Consume?", in *Perspectives on Conservation,* ed. Henry Jarret (Baltimore: Johns Hopkins University Press, 1958), pp. 91–2.

4. Carl Sauer, "Theme of Plant and Animal Destruction in Economic History" (1938), in his *Land and Life* (Berkeley: University of California Press, 1963), p. 154.

5. Galbraith, "How Much Should a Country Consume?", p. 97.

6. Samuel Hays, "From Conservation to Environment: Environmental Politics in the United States since World War Two", *Environmental Review* 6, no. 1 (1982), p. 21.

7. For details, see my essays, "Radical American Environmentalism and Wilderness Preservation: A Third World Critique", *Environmental Ethics* 11, no. 1 (spring 1989), and "The Two Phases of American Environmentalism: A Critical History", in *Decolonizing Knowledge,* ed. Stephen A. Marglin and Frederique Apffel-Marglin (Oxford: Clarendon Press, 1996).

8. Paul R. Ehrlich, *The Population Bomb* (New York: Ballantine Books, 1968), p. 15.

9. See Satyajit Singh, "Environment, Class and State in India: A Perspective on Sustainable Irrigation" (PhD dissertation, Delhi University, 1994).

10. Galbraith, "How Much Should a Country Consume?", p. 90.

11. Helmut Lippelt, "Green Politics in Progress: Germany", *International Journal of Sociology and Social Policy* 12, nos. 4–7 (1992), p. 197.

12. Rudolf Bahro, *From Red to Green: Interviews with New Left Review* (London: Verso, 1984), p. 184.

13. This paragraph is based on Werner Hülsberg, *The German Greens: A Social and Political Profile* (London: Verso, 1988); but see also Margit Mayer and John Ely, eds., *Between Movement and Party: The Paradox of the German Greens* (Philadelphia: Temple University Press, 1997), and Saral Sarkar, "The Green Movement in West Germany", *Alternatives* 11, no. 2 (1986).

14. Mahatma Gandhi, "Discussion with a Capitalist", *Young India,* 20 December 1928, in the *Collected Works of Mahatma Gandhi* (hereafter *CWMG*) [New Delhi: Publications Division, n.d.], vol. 38, p. 243.

15. "The Same Old Argument", *Young India,* 7 October 1926 (*CWMG,* vol. 31, p. 478).

16. "Choice before Us", *Young India,* 2 June 1927 (*CWMG,* vol. 33, pp. 417–8).

17. "No and Yes", *Young India,* 17 March 1927 (*CWMG,* vol. 33, p. 163).

18. See *Harijan,* 23 June 1946.

19. See especially the two *Citizens Reports on the Indian Environment,* published in 1982 and 1985 by the New Delhi–based Centre for Science and Environment. See also the magisterial essay by the centre's director, Anil Agarwal, "Human–Nature Interactions in a Third World Country", *Environmentalist* 6, no. 3 (1986).

20. The following paragraphs expand and elaborate on some ideas first presented in Madhav Gadgil and Ramachandra Guha, *Ecology and Equity: The Use and Abuse of Nature in Contemporary India* (London: Routledge, 1995).

21. See Pranab Bardhan, *The Political Economy of India's Development* (Oxford: Clarendon Press, 1984).

22. See Eenakshi Ganguly-Thukral, ed., *Big Dams, Displaced People* (New Delhi: Sage Publishers, 1992).

23. The environmentalist and social critic Wolfgang Sachs.

24. See, for example, Ashis Nandy, ed., *Science, Hegemony and Violence: A Requiem for Modernity* (New Delhi: Oxford University Press, 1989); Vandana Shiva, *Staying Alive: Women, Ecology and Development* (London: Zed Books, 1989).

25. See, for example, Arturo Escobar, *Encountering Development: The Making and Unmaking of the Third World* (Princeton, N.J.: Princeton University Press, 1995); James Scott, *Seeing Like a State: How Certain Schemes to Improve the Human Condition Have Failed* (New Haven: Yale University Press, 1998); and especially Wolfgang Sachs, ed., *The Development Dictionary: A Guide to Knowledge as Power* (London: Zed Books, 1992).

26. Majid Rahnema, ed., *The Post-Development Reader* (London: Zed Books, 1998).

27. And elaborated in more detail in Gadgil and Guha, *Ecology and Equity.*

28. Wolfgang Sachs et al., *Greening the North: A Post-Industrial Blueprint for Ecology and Equity* (London: Zed Books, 1998). on which the rest of this section is based. See also F. Schmidt-Beek, ed., *Carnoules Declaration: Factor 10 Club* (Wüppertal: WIKUE, 1994), which sets the target of a 90 percent reduction in material use by the industrialised countries.

29. Sachs et al., *Greening the North,* p. 159.

30. See Martinez-Alier's essay, "Poverty and the Environment", in Ramachandra Guha and Juan Martinez-Alier, *Varieties of Environmentalism: Essays North and South* (London: Earthscan, 1997). See also Juan Martinez-Alier, *Ecological Economics: Energy, Environment, Society,* rev. ed. (London: Basil Blackwell, 1991).

31. In this connection, see Anil Agarwal and Sunita Narain, *Global Warming in an Unequal World: A Case of Environmental Colonialism?* (New Delhi: Centre for Science and Environment, 1992).

Critical Thinking

1. List five "facts" or "data" the author provides that supports Article 36.

2. Based on your readings in this text, describe some of the potential global consequences of the "consumer society."

3. What would be the global environmental consequences (potential) of both India and China "developing" the way the United States has?

4. With regard to patterns of consumption, what is the difference between "omnivores" and "ecosystem people"?

5. Explain the idea of "inequalities of consumption between nations."

6. What are some implications this article has for global development? For feeding humanity? For a thirsty planet? For our quest for energy? Pick one and discuss.

7. Identify in this article three key terms, concepts, or principles that are used in your textbook (environmental science, economics, sociology, history, geography, etc,) or employed in the discipline you are currently studying. (Note: The terms, concepts. or principles may be implicit, explicit, implied, or inferred.)

RAMACHANDRA GUHA is a historian and anthropologist who lives in Bangalore. His paper is the product of a research and writing grant from the John D. and Catherine T. MacArthur Foundation.

Reversal of Fortune

The formula for human well-being used to be simple: Make money, get happy. So why is the old axiom suddenly turning on us?

Bill McKibben

For most of human history, the two birds More and Better roosted on the same branch. You could toss one stone and hope to hit them both. That's why the centuries since Adam Smith launched modern economics with his book *The Wealth of Nations* have been so single-mindedly devoted to the dogged pursuit of maximum economic production. Smith's core ideas—that individuals pursuing their own interests in a market society end up making each other richer; and that increasing efficiency, usually by increasing scale, is the key to increasing wealth—have indisputably worked. They've produced more More than he could ever have imagined. They've built the unprecedented prosperity and ease that distinguish the lives of most of the people reading these words. It is no wonder and no accident that Smith's ideas still dominate our politics, our outlook, even our personalities.

But the distinguishing feature of our moment is this: Better has flown a few trees over to make her nest. And that changes everything. Now, with the stone of your life or your society gripped in your hand, you have to choose. It's More or Better.

Which means, according to new research emerging from many quarters, that our continued devotion to growth above all is, on balance, making our lives worse, both collectively and individually. Growth no longer makes most people wealthier, but instead generates inequality and insecurity. Growth is bumping up against physical limits so profound—like climate change and peak oil—that trying to keep expanding the economy may be not just impossible but also dangerous. And perhaps most surprisingly, growth no longer makes us happier. Given our current dogma, that's as bizarre an idea as proposing that gravity pushes apples skyward. But then, even Newtonian physics eventually shifted to acknowledge Einstein's more complicated universe.

[I] "We Can Do It If We Believe It": FDR, LBJ, and the Invention of Growth

It was the great economist John Maynard Keynes who pointed out that until very recently, "there was no very great change in the standard of life of the average man living in the civilized centers of the earth." At the utmost, Keynes calculated, the standard of living roughly doubled between 2000 B.C. and the dawn of the 18th century—four millennia during which we basically didn't learn to do much of anything new. Before history began, we had already figured out fire, language, cattle, the wheel, the plow, the sail, the pot. We had banks and governments and mathematics and religion.

And then, something new finally did happen. In 1712, a British inventor named Thomas Newcomen created the first practical steam engine. Over the centuries that followed, fossil fuels helped create everything we consider normal and obvious about the modern world, from electricity to steel to fertilizer; now, a 100 percent jump in the standard of living could suddenly be accomplished in a few decades, not a few millennia.

In some ways, the invention of the idea of economic growth was almost as significant as the invention of fossil-fuel power. But it took a little longer to take hold. During the Depression, even FDR routinely spoke of America's economy as mature, with no further expansion anticipated. Then came World War II and the postwar boom—by the time Lyndon Johnson moved into the White House in 1963, he said things like: "I'm sick of all the people who talk about the things we can't do. Hell, we're the richest country in the world, the most powerful. We can do it all. . . . We can do it if we believe it." He wasn't alone in thinking this way. From Moscow, Nikita Khrushchev thundered, "Growth of industrial and agricultural production is the battering ram with which we shall smash the capitalist system."

Yet the bad news was already apparent, if you cared to look. Burning rivers and smoggy cities demonstrated the dark side of industrial expansion. In 1972, a trio of MIT researchers released a series of computer forecasts they called "limits to growth," which showed that unbridled expansion would eventually deplete our resource base. A year later the British economist E. F. Schumacher wrote the best-selling *Small Is Beautiful.* (Soon after, when Schumacher came to the United States on a speaking tour, Jimmy Carter actually received him at the White House—imagine the current president making time for any economist.) By 1979, the sociologist Amitai Etzioni reported to President Carter that only 30 percent of Americans were "progrowth," 31 percent were "anti-growth," and 39 percent were "highly uncertain."

Such ambivalence, Etzioni predicted, "is too stressful for societies to endure," and Ronald Reagan proved his point. He convinced us it was "Morning in America"—out with limits, in with Trump. Today, mainstream liberals and conservatives compete mainly on the question of who can flog the economy harder. Larry Summers, who served as Bill Clinton's secretary of the treasury, at one point declared that the Clinton administration "cannot and will not accept any 'speed limit' on American economic growth. It is the task of economic policy to grow the economy as rapidly, sustainably, and inclusively as possible." It's the economy, stupid.

[2] Oil Bingeing, Chinese Cars, and the End of the Easy Fix

Except there are three small things. The first I'll mention mostly in passing: Even though the economy continues to grow, most of us are no longer getting wealthier. The average wage in the United States is less now, in real dollars, than it was 30 years ago. Even for those with college degrees, and although productivity was growing faster than it had for decades, between 2000 and 2004 earnings fell 5.2 percent when adjusted for inflation, according to the most recent data from White House economists. Much the same thing has happened across most of the globe. More than 60 countries around the world, in fact, have seen incomes per capita fall in the past decade.

For the second point, it's useful to remember what Thomas Newcomen was up to when he helped launch the Industrial Revolution—burning coal to pump water out of a coal mine. This revolution both depended on, and revolved around, fossil fuels. "Before coal," writes the economist Jeffrey Sachs, "economic production was

limited by energy inputs, almost all of which depended on the production of biomass: food for humans and farm animals, and fuel wood for heating and certain industrial processes." That is, energy depended on how much you could grow. But fossil energy depended on how much had grown eons before—all those billions of tons of ancient biology squashed by the weight of time till they'd turned into strata and pools and seams of hydrocarbons, waiting for us to discover them.

To understand how valuable, and irreplaceable, that lake of fuel was, consider a few other forms of creating usable energy. Ethanol can perfectly well replace gasoline in a tank; like petroleum, it's a way of using biology to create energy, and right now it's a hot commodity, backed with billions of dollars of government subsidies. But ethanol relies on plants that grow anew each year, most often corn; by the time you've driven your tractor to tend the fields, and your truck to carry the crop to the refinery, and powered your refinery, the best-case "energy output-to-input ratio" is something like 1.34-to-1. You've spent 100 Btu of fossil energy to get 134 Btu. Perhaps that's worth doing, but as Kamyar Enshayan of the University of Northern Iowa points out, "it's not impressive" compared to the ratio for oil, which ranges from 30-to-1 to 200-to-1, depending on where you drill it. To go from our fossil-fuel world to a biomass world would be a little like leaving the Garden of Eden for the land where bread must be earned by "the sweat of your brow."

And east of Eden is precisely where we may be headed. As everyone knows, the past three years have seen a spate of reports and books and documentaries suggesting that humanity may have neared or passed its oil peak—that is, the point at which those pools of primeval plankton are half used up, where each new year brings us closer to the bottom of the barrel. The major oil companies report that they can't find enough new wells most years to offset the depletion in the old ones; rumors circulate that the giant Saudi fields are dwindling faster than expected; and, of course, all this is reflected in the cost of oil.

The doctrinaire economist's answer is that no particular commodity matters all that much, because if we run short of something, it will pay for someone to develop a substitute. In general this has proved true in the past: Run short of nice big sawlogs and someone invents plywood. But it's far from clear that the same precept applies to coal, oil, and natural gas. This time, there is no easy substitute: I like the solar panels on my roof, but they're collecting diffuse daily energy, not using up eons of accumulated power. Fossil fuel was an exception to the rule, a one-time gift that underwrote a one-time binge of growth.

This brings us to the third point: If we do try to keep going, with the entire world aiming for an economy structured like America's, it won't be just oil that we'll run short of. Here are the numbers we have to contend with: Given current rates of growth in the Chinese economy, the 1.3 billion residents of that nation alone will, by 2031, be about as rich as we are. If they then eat meat, milk, and eggs at the rate that we do, calculates ecostatistician Lester Brown, they will consume 1,352 million tons of grain each year—equal to two-thirds of the world's entire 2004 grain harvest. They will use 99 million barrels of oil a day, 15 million more than the entire world consumes at present. They will use more steel than all the West combined, double the world's production of paper, and drive 1.1 billion cars—1.5 times as many as the current world total. And that's just China; by then, India will have a bigger population, and its economy is growing almost as fast. And then there's the rest of the world.

Trying to meet that kind of demand will stress the earth past its breaking point in an almost endless number of ways, but let's take just one. When Thomas Newcomen fired up his pump on that morning in 1712, the atmosphere contained 275 parts per million of carbon dioxide. We're now up to 380 parts per million, a level higher than the earth has seen for many millions of years, and climate change has only just begun. The median predictions of the world's climatologists—by no means the worst-case scenario—show that unless we take truly enormous steps to rein in our use of fossil fuels, we can expect average temperatures to rise another four or five degrees before the century is out, making the globe warmer than it's been since long before primates appeared. We might as well stop calling it earth and have a contest to pick some new name, because it will be a different planet. Humans have never done anything more profound, not even when we invented nuclear weapons.

How does this tie in with economic growth? Clearly, getting rich means getting dirty—that's why, when I was in Beijing recently, I could stare straight at the sun (once I actually figured out where in the smoggy sky it was). But eventually, getting rich also means wanting the "luxury" of clean air and finding the technological means to achieve it. Which is why you can once again see the mountains around Los Angeles; why more of our rivers are swimmable every year. And economists have figured out clever ways to speed this renewal: Creating markets for trading pollution credits, for instance, helped cut those sulfur and nitrogen clouds more rapidly and cheaply than almost anyone had imagined.

But getting richer doesn't lead to producing less carbon dioxide in the same way that it does to less smog—in fact, so far it's mostly the reverse. Environmental destruction of the old-fashioned kind—dirty air, dirty water—results from something going wrong. You haven't bothered to stick the necessary filter on your pipes, and so the crud washes into the stream; a little regulation, and a little money, and the problem disappears. But the second, deeper form of environmental degradation comes from things operating exactly as they're supposed to, just too much so. Carbon dioxide is an inevitable byproduct of burning coal or gas or oil—not something going wrong. Researchers are struggling to figure out costly and complicated methods to trap some CO_2 and inject it into underground mines—but for all practical purposes, the vast majority of the world's cars and factories and furnaces will keep belching more and more of it into the atmosphere as long as we burn more and more fossil fuels.

True, as companies and countries get richer, they can afford more efficient machinery that makes better use of fossil fuel, like the hybrid Honda Civic I drive. But if your appliances have gotten more efficient, there are also far more of them: The furnace is better than it used to be, but the average size of the house it heats has doubled since 1950. The 60-inch TV? The always-on cable modem? No need for you to do the math—the electric company does it for you, every month. Between 1990 and 2003, precisely the years in which we learned about the peril presented by global warming, the United States' annual carbon dioxide emissions increased by 16 percent. And the momentum to keep going in that direction is enormous. For most of us, growth has become synonymous with the economy's "health," which in turn seems far more palpable than the health of the planet. Think of the terms we use—the economy, whose temperature we take at every newscast via the Dow Jones average, is "ailing" or it's "on the mend." It's "slumping" or it's "in recovery." We cosset and succor its every sniffle with enormous devotion, even as we more or less ignore the increasingly urgent fever that the globe is now running. The ecological economists have an enormous task ahead of them—a nearly insurmountable task, if it were "merely" the environment that is in peril. But here is where things get really interesting. It turns out that the economics of environmental destruction are closely linked to another set of leading indicators—ones that most humans happen to care a great deal about.

[3] "It Seems That Well-Being Is a Real Phenomenon": Economists Discover Hedonics

Traditionally, happiness and satisfaction are the sort of notions that economists wave aside as poetic irrelevance, the kind of questions that occupy people with no head for numbers who had to major in liberal arts. An orthodox economist has a simple happiness formula: If you buy a Ford Expedition, then ipso facto a Ford Expedition is what makes you happy. That's all we need to know. The economist would call this idea "utility maximization," and in the words of the economic historian Gordon Bigelow, "the theory holds that every time a person buys something, sells something, quits a job, or invests, he is making a rational decision about what will . . . provide him 'maximum utility.' If you bought a Ginsu knife at 3 A.M. a neoclassical economist will tell you that, at that time, you calculated that this purchase would optimize your resources." The beauty of this principle lies in its simplicity. It is perhaps the central assumption of the world we live in: You can tell who I really am by what I buy.

Yet economists have long known that people's brains don't work quite the way the model suggests. When Bob Costanza, one of the fathers of ecological economics and now head of the Gund Institute at the University of Vermont, was first edging into economics in the early 1980s, he had a fellowship to study "social traps"—the nuclear arms race, say—in which "short-term behavior can get out of kilter with longer broad-term goals."

It didn't take long for Costanza to demonstrate, as others had before him, that, if you set up an auction in a certain way, people will end up bidding $1.50 to take home a dollar. Other economists have shown that people give too much weight to "sunk costs"—that they're too willing to throw good money after bad, or that they value items more highly if they already own them than if they are considering acquiring them. Building on such insights, a school of "behavioral economics" has emerged in recent years and begun plumbing how we really behave.

The wonder is that it took so long. We all know in our own lives how irrationally we are capable of acting, and how unconnected those actions are to any real sense of joy. (I mean, there you are at 3 A.M. thinking about the Ginsu knife.) But until fairly recently, we had no alternatives to relying on Ginsu knife and Ford Expedition purchases as the sole measures of our satisfaction. How else would we know what made people happy?

That's where things are now changing dramatically: Researchers from a wide variety of disciplines have started to figure out how to assess satisfaction, and economists have begun to explore the implications. In 2002 Princeton's Daniel Kahneman won the Nobel Prize in economics even though he is trained as a psychologist. In the book *Well-Being,* he and a pair of coauthors announce a new field called "hedonics," defined as "the study of what makes experiences and life pleasant or unpleasant. . . . It is also concerned with the whole range of circumstances, from the biological to the societal, that occasion suffering and enjoyment." If you are worried

that there might be something altogether too airy about this, be reassured—Kahneman thinks like an economist. In the book's very first chapter, "Objective Happiness," he describes an experiment that compares "records of the pain reported by two patients undergoing colonoscopy," wherein every 60 seconds he insists they rate their pain on a scale of 1 to 10 and eventually forces them to make "a hypothetical choice between a repeat colonoscopy and a barium enema." Dismal science indeed.

As more scientists have turned their attention to the field, researchers have studied everything from "biases in recall of menstrual symptoms" to "fearlessness and courage in novice paratroopers." Subjects have had to choose between getting an "attractive candy bar" and learning the answers to geography questions; they've been made to wear devices that measured their blood pressure at regular intervals; their brains have been scanned. And by now that's been enough to convince most observers that saying "I'm happy" is more than just a subjective statement. In the words of the economist Richard Layard, "We now know that what people say about how they feel corresponds closely to the actual levels of activity in different parts of the brain, which can be measured in standard scientific ways." Indeed, people who call themselves happy, or who have relatively high levels of electrical activity in the left prefrontal region of the brain, are also "more likely to be rated as happy by friends," "more likely to respond to requests for help," "less likely to be involved in disputes at work," and even "less likely to die prematurely." In other words, conceded one economist, "it seems that what the psychologists call subjective well-being is a real phenomenon. The various empirical measures of it have high consistency, reliability, and validity."

The idea that there is a state called happiness, and that we can dependably figure out what it feels like and how to measure it, is extremely subversive. It allows economists to start thinking about life in richer (indeed) terms, to stop asking "What did you buy?" and to start asking "Is your life good?" And if you can ask someone "Is your life good?" and count on the answer to mean something, then you'll be able to move to the real heart of the matter, the question haunting our moment on the earth: Is more better?

[4] If We're So Rich, How Come We're So Damn Miserable?

In some sense, you could say that the years since World War II in America have been a loosely controlled experiment designed to answer this very question. The environmentalist Alan Durning found that in 1991 the average American family owned twice as many cars as it did in

1950, drove 2.5 times as far, used 21 times as much plastic, and traveled 25 times farther by air. Gross national product per capita tripled during that period. Our houses are bigger than ever and stuffed to the rafters with belongings (which is why the storage-locker industry has doubled in size in the past decade). We have all sorts of other new delights and powers—we can send email from our cars, watch 200 channels, consume food from every corner of the world. Some people have taken much more than their share, but on average, all of us in the West are living lives materially more abundant than most people a generation ago.

What's odd is, none of it appears to have made us happier. Throughout the postwar years, even as the GNP curve has steadily climbed, the "life satisfaction" index has stayed exactly the same. Since 1972, the National Opinion Research Center has surveyed Americans on the question: "Taking all things together, how would you say things are these days—would you say that you are very happy, pretty happy, or not too happy?" (This must be a somewhat unsettling interview.) The "very happy" number peaked at 38 percent in the 1974 poll, amid oil shock and economic malaise; it now hovers right around 33 percent.

And it's not that we're simply recalibrating our sense of what happiness means—we are actively experiencing life as grimmer. In the winter of 2006 the National Opinion Research Center published data about "negative life events" comparing 1991 and 2004, two data points bracketing an economic boom. "The anticipation would have been that problems would have been down," the study's author said. Instead it showed a rise in problems—for instance, the percentage who reported breaking up with a steady partner almost doubled. As one reporter summarized the findings, "There's more misery in people's lives today."

This decline in the happiness index is not confined to the United States; as other nations have followed us into mass affluence, their experiences have begun to yield similar results. In the United Kingdom, real gross domestic product per capita grew two-thirds between 1973 and 2001, but people's satisfaction with their lives changed not one whit. Japan saw a fourfold increase in real income per capita between 1958 and 1986 without any reported increase in satisfaction. In one place after another, rates of alcoholism, suicide, and depression have gone up dramatically, even as we keep accumulating more stuff. Indeed, one report in 2000 found that the average American child reported higher levels of anxiety than the average child under psychiatric care in the 1950s—our new normal is the old disturbed.

If happiness was our goal, then the unbelievable amount of effort and resources expended in its pursuit since 1950 has been largely a waste. One study of life satisfaction and mental health by Emory University professor Corey Keyes found just 17 percent of Americans "flourishing" in mental health terms, and 26 percent either "languishing" or out-and-out depressed.

[5] Danes (and Mexicans, the Amish, and the Masai) Just Want to Have Fun

How is it, then, that we became so totally, and apparently wrongly, fixated on the idea that our main goal, as individuals and as nations, should be the accumulation of more wealth? The answer is interesting for what it says about human nature. Up to a certain point, more really does equal better. Imagine briefly your life as a poor person in a poor society—say, a peasant farmer in China. (China has one-fourth of the world's farmers, but one-fourteenth of its arable land; the average farm in the southern part of the country is about half an acre, or barely more than the standard lot for a new American home.) You likely have the benefits of a close and connected family, and a village environment where your place is clear. But you lack any modicum of security for when you get sick or old or your back simply gives out. Your diet is unvaried and nutritionally lacking; you're almost always cold in winter.

In a world like that, a boost in income delivers tangible benefits. In general, researchers report that money consistently buys happiness right up to about $10,000 income per capita. That's a useful number to keep in the back of your head—it's like the freezing point of water, one of those random figures that just happens to define a crucial phenomenon on our planet. "As poor countries like India, Mexico, the Philippines, Brazil, and South Korea have experienced economic growth, there is some evidence that their average happiness has risen," the economist Layard reports. Past $10,000 (per capita, mind you—that is, the average for each man, woman, and child), there's a complete scattering: When the Irish were making two-thirds as much as Americans they were reporting higher levels of satisfaction, as were the Swedes, the Danes, the Dutch. Mexicans score higher than the Japanese; the French are about as satisfied with their lives as the Venezuelans. In fact, once basic needs are met, the "satisfaction" data scrambles in mind-bending ways. A sampling of *Forbes* magazine's "richest Americans" have identical happiness scores with Pennsylvania Amish, and are only a whisker above Swedes taken as a whole, not to mention the Masai. The "life satisfaction" of pavement dwellers—homeless people—in Calcutta is among the lowest recorded, but it almost doubles when they move into a slum, at which point they are basically as satisfied with their lives as a sample of college students drawn from 47 nations. And so on.

On the list of major mistakes we've made as a species, this one seems pretty high up. Our single-minded focus on increasing wealth has succeeded in driving the planet's ecological systems to the brink of failure, even as it's failed to make us happier. How did we screw up?

The answer is pretty obvious—we kept doing something past the point that it worked. Since happiness had increased with income in the past, we assumed it would inevitably do so in the nature. We make these kinds of mistakes regularly: Two beers made me feel good, so ten will make me feel five times better. But this case was particularly extreme—in part because as a species, we've spent so much time simply trying to survive. As the researchers Ed Diener and Martin Seligman—both psychologists—observe, "At the time of Adam Smith, a concern with economic issues was understandably primary. Meeting simple human needs for food, shelter and clothing was not assured, and satisfying these needs moved in lockstep with better economics." Freeing people to build a more dynamic economy was radical and altruistic.

Consider Americans in 1820, two generations after Adam Smith. The average citizen earned, in current dollars, less than $1,500 a year, which is somewhere near the current average for all of Africa. As the economist Deirdre McCloskey explains in a 2004 article in the magazine *Christian Century,* "Your great-great-great-grandmother had one dress for church and one for the week, if she were not in rags. Her children did not attend school, and probably could not read. She and her husband worked eighty hours a week for a diet of bread and milk—they were four inches shorter than you." Even in 1900, the average American lived in a house the size of today's typical garage. Is it any wonder that we built up considerable velocity trying to escape the gravitational pull of that kind of poverty? An object in motion stays in motion, and our economy— with the built-up individual expectations that drive it—is a mighty object indeed.

You could call it, I think, the Laura Ingalls Wilder effect. I grew up reading her books—*Little House on the Prairie, Little House in the Big Woods*—and my daughter grew up listening to me read them to her, and no doubt she will read them to her children. They are the ur-American story. And what do they tell? Of a life rich in family, rich in connection to the natural world, rich in adventure—but materially deprived. That one dress, that same bland dinner. At Christmastime, a penny—a penny! And a stick of candy, and the awful deliberation about whether to stretch it out with tiny licks or devour it in an orgy of happy greed. A rag doll was the zenith of aspiration. My daughter likes dolls too, but her bedroom boasts a density of Beanie Babies that mimics the manic biodiversity of the deep rainforest. Another one? Really, so what? Its marginal utility, as an economist might say, is low. And so it is with all of us. We just haven't figured that out because the momentum of the past is still with us—we still imagine we're in that little house on the big prairie.

[6] This Year's Model Home: "Good for the Dysfunctional Family"

That great momentum has carried us away from something valuable, something priceless: It has allowed us to become (very nearly forced us to become) more thoroughly individualistic than we really wanted to be. We left behind hundreds of thousands of years of human community for the excitement, and the isolation, of "making something of ourselves," an idea that would not have made sense for 99.9 percent of human history. Adam Smith's insight was that the interests of each of our individual selves could add up, almost in spite of themselves, to social good—to longer lives, fuller tables, warmer houses. Suddenly the community was no longer necessary to provide these things; they would happen as if by magic. And they did happen. And in many ways it was good.

But this process of liberation seems to have come close to running its course. Study after study shows Americans spending less time with friends and family, either working longer hours, or hunched over their computers at night. And each year, as our population grows by 1 percent we manage to spread ourselves out over 6 to 8 percent more land. Simple mathematics says that we're less and less likely to bump into the other inhabitants of our neighborhood, or indeed of our own homes. As the *Wall Street Journal* reported recently, "Major builders and top architects are walling people off. They're touting one-person 'Internet alcoves,' locked-door 'away rooms,' and his-and-her offices on opposite ends of the house. The new floor plans offer so much seclusion, they're 'good for the dysfunctional family,' says Gopal Ahluwahlia, director of research for the National Association of Home Builders." At the building industry's annual Las Vegas trade show, the "showcase 'Ultimate Family Home' hardly had a family room," noted the *Journal*. Instead, the boy's personal playroom had its own 42-inch plasma TV, and the girl's bedroom had a secret mirrored door leading to a "hideaway karaoke room." "We call this the ultimate home for families who don't want anything to do with one another," said Mike McGee, chief executive of Pardee Homes of Los Angeles, builder of the model.

This transition from individualism to hyper-individualism also made its presence felt in politics. In the 1980s, British prime minister Margaret Thatcher asked, "Who is society? There is no such thing. There are individual men and women, and there are families." Talk about everything solid melting into air—Thatcher's maxim would have spooked Adam Smith himself. The "public realm"—things like parks and schools and Social Security, the last reminders of the communities from which we came—is under steady and increasing attack. Instead of contributing to the shared risk of health insurance, Americans are encouraged to go it alone with "health savings accounts." Hell, even the nation's most collectivist institution, the U.S. military, until recently recruited under the slogan an "Army of One." No wonder the show that changed television more than any other in the past decade was Survivor, where the goal is to end up alone on the island, to manipulate and scheme until everyone is banished and leaves you by yourself with your money.

It's not so hard, then, to figure out why happiness has declined here even as wealth has grown. During the same decades when our lives grew busier and more isolated, we've gone from having three confidants on average to only two, and the number of people saying they have no one to discuss important matters with has nearly tripled. Between 1974 and 1994, the percentage of Americans who said they visited with their neighbors at least once a month fell from almost two-thirds to less than half, a number that has continued to fall in the past decade. We simply worked too many hours earning, we commuted too far to our too-isolated homes, and there was always the blue glow of the tube shining through the curtains.

[7] New Friend or New Coffeemaker? Pick One

Because traditional economists think of human beings primarily as individuals and not as members of a community, they miss out on a major part of the satisfaction index. Economists lay it out almost as a mathematical equation: Overall, "evidence shows that companionship . . . contributes more to well-being than does income," writes Robert E. Lane, a Yale political science professor who is the author of The Loss of Happiness in Market Democracies. But there is a notable difference between poor and wealthy countries: When people have lots of companionship but not much money, income "makes more of a contribution to subjective well-being." By contrast, "where money is relatively plentiful and companionship relatively scarce, companionship will add more to subjective well-being." If you are a poor person in China, you have plenty

of friends and family around all the time—perhaps there are four other people living in your room. Adding a sixth doesn't make you happier. But adding enough money so that all five of you can eat some meat from time to time pleases you greatly. By contrast, if you live in a suburban American home, buying another coffeemaker adds very little to your quantity of happiness—trying to figure out where to store it, or wondering if you picked the perfect model, may in fact decrease your total pleasure. But a new friend, a new connection, is a big deal. We have a surplus of individualism and a deficit of companionship, and so the second becomes more valuable.

Indeed, we seem to be genetically wired for community. As biologist Edward O. Wilson found, most primates live in groups and get sad when they're separated—"an isolated individual will repeatedly pull a lever with no reward other than the glimpse of another monkey." Why do people so often look back on their college days as the best years of their lives? Because their classes were so fascinating? Or because in college, we live more closely and intensely with a community than most of us ever do before or after? Every measure of psychological health points to the same conclusion: People who "are married, who have good friends, and who are close to their families are happier than those who are not," says Swarthmore psychologist Barry Schwartz. "People who participate in religious communities are happier than those who do not." Which is striking, Schwartz adds, because social ties "actually decrease freedom of choice"—being a good friend involves sacrifice.

Do we just think we're happier in communities? Is it merely some sentimental good-night-John-Boy affectation? No—our bodies react in measurable ways. According to research cited by Harvard professor Robert Putnam in his classic book Bowling Alone, if you do not belong to any group at present, joining a club or a society of some kind cuts in half the risk that you will die in the next year. Check this out: When researchers at Carnegie Mellon (somewhat disgustingly) dropped samples of cold virus directly into subjects' nostrils, those with rich social networks were four times less likely to get sick. An economy that produces only individualism undermines us in the most basic ways.

Here's another statistic worth keeping in mind: Consumers have 10 times as many conversations at farmers' markets as they do at supermarkets—an order of magnitude difference. By itself, that's hardly life-changing, but it points at something that could be: living in an economy where you are participant as well as consumer, where you have a sense of who's in your universe and how it fits together. At the same time, some studies show local agriculture using less energy (also by an order of magnitude) than the "it's always summer somewhere" system we operate on now. Those are big numbers, and it's worth thinking about what they suggest—especially since, between peak oil and climate change, there's no longer really a question that we'll have to wean ourselves of the current model.

So as a mental experiment, imagine how we might shift to a more sustainable kind of economy. You could use government policy to nudge the change—remove subsidies from agribusiness and use them instead to promote farmer-entrepreneurs; underwrite the cost of windmills with even a fraction of the money that's now going to protect oil flows. You could put tariffs on goods that travel long distances, shift highway spending to projects that make it easier to live near where you work (and, by cutting down on commutes, leave some time to see the kids). And, of course, you can exploit the Net to connect a lot of this highly localized stuff into something larger. By way of example, a few of us are coordinating the first nationwide global warming demonstration—but instead of marching on Washington, we're rallying in our local areas, and then fusing our efforts, via the website stepitup07.org, into a national message.

It's easy to dismiss such ideas as sentimental or nostalgic. In fact, economies can be localized as easily in cities and suburbs as rural villages (maybe more easily), and in ways that look as much to the future as the past, that rely more on the solar panel and the Internet than the white picket fence. In fact, given the trendlines for phenomena such as global warming and oil supply, what's nostalgic and sentimental is to keep doing what we're doing simply because it's familiar.

[8] The Oil-For-People Paradox: Why Small Farms Produce More Food

To understand the importance of this last point, consider the book *American Mania* by the neuroscientist Peter Whybrow. Whybrow argues that many of us in this country are predisposed to a kind of dynamic individualism—our gene pool includes an inordinate number of people who risked everything to start over. This served us well in settling a continent and building our prosperity. But it never got completely out of control, says Whybrow, because "the marketplace has always had its natural constraints. For the first two centuries of the nation's existence, even the most insatiable American citizen was significantly leashed by the checks and balances inherent in a closely knit community, by geography, by the elements of weather, or, in some cases, by religious practice."

You lived in a society—a habitat—that kept your impulses in some kind of check. But that changed in the past few decades as the economy nationalized and then globalized. As we met fewer actual neighbors in the course of a day, those checks and balances fell away. "Operating in a world of instant communication with minimal social tethers," Whybrow observes, "America's engines of commerce and desire became turbocharged."

Adam Smith himself had worried that too much envy and avarice would destroy "the empathic feeling and neighborly concerns that are essential to his economic model," says Whybrow, but he "took comfort in the fellowship and social constraint that he considered inherent in the tightly knit communities characteristic of the 18th century." Businesses were built on local capital investment, and "to be solicitous of one's neighbor was prudent insurance against future personal need." For the most part, people felt a little constrained about showing off wealth; indeed, until fairly recently in American history, someone who was making tons of money was often viewed with mixed emotions, at least if he wasn't giving back to the community. "For the rich," Whybrow notes, "the reward system would be balanced between the pleasure of self-gain and the civic pride of serving others. By these mechanisms the most powerful citizens would be limited in their greed."

Once economies grow past a certain point, however, "the behavioral contingencies essential to promoting social stability in a market-regulated society—close personal relationships, tightly knit communities, local capital investment, and so on—are quickly eroded." So re-localizing economies offers one possible way around the gross inequalities that have come to mark our societies. Instead of aiming for growth at all costs and hoping it will trickle down, we may be better off living in enough contact with each other for the affluent to once again feel some sense of responsibility for their neighbors. This doesn't mean relying on noblesse oblige; it means taking seriously the idea that people, and their politics, can be changed by their experiences. It's a hopeful sign that more and more local and state governments across the country have enacted "living wage" laws. It's harder to pretend that the people you see around you every day should live and die by the dictates of the market.

Right around this time, an obvious question is doubtless occurring to you. Is it foolish to propose that a modern global economy of 6 (soon to be 9) billion people should rely on more localized economies? To put it more bluntly, since for most people "the economy" is just a fancy way of saying "What's for dinner?" and "Am I having any?", doesn't our survival depend on economies that function on a massive scale—such as highly industrialized agriculture? Turns out the answer is no—and the reasons why offer a template for rethinking the rest of the economy as well.

We assume, because it makes a certain kind of intuitive sense, that industrialized farming is the most productive farming. A vast Midwestern field filled with high-tech equipment ought to produce more food than someone with a hoe in a small garden. Yet the opposite is true. If you are after getting the greatest yield from the land, then smaller farms in fact produce more food.

If you are one guy on a tractor responsible for thousands of acres, you grow your corn and that's all you can do—make pass after pass with the gargantuan machine across a sea of crop. But if you're working 10 acres, then you have time to really know the land, and to make it work harder. You can intercrop all kinds of plants—their roots will go to different depths, or they'll thrive in each other's shade, or they'll make use of different nutrients in the soil. You can also walk your fields, over and over, noticing. According to the government's most recent agricultural census, smaller farms produce far more food per acre, whether you measure in tons, calories, or dollars. In the process, they use land, water, and oil much more efficiently; if they have animals, the manure is a gift, not a threat to public health. To feed the world, we may actually need lots more small farms.

But if this is true, then why do we have large farms? Why the relentless consolidation? There are many reasons, including the way farm subsidies have been structured, the easier access to bank loans (and politicians) for the big guys, and the convenience for food-processing companies of dealing with a few big suppliers. But the basic reason is this: We substituted oil for people. Tractors and synthetic fertilizer instead of farmers and animals. Could we take away the fossil fuel, put people back on the land in larger numbers, and have enough to eat?

The best data to answer that question comes from an English agronomist named Jules Pretty, who has studied nearly 300 sustainable agriculture projects in 57 countries around the world. They might not pass the U.S. standards for organic certification, but they're all what he calls "low-input." Pretty found that over the past decade, almost 12 million farmers had begun using sustainable practices on about 90 million acres. Even more remarkably, sustainable agriculture increased food production by 79 percent per acre. These were not tiny isolated demonstration farms—Pretty studied 14 projects where 146,000 farmers across a broad swath of the developing world were raising potatoes, sweet potatoes, and cassava, and he found that practices such as cover-cropping and fighting pests with natural adversaries had increased production 150 percent—17 tons per household. With 4.5 million small

Asian grain farmers, average yields rose 73 percent. When Indonesian rice farmers got rid of pesticides, their yields stayed the same but their costs fell sharply.

"I acknowledge," says Pretty, "that all this may sound too good to be true for those who would disbelieve these advances. Many still believe that food production and nature must be separated, that 'agroecological' approaches offer only marginal opportunities to increase food production, and that industrialized approaches represent the best, and perhaps only, way forward. However, prevailing views have changed substantially in just the last decade."

And they will change just as profoundly in the decades to come across a wide range of other commodities. Already I've seen dozens of people and communities working on regional-scale sustainable timber projects, on building energy networks that work like the Internet by connecting solar rooftops and backyard windmills in robust mini-grids. That such things can begin to emerge even in the face of the political power of our reigning economic model is remarkable; as we confront significant change in the climate, they could speed along the same kind of learning curve as Pretty's rice farmers and wheat growers. And they would not only use less energy; they'd create more community. They'd start to reverse the very trends I've been describing, and in so doing rebuild the kind of scale at which Adam Smith's economics would help instead of hurt.

In the 20th century, two completely different models of how to run an economy battled for supremacy. Ours won, and not only because it produced more goods than socialized state economies. It also produced far more freedom, far less horror. But now that victory is starting to look Pyrrhic; in our overheated and underhappy state, we need some new ideas.

We've gone too far down the road we're traveling. The time has come to search the map, to strike off in new directions. Inertia is a powerful force; marriages and corporations and nations continue in motion until something big diverts them. But in our new world we have much to fear, and also much to desire, and together they can set us on a new, more promising course.

Critical Thinking

1. Why is trying to keep expanding the economy bad for our environment?

2. Explain what it means to have to pick between more or better. Is one selection more compatible with "environmental sustainability"? Why?

3. Locate on your text's world map, the region of the world's peoples who could "relate" to this article.

4. The author states: "Our single-minded focus on increasing wealth has succeeded in driving the planet's ecological systems to the brink of failure, even as it's failed to make

us happier." Beyond the author's evidence, what evidence supporting this statement do you see?

5. The author offers eight ideas to support his thesis. Locate one article in this text for each idea that you feel complements or conflicts with the author's view, and discuss.

6. Identify in this article three key terms, concepts, or principles that are used in your textbook (environmental science, economics, sociology, history, geography, etc.) or employed in the discipline you are currently studying. (Note: The terms, concepts, or principles may be implicit, explicit, implied, or inferred.)

7. Is it possible to be "happy" without being wealthy or owning material things? Explain. What do you need to be happy?

8. Do you think your definition of happiness is compatible with environmental sustainability? Discuss.

Approaching Environmental Issues Paradigm Worksheet

This worksheet can be photocopied and used alongside any of the articles to help you better conceptualize and understand the connections and linkages between an **Environmental Issue** and its three critical components: **Environment** (natural science concepts / principles); **People** (social science concepts / principles); and **Place / Patterns** (geographical science concepts / principles).

Environment
(Natural Science)

List aspects of the "natural sciences" (biology, chemistry, ecology, botany, physics, etc.) linked to the issue:

_____

_____

_____

_____

_____

_____

_____

_____

Environmental Issue / Concern:

_____

People
(Social Science)

List aspects of the "social sciences" (sociology, economics, politics, history, culture, etc.) linked to the issue:

_____

_____

_____

_____

_____

_____

_____

Places and Patterns
(Geographical Science)
Describe *geographic places*, movements, patterns linked to the issue:

_____

_____

_____

_____

_____

_____

_____

_____

_____

_____

_____

_____

_____

_____

_____

_____

_____

_____

Test-Your-Knowledge Form

We encourage you to photocopy and use this page as a tool to assess how the articles in *Annual Editions* expand on the information in your textbook. By reflecting on the articles you will gain enhanced text information. You can also access this useful form on a product's book support website at www.mhhe.com/cls

NAME: DATE:

TITLE AND NUMBER OF ARTICLE:

BRIEFLY STATE THE MAIN IDEA OF THIS ARTICLE:

LIST THREE IMPORTANT FACTS THAT THE AUTHOR USES TO SUPPORT THE MAIN IDEA:

WHAT INFORMATION OR IDEAS DISCUSSED IN THIS ARTICLE ARE ALSO DISCUSSED IN YOUR TEXTBOOK OR OTHER READINGS THAT YOU HAVE DONE? LIST THE TEXTBOOK CHAPTERS AND PAGE NUMBERS:

LIST ANY EXAMPLES OF BIAS OR FAULTY REASONING THAT YOU FOUND IN THE ARTICLE:

LIST ANY NEW TERMS/CONCEPTS THAT WERE DISCUSSED IN THE ARTICLE, AND WRITE A SHORT DEFINITION:

NOTES

NOTES

NOTES

NOTES

NOTES

NOTES

NOTES